PrP^{Sc} Prions: State of the Art

PrP^Sc Prions: State of the Art

Special Issue Editors

Joaquín Castilla
Jesús R. Requena

MDPI • Basel • Beijing • Wuhan • Barcelona • Belgrade

MDPI

Special Issue Editors

Joaquín Castilla
CIC bioGUNE
Spain

Jesús R. Requena
University of Santiago de Compostela
Spain

Editorial Office
MDPI
St. Alban-Anlage 66
Basel, Switzerland

This is a reprint of articles from the Special Issue published online in the open access journal *Pathogens* (ISSN 2076-0817) from 2017 to 2018 (available at: https://www.mdpi.com/journal/pathogens/special_issues/prions_study)

For citation purposes, cite each article independently as indicated on the article page online and as indicated below:

LastName, A.A.; LastName, B.B.; LastName, C.C. Article Title. *Journal Name* **Year**, *Article Number*, Page Range.

ISBN 978-3-03897-308-9 (Pbk)
ISBN 978-3-03897-309-6 (PDF)

Cover image courtesy of Jesús R. Requena and Joaquín Castilla.

Contents

About the Special Issue Editors

Joaquín Castilla, Ph.D., obtained his Ph.D at the Autonomous University of Madrid. His expertise since 1998 is based on in vitro and in vivo replication of prions. In particular, his group studies the strain and species barrier phenomena in a cell-free system, trying to dissect the molecular mechanisms by which prions propagate and focusing on new anti-prion therapeutic approaches. His main achievements are: (i) the development of the most sensitive method for prion detection; (ii) generation of prion infectivity in a test tube, contributing to validation of the protein-only hypothesis; (iii) prion detection in blood for the first time in pre-symptomatic and symptomatic animals; (iv) confirmation that the in vitro prion propagation faithfully mimics the three major phenomena governing transmissible spongiform diseases: infectivity, strain phenomenon and transmission barrier; and (v) establishing a reliable in vitro method for generating a diversity of recombinant prions with an invaluable value to understand the structure and other keys in the prion field.

Jesús R. Requena, Ph.D., obtained his Ph.D. at the University of Santiago de Compostela. He conducted postdoctoral research on the role of post-translational modification of proteins in disease and ageing at the University of South Carolina, with John Baynes, and at the Protein Biochemistry Lab, NHLBI, NIH, with Earl Stadtman and Rod Levine. There, he started to work on the biochemistry of prions and has since devoted his research activities to trying to understand the structure of PrP^{Sc}, which he believes holds the key to understanding how prions propagate. Recently, in collaboration with Holger Wille and Howard Young from the University of ASlberta, he has revealed the basic architecture of PrP^{Sc} which conforms to a 4-rung beta-solenoid, using cryo-electron microscopy. In May, 2018, he chaired the Prion 2018 conference in Santiago de Compostela.

Preface to "PrP^{Sc} Prions: State of the Art"

Prions were defined by Stanley Prusiner, in 1982, as "proteinaceous infectious particles" and more recently redefined as "alternative protein conformations that exhibit self-propagation" (Watts & Prusiner, Cold Spring Harb. Perspect. Med. 2018; 8:a023507). The scrapie isoform of the prion protein, PrP^{Sc}, was the first prion to be identified. Identification of several fungal and yeast prions followed. More recently, a strong debate has erupted about whether Aβ, tau, alpha-synuclein, and other amyloids that propagate throughout the brain through templating mechanisms are prions too. From a purely biochemical point of view it seems that they are. However, the fact that only the PrP^{Sc} prion is known to date to have caused epidemic and epizootic events puts it in a unique position within biology. In this special issue of Pathogens, a number of reviews summarize the state of the art of our knowledge of PrP^{Sc} prions. Reviews are presented on what we know about their structure and propagation, the basis of strains and transmission barriers, the mechanisms of PrP^{Sc} toxicity, the possible function of PrP^{Sc}'s properly folded precursor, PrPC and its evolutionary history, and recent technical breakthroughs in diagnostics and therapy development, among other key aspects of PrP^{Sc} prion biology.

Joaquín Castilla, Jesús R. Requena
Special Issue Editors

pathogens

MDPI

Review

What Is Our Current Understanding of PrP^{Sc}-Associated Neurotoxicity and Its Molecular Underpinnings?

Daniel Hughes [1] and Mark Halliday [2,*]

[1] MRC Toxicology Unit, Hodgkin Building, University of Leicester, Lancaster Road, Leicester LE1 9HN, UK; dh232@le.ac.uk
[2] Department of Clinical Neurosciences, University of Cambridge, Cambridge Biomedical Campus, Cambridge CB2 0AH, UK
*   Correspondence: mh931@medschl.cam.ac.uk; Tel.: +44-(0)122-376-2116

Received: 29 September 2017; Accepted: 27 November 2017; Published: 1 December 2017

Abstract: The prion diseases are a collection of fatal, transmissible neurodegenerative diseases that cause rapid onset dementia and ultimately death. Uniquely, the infectious agent is a misfolded form of the endogenous cellular prion protein, termed PrP^{Sc}. Despite the identity of the molecular agent remaining the same, PrP^{Sc} can cause a range of diseases with hereditary, spontaneous or iatrogenic aetiologies. However, the link between PrP^{Sc} and toxicity is complex, with subclinical cases of prion disease discovered, and prion neurodegeneration without obvious PrP^{Sc} deposition. The toxic mechanisms by which PrP^{Sc} causes the extensive neuropathology are still poorly understood, although recent advances are beginning to unravel the molecular underpinnings, including oxidative stress, disruption of proteostasis and induction of the unfolded protein response. This review will discuss the diseases caused by PrP^{Sc} toxicity, the nature of the toxicity of PrP^{Sc}, and our current understanding of the downstream toxic signaling events triggered by the presence of PrP^{Sc}.

Keywords: prion disease; neurodegeneration; neurotoxicity; proteostasis; PrP^{Sc}

1. Introduction

The Prion diseases, or transmissible spongiform encephalopathies, are a group of fatal neurodegenerative diseases characterized by extensive neuronal death, spongiform change and gliosis. Uniquely among neurodegenerative disease, the prion diseases are transmissible, with infection between members of the same species, and in some cases between different species, possible. Since the discovery that the infectious component of prions is comprised solely of protein, more specifically a misfolded form of the cellular prion protein (PrP^C), a great deal of research has focused on how this misfolded protein can cause such extensive and catastrophic damage to neurons. The major histopathological feature is the accumulation of extracellular amyloid plaques, comprised of misfolded PrP^C (termed PrP^{Sc} for PrP scrapie). However, it is not known how PrP^{Sc} causes neurodegeneration, or indeed what the exact toxic species is, as the link between PrP^{Sc}, infectivity and toxicity is not sharply defined. The role of cellular PrP^C also remains elusive, further clouding the investigation into disease processes. Due to its insolubility, the structure of PrP^{Sc} has not been definitively proven, hampering rational drug discovery efforts, although recent advances have allowed various models to be proposed [1]. Unfortunately, and largely due to our poor understanding of these molecular mechanisms, therapeutic treatments for prion disease remain elusive.

Recent research has begun to unravel the role PrP^{Sc} plays in the neurodegeneration associated with prion disease. This review will discuss the diseases caused by PrP^{Sc} toxicity, the nature of the toxicity of PrP^{Sc}, and our current understanding of the downstream toxic signaling events triggered by the presence of PrP^{Sc}.

1.1. PrP^C

The human prion protein is highly conserved in mammals, suggesting an essential role for the protein [2]. It is a small glycoprotein found mainly on the outer leaflet of the plasma membrane, held in place by a C-terminal glycosylphosphatidylinositol (GPI) anchor [3]. Although highly expressed in the tissues of the central nervous system (CNS), PrP^C is also expressed to varying degrees in most tissues in the body. Expression begins early in embryogenesis, and in adults the highest levels are in neurons, with moderate expression observed in glial cells and the peripheral nervous system [4]. Human PrP^C is synthesized as a 231 amino acid polypeptide (after removal of a 22 residue signal peptide [5]), which is processed through the endoplasmic reticulum (ER) and golgi apparatus. Post-translational modifications, including the removal of a signal sequence from the C-terminal end of PrP^C, result in a mature protein of 208 amino acids in length. The main structural features are a globular C-terminal domain made up of three alpha helices with a small antiparallel beta sheet composed of two separate strands, and a largely unstructured flexible N-terminal tail [6].

The exact cellular function of PrP^C remains unclear, with several distinct and overlapping roles suggested. One of the most important proposed roles of PrP^C is the maintenance of myelination in the peripheral nervous system [7]. Interestingly, neuron-specific PrP^C expression is enough to maintain myelination, as although PrP^C is expressed in Schwann cells, it appears to not be essential there [7]. It is unknown what effect PrP^C has on CNS myelin. The first proposed role for PrP^C was in Cu^{2+} homeostasis due it its high affinity binding [8], although a functional physiological role for this affinity remains elusive. PrP^C has a putative role in protecting against stress, especially oxidative and some apoptotic stresses [9–11]. There is also evidence it helps to regulate neuronal excitability and memory [12,13]. Interestingly, it is also involved in the regulation of the circadian rhythm [14] and cellular differentiation [15,16]. The wide variety of biological roles has led to the suggestion that PrP^C is a scaffold protein that regulates the formation of a number of multi-protein complexes, but it is unlikely that the entirety of its physiological roles have been discovered [2].

1.2. PrP^{Sc}

The crucial event in the development of prion disease is the structural and conformational change of PrP^C to the disease associated misfolded form, PrP^{Sc}. This conversion changes PrP^C from a protein characterized by alpha-helices to a partially protease-resistant misfolded protein categorized by beta sheets [17]. Proteinase K (PK) partially digests PrP^{Sc} and is often used to determine the presence of misfolded PrP^{Sc} [18]. Despite this conversion being essential for the pathogenesis of prion disease, the molecular underpinnings are still not understood. The transformation is thought to be a post-translational change in conformation which initiates the catalytic conversion of PrP^C into more PrP^{Sc}, by the interaction of existing PrP^{Sc} molecules (Figure 1). Continuous synthesis of PrP^C in the brain only provides more substrate for the pathological conversion to PrP^{Sc}. While this mainly occurs after exposure to already misfolded PrP^{Sc}, conversion can occur spontaneously in rare cases without exposure or a genetic basis. There are no primary sequence differences between PrP^C and PrP^{Sc}, so the change is mediated by different secondary structures and a propensity to aggregate. This pathological change involving only the prion protein is summarized in the protein only hypothesis [19]. Strong evidence supporting PrP^{Sc} as being the main cause of prion disease comes from the production of infectious PrP^{Sc} in vitro [20]. Whatever the mechanism, this conversion is the basis for all the prion diseases.

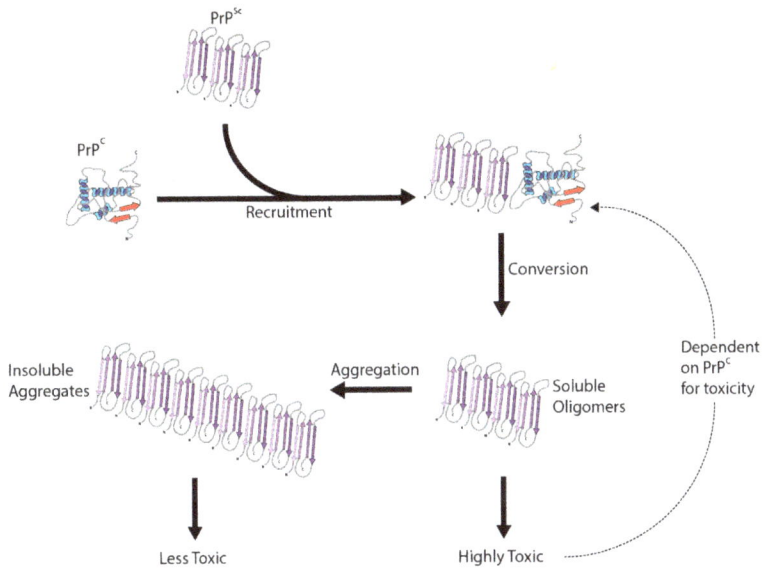

Figure 1. The conversion of PrP^C to PrP^{Sc}. The protein only hypothesis of prion conversion posits that misfolded PrP^{Sc} acts a catalyst, directly binding to PrP^C and causing its conversion to PrP^{Sc}. This self-perpetuating recruitment leads to large aggregates of PrP^{Sc}, and underlies its infectious potential. Surprisingly, aggregated PrP^{Sc} appears to be minimally toxic, with smaller, soluble oligomers of PrP^{Sc} likely mediating the majority of the observed neurodegeneration. Importantly, PrP^C is required for both the conversion process and for the toxicity of PrP^{Sc} to manifest.

1.3. The PRNP Gene

PrP^C is encoded by the *PRNP* gene found on chromosome 20 in humans, and chromosome 2 in mice. It is significantly conserved across mammalian species and even vertebrates as a whole. It contains three exons, but the entire open reading frame lies within exon 3 [21,22], with all of the disease-associated mutations discovered so far located within exon 3 [23]. The *PRNP* gene encodes a nonapeptide region followed by four octarepeats; this motif is thought to be important for its copper binding ability. More than 30 disease-causing mutations in *PRNP* have been discovered, leading to a single amino acid substitution, the addition of superfluous residues or an early truncation of the protein [24]. A number of insertion mutations have also been discovered in the octarepeat region. Many of these mutations are believed to facilitate the conversion of PrP^C to PrP^{Sc}, linking these mutations to disease. There are also polymorphisms in the *PRNP* gene that can influence the risk of developing prion disease. The most important is at codon 129, as it predisposes to sporadic, iatrogenic and variant Creutzfeldt–Jakob Disease (CJD-see below) [25]. Codon 129 codes for either methionine (M) or valine (V), and M/M homozygosity predisposes to an earlier and more rapid onset of disease, while heterozygosity is protective. A glutamate to lysine substitution at codon 219 also appears to confer a protective effect against prion disease [26]. The shortest incubation times for prion disease occur when PrP^{Sc} and the host PrP^C share the same sequence, and when inoculation occurs intracerebrally instead of peripherally [27]. If the inoculating prion differs to the host PrP^C, incubation times can be greatly increased, or clinical signs of disease never develop. This can prevent transmission between species, and is known as the species barrier.

1.4. Human Prion Disease

Human prion diseases are characterized by the presence of spongiform change, gliosis, amyloidosis and neuronal loss. Spongiosis appears as a series of vacuoles in fixed brain tissue. Astrocyte proliferation and neuronal cell death are other common features, and insoluble amyloid plaques containing aggregates of protease resistant prion protein (PrP^{Sc}) are often correlated with prion diseases. Uniquely in the field of neurodegeneration, prion diseases are transmissible between members of the same species, and often between (mammalian) species, although not freely as species barriers do exist. They can be sporadic, familial or acquired in origin. The most common is CJD; others include Kuru, Fatal Familial Insomnia (FFI) and Gerstmann–Straussler–Scheinker (GSS) disease. Although all are caused by the misfolding of PrP^C, these diseases often display startlingly different pathological and biochemical characteristics. These diseases can also affect different regions of the brain, causing further differences in disease course and symptoms.

Mutations in *PRNP* cause inherited prion disease that accounts for approximately 15% of prion disease cases, producing a wide spectrum of clinical phenotypes [28]. Inherited prion diseases generally have an earlier onset, but slower disease progression than sporadic cases. These mutations are autosomal dominant, and can result in either an expanded octapeptide repeat in the normal sequence of the prion protein, a non-conservative point mutation or a stop codon insertion in the *PRNP* open reading frame (ORF). This can lead to familial CJD (fCJD), GSS and FFI. fCJD causes a rapidly progressive dementia with myoclonus and abnormal electroencephalogram (EEG) recordings, GSS is characterized by a slow progression of ataxia and late onset dementia, and FFI is unique with its refractory insomnia, dysautonomia and motor dysfunction. These disease syndromes are not absolute; however, the same mutation can lead to highly divergent phenotypic and pathological variation between individuals [29].

Sporadic CJD (sCJD) accounts for 85% of cases of human prion disease, occurring in around one in a million people over the age of 65. Early onset cases are extremely rare. The disease presents with a rapidly progressive dementia with myoclonus and development of movement disorders such as tremor and rigidity. Associated neurological symptoms include cerebellar ataxia, pyramidal and extra pyramidal signs, and cortical blindness. Most cases have a characteristic EEG that includes periodic sharp-wave complexes. Death occurs after an average of 4 months from diagnosis, making it one the most aggressive forms of neurodegeneration [30].

Acquired prion diseases include Kuru, iatrogenic CJD (iCJD) and vCJD. Kuru is caused by the eating of infected brain tissue, and is characterized by progressive cerebellar ataxia, mood and personality changes, and a late onset dementia [31]. Death occurs approximately one year after the emergence of clinical symptoms. iCJD is rare, and has occurred after the exposure of patients to contaminated medical treatments or equipment. Contaminated dura matter and corneal grafts, inoculation with human pituitary-derived growth hormone and gonadotrophins have all been reported [32]. Improperly sterilized surgical equipment has also led to iCJD after brain surgery. iCJD caused by intracerebral infection is relatively rapid in onset and duration, with prominent early dementia. In contrast, peripheral inoculation is associated with a prolonged incubation time and late onset dementia.

In the mid-1990s, in the wake of the Bovine Spongiform Encephalopathy (BSE) epidemic, a new neurodegenerative illness emerged in the UK. Clinically and pathologically, it resembled sCJD, but the disease had a longer duration with a protracted neuropsychiatric syndrome, and critically, mainly affected young people [33,34]. After the realization, it was a new disease, it was termed new variant, or variant CJD (vCJD) [35]. The age of onset was much earlier than sporadic CJD, with a mean age of 29, and patients as young as 16. The initial symptoms are mainly behavioural, followed by ataxia and movement disorders. Dementia occurs at a much later point in the disease than CJD, with EEG abnormalities frequently absent. It also progresses slower than sporadic CJD, with a mean duration of 14 months. As none of the patients had *PRNP* mutations and were at a very low risk of iatrogenic exposure, BSE was considered to be the most likely cause. Molecular studies on vCJD tissue showed that the biochemical properties of the protease-resistant prion protein found in these patients were distinct from other human prion diseases [36], but similar to that of BSE [37], leading to the acceptance that BSE exposure causes vCJD.

1.5. Animal Prion Diseases

Several mammalian species are also afflicted by prion disease, including scrapie in sheep, Chronic Wasting Disease (CWD), which mainly affects deer and elk in North America and transmissible mink encephalopathy, which affects mink feeding on infected livestock. Unlike in humans, most cases are infectious in origin, although the increased surveillance for prion disease after the BSE outbreak is identifying increasing numbers of spontaneous cases. The histopathological features are grossly similar between human and animal prion disease. BSE affects the brainstem of cattle causing ataxia, with a presymptomatic incubation time of 5 years [38]. CWD is thought to be highly contagious due to the high number of animals infected despite the free-ranging habits of deer and elk. It has spread through much of North America and been detected in South Korea [39,40]. CWD has recently been observed in free ranging reindeer in Norway, with the origin of this outbreak currently unknown [41]. Despite not being natural carriers of prion disease, mice have been used extensively to model prion disease. They have been infected with sheep scrapie, and genetic modification of the carried *Prnp* gene allows prion disease from other hosts to be replicated (Figure 2).

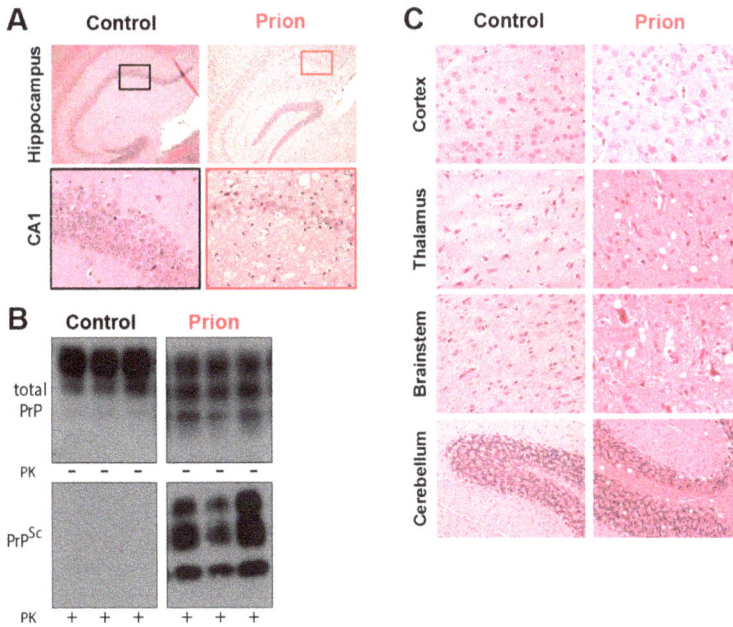

Figure 2. The neuropathology of prion disease. (**A**). The Rocky Mountain Laboratory (RML) strain of prion disease induces extensive neurodegeneration in mice, especially in the hippocampus where considerable neuronal death is observed in the CA1 region (hematoxylin and eosin stained hippocampal sections from uninfected control mice, or prion-infected terminal tg37$^{+/-}$ mice) (**B**). Prion infection is associated with the accumulation of PrP^{Sc}, which is often detected by its partial proteinase resistance to digestion by proteinase K (total PrP and PrP^{Sc} levels, detected with and without proteinase K digestion) (**C**). Spongiosis, an intracellular oedema that appears as holes in histological slices after fixation, is observed throughout the brain in both human and animal cases of prion disease (hematoxylin and eosin stained hippocampal sections from uninfected control mice, or prion-infected terminal tg37$^{+/-}$ mice).

1.6. Selective Neuronal Vulnerability in Prion Diseases

The same disease agent, PrP^{Sc}, is associated with all the prion diseases, however, the signs and symptoms of each disease can differ dramatically. This may be due to the regional differences of PrP^{Sc} accumulation in the brain and the neurons affected. This is thought to be a result of many factors including specific interactions of different protein conformers as well as region-specific micro-environments which contain a different combination of metals, chaperone proteins and translational machinery [42].

For instance, in FFI, neurodegeneration occurs in the thalamus, accounting for the insomnia associated with this prion disease due to the involvement of the thalamus in sleep regulation [30,43]. In Kuru, the damage often occurs in the cerebellum, leading to defects in coordination, while GSS has a wider range of clinical manifestations ranging from cerebellar ataxia to spastic parapesis, often in combination with dementia [43]. In CJD, the cerebral cortex is the main affected brain region [44], which results in mental impairments, mood change and various visual disturbances.

2. Is the Prion Protein Directly Responsible for Prion Disease?

Despite initial resistance, the protein only hypothesis of prion disease is now widely accepted. However, many unanswered questions remain. The most pressing, and to this day still the most elusive, is how exactly does the conversion of PrP^{C} to PrP^{Sc} cause prion disease? There are several possibilities that have been suggested to be a cause: the conversion of PrP^{C} to PrP^{Sc} causes a toxic loss of function in the PrP^{C} protein; PrP^{Sc}, or aggregates of PrP^{Sc}, are directly toxic to neurons; the conversion process itself is somehow toxic, or there are transient intermediaries formed that mediate the toxicity. In addition to the main cause, multiple toxic downstream processes are likely to be engaged, with neuroprotective responses failing or behaving ineffectively.

2.1. PrP^{C} is Essential for Prion Disease

It is now known that PrP^{C} loss of function is not the main cause of prion disease. Knockout mice models are grossly normal and display no obvious phenotypes [12,45,46]. In species where prion disease is naturally occurring, such as cows and goats, PrP knockout is again non-toxic [47,48]. Therefore, loss of PrP^{C} function does not cause prion-induced neurodegeneration. However, since the first reports of PrP^{C} knockout, many subtle phenotypes have been described, including some related to neuroprotective pathways such as neurogenesis and stress protection [49]. Although not a direct cause, impairment of some of these processes might contribute to disease under specific stress insults.

PrP^{C} does appear to have some directly neuroprotective abilities. PrP^{C} is upregulated in neurons after ischaemic stroke in humans, and knocking out PrP^{C} was shown to greatly increase infarct size in an animal model [50]. PrP^{C} is involved in the maintenance of myelin in the mouse peripheral nervous system [7]. PrP^{C} is also involved in protecting against the neurotoxicity induced by the artificial expression of its closest homologue, doppel (Dpl). This was first discovered due to the accidental expression in the brain of Dpl by a group trying to delete the Prnp gene in mice [51], caused by the fusion of the PrP promoter to the Dpl open reading frame and its subsequent ectopic expression in the brain [52]. Dpl expression caused Purkinje cell death [51] and a late onset ataxia [53] that could be rescued by PrP^{C} expression in a dose-dependent manner [54]. However, Dpl is not normally expressed in the CNS in any significant quantities, and levels are not increased during prion disease [55], so the physiological extent of neuroprotection by PrP^{C} cannot be inferred.

The most important observation from PrP^{C} knock out experiments is the absolute and total requirement of the presence of PrP^{C} for any PrP^{Sc} induced pathology. This was first demonstrated in mice lacking PrP^{C}, which were resistant to prion infection [56]. A set of elegant experiments further proved this effect. Brandner and colleagues grafted neural tissue overexpressing PrP^{C} into the brain of PrP null mice [57]. After inoculation with prions, the grafts accumulated high levels of PrP^{Sc} and developed the severe histopathological changes characteristic of prion disease. Substantial amounts of graft-derived PrP^{Sc} migrated into the surrounding areas of the host brain, but even 16 months after inoculation no pathological changes were seen in PrP null tissue [57]. Therefore, in addition to being resistant to scrapie infection, brain tissue devoid of PrP^C is not damaged by exogenous PrP^{Sc}, providing further evidence that PrP^{Sc} is not directly toxic in vivo. This was further demonstrated in experiments in which PrP^C was depleted during the course of prion infection [58]. Double transgenic mice were generated that had floxed PrP transgenes, from which the PrP coding sequence is deleted by neuronal Cre expression at 9 weeks of age. When these mice are inoculated with prions before PrP knockout, they develop the initial stages of prion disease, including spongiosis and hippocampal shrinkage. When the Cre-mediated excision of the Prnp gene occurred, prion disease was prevented from developing and the early spongiform changes were reversed, despite continued prion replication in non-neuronal cells and further astrocytic extra-neuronal PrP^{Sc} deposition. The mice lived for the normal lifespan, and remarkably, they never developed further clinical disease [58]. These results also argue against direct neurotoxicity of PrP^{Sc}, because the continued non-neuronal replication and accumulation of PrP^{Sc} throughout the brains of scrapie-infected mice was not pathogenic.

2.2. The Weak Links between PrP^{Sc} and Neurotoxicity

A number of other experiments have also demonstrated the complicated relationship between PrP^{Sc} and toxicity. Interestingly, the GPI-anchor has been demonstrated to be required for PrP^{Sc} induced toxicity. Prion inoculated mice expressing anchor-less PrP which is released into the extracellular space instead of being tethered to the plasma membrane, freely replicate PrP^{Sc} but do not develop disease [59]. This again suggests that PrP^{Sc} is not directly toxic to neurons, and also that either conversion at the membrane or subsequent internalization mediated by the GPI anchor is required for toxicity. In concert with this, the levels of PrP^{Sc} deposition are poorly correlated with disease progression, with subclinical cases of prion infection and prion disease observed with low prion titers observed in both animal and human cases [60]. Several studies have noted neurodegeneration without amyloidgenic PrP^{Sc} when passaging BSE prions into mice or rats [61,62]. In humans, several inherited mutations cause neurodegeneration without plaques [63–65]. A transmembrane form of the prion protein has been demonstrated to cause GSS without any detectable PrP^{Sc} [66]. Studies using hamster prions injection into mice have demonstrated cases of substantial PrP^{Sc} replication without the emergence of clinical signs [67,68]. These experiments have profound implications for the development, diagnosis and treatment of prion disease. It may be that PrP^{Sc} is a better marker for prion infection rather than prion-induced neurodegeneration, and again demonstrates the relative lack of neurotoxicity of PrP^{Sc}. Subclinical infection also poses a public health risk, as PrP^{Sc} from non-symptomatic individuals can still be infectious to other, perhaps more susceptible individuals [69].

Unfortunately, the identity of the actual infective agent even in purified scrapie infectious fractions remains a source for debate. Only 1 in 10^5 particles appears to be infectious [70], so the structure of the infectious agent cannot be definitively inferred. The most infectious particles have been shown to be non-fibrillar in nature, and comprised of 14–28 PrP molecules, with infectivity significantly reduced in oligomers larger or smaller than this [71]. PrP^{Sc} is partially protease resistant by definition, but infectious PK-sensitive forms of PrP^{Sc} have also been detected [72–74]. These PK-sensitive forms share structural features with PK-resistant PrP^{Sc}, and are similarly infective [74]. In contrast, high levels of PK-resistant PrP^{Sc} are not always correlated with disease [75], or infectivity [76]. Partial protease resistance is not consistently correlated with infectivity, and non-infective protease resistant PrP can be produced [77]. Furthermore, infectious PrP^{Sc} can be comprised of both protease sensitive and resistant fractions. All of these studies highlight the complexity and heterogeneity of the toxic agent in prion disease, obfuscating a better understanding of the mechanisms of disease.

2.3. PrP^{Sc} Structure and Toxicity

A puzzling characteristic of prion disease is the existence of different strains/isolates of PrP^{Sc} that when inoculated into model organisms can produce drastically different incubation times, clinical signs and pathology. Biochemically, the strains display different immunohistopathological characteristics and protease sensitivities [78]. As prions are comprised only of protein and are generated by the conversion of host PrP^C, the prion strain phenomenon cannot be attributed to genetic variability. Instead, prion strains are likely to result from distinct conformational changes, that are maintained during the conversion process [79]. This suggests that the structure of PrP^{Sc} may help to explain the associated neurotoxicity. Unfortunately, difficulties in purifying and determining the structure of PrP^{Sc} have hindered investigations.

The toxic conversion results in PrP^{Sc} characterized by extensive β-sheet secondary structure [17], a protease resistant core [80] and new epitopes not shared with PrP^C [81]. From here, monomers, oligomers, protofibrils and insoluble fibrils of PrP^{Sc} then accumulate, creating a heterogeneous assortment of structures. It is believed the β-sheet is essential for the aggregation of PrP^{Sc} into amyloid fibrils [82]. Several models of PrP^{Sc} amyloid have been suggested, including short compact fibrils or parallel β-sheets (see [1] for review). It is believed that fibrils dictate infectivity and the species barrier effect [1], as fibril load correlates with infectivity, but not toxicity [83]. Even if they are not the toxic species, their presence may catalyze the formation of a more toxic species [84], with growing evidence pointing towards oligomers of PrP^{Sc} as being the culprit. They display increased toxicity compared to fibrils both in vitro and in vivo [85–87]. These results are in agreement with the current evidence from other protein misfolding neurodegenerative diseases, where smaller oligomer molecules are thought to be the most toxic [88]. Determining the exact structure of the various PrP^{Sc} species remains elusive, and thus so does the exact relationship between PrP^{Sc} structure and toxicity. There is evidence that the disordered N-terminal domain of PrP^C mediates the toxicity of PrP^{Sc} [89]; it is likely that a better understanding of similar relationships between structure and function will improve our understanding of PrP^{Sc} neurotoxicity.

3. The Molecular Underpinnings of PrP^{Sc} Toxicity

Despite the weak evidence for the direct neurotoxicity of PrP^{Sc}, there are numerous reported detrimental effects of PrP^{Sc} formation/aggregation that explain at least some of the toxicity of prion disease. This includes the induction or inhibition of a range of cellular processes, discussed below.

3.1. Autophagy

Autophagy is the cell's main clearance mechanism for aggregated or dysfunctional proteins, which delivers cytoplasmic macromolecules or organelles to be degraded to the lysosomes. The structure to be degraded is enclosed in a double membrane structure termed the autophagic vacuole (or autophagosome), which then fuses with a lysosome containing hydrolases and digestive enzymes. Autophagic processes can ultimately initiate a form of cell death similar to apoptosis if levels of aggregated protein are deemed to be too high. Due to the large buildup of aggregated proteins in the protein misfolding neurodegenerative diseases, it is no surprise that dysfunction of this pathway is often observed, and increasingly autophagy is being explored as a possible therapeutic target. The links between autophagy and prion disease were first described in a hamster model of prion disease [90], where large autophagy vacuoles were observed, which increased in size as the disease progressed. Since then, autophagic vacuoles have been observed in many experimental models and human patients [91–94].

Pharmacological treatments that induce autophagy have conferred neuroprotection in various models of prion disease. Lithium, astemizole and the experimental compound FK506 have all been reported to induce autophagy and prolong survival in prion infected mice, with the authors attributing the neuroprotective effects to an increased removal and degradation of PrP^{Sc} [95–97]. The mTOR inhibitor rapamycin, which also induces autophagy, was shown to be neuroprotective in a mouse model of GSS [98]. Despite these results, it is still unclear if reduced or dysfunctional autophagy is a direct toxic effect of PrP^{Sc}, or if inducing autophagy confers neuroprotection by increasing clearance of PrP^{Sc} preventing other toxic downstream effects from occurring.

3.2. The Induction of Apoptosis

Apoptosis is the programmed death of cells, and is characterized by cell shrinkage, condensation of the chromatin and fragmentation of the nucleus. It is an active process requiring gene transcription and protein translation, and markedly different from the uncontrolled necrosis [99]. Apoptosis occurs in both experimental prion disease and human patients [100–104]. However, it is likely that the apoptosis observed in prion disease is a downstream effect of the prion infection, rather than the direct cause of PrP^{Sc} toxicity. Genetic deletion of the major pro-apoptotic protein Bax has no effect on prion disease progression [105], and over expression of the anti-apoptotic Bcl-2 again has no effect [106]. Deletion of caspase-12, a pro-apoptotic protein induced by ER stress, also failed to protect neurons in prion diseased mice [107].

3.3. Proteasome/Ubiquitin Inhibition

The Ubiquitin Proteasome (UPS) system is the main route for targeted protein degradation in mammalian cells. Degradation of proteins by the UPS occurs over two stages; firstly, the protein to be degraded is conjugated with multiple ubiquitin (Ub) molecules, targeting it for destruction. The Ub tagged protein is then degraded by the 26S proteasome [108] and broken down into its constituent peptides, which can be recycled for future protein synthesis.

Numerous studies suggest that disruption of the UPS by PrP^{Sc} in prion disease is a contributing factor to pathogenesis. An increase in ubiquitin immuno-reactivity, which is indicative of UPS dysregulation, has been reported in prion-infected mice in the early stages of disease and precedes behavioural deficits as well as correlating with PrP^{Sc} deposition [109].

PrP isoforms have been shown to directly inhibit the 26S proteasome [109–113]. In vitro data using purified proteasomes and three different cell lines show that PrP oligomers directly inhibit the 26S proteasome [112]. It is thought that PrP molecules rich in β-sheet conformations such as PrP^{Sc} inhibit the 26S proteasome by reducing gate opening of the 20S subunit and hence limiting substrate entry [111]. This has been inferred because β-sheet rich PrP does not inhibit mutant proteasomes with a constitutively open gate [110]. Inhibition of the proteasome can also increase aggregation of PrP^{Sc}, in addition to affecting the processing of PrP^{C} [114,115]. This could result in a positive feedback mechanism where PrP^{Sc} induces proteasomal malfunction, which in turn leads to increased aggregation of PrP^{Sc} and a backlog of poly-ubiquitinated proteins (Figure 3).

Proteasomal dysfunction is a common feature of many neurodegenerative diseases and given that PrP^{Sc} can directly inhibit the UPS, this could make activation of the UPS a possible therapeutic target in prion disease. McKinnon et al., found that activation of the UPS by a small molecule inhibitor resulted in enhanced clearance of poly-ubiquitinated substrates and reduced PrP^{Sc} load in prion infected cells [109]. Reducing PrP^{Sc} and its effects on proteostasis in this manner could be doubly beneficial.

3.4. The Unfolded Protein Response

Recent studies have implicated aberrant signaling by the unfolded protein response (UPR) as a major pathological player in prion disease progression and neuronal cell death [116–122]. The UPR is made up of three signaling cascades all beginning in the endoplasmic reticulum membrane. They are controlled by PKR-like endoplasmic reticulum kinase (PERK), Inositol-requiring enzyme 1 (IRE1) and Activating transcription factor 6 (ATF6) [123–125], and all are activated in response to an unfolded protein load within the ER. PERK phosphorylation attenuates protein translation via the phosphorylation of eIF2α at serine 51, in an attempt to prevent additional unfolded protein production, and a number of stress response genes are upregulated by ATF6, XBP1 (a transcription factor downstream of IRE1) and ATF4 (a transcription factor induced by eIF2α phosphorylation) [126]. After the resolution of unfolded protein stress, GADD34, the eIF2α-targeting component of protein phosphatase 1, reduces eIF2α phosphorylation and restores translation [127].

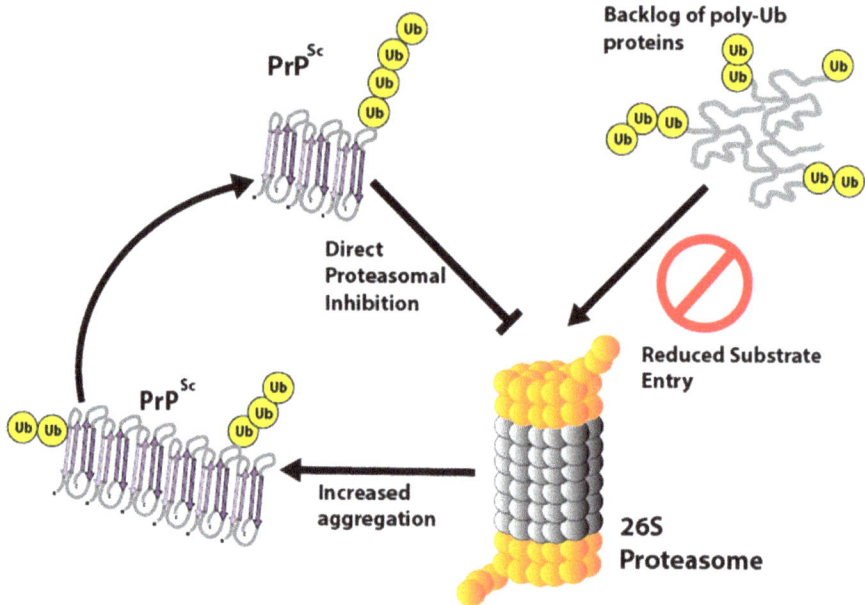

Figure 3. Disruption of the ubiquitin proteasome system by PrP^{Sc}. PrP^{Sc} can directly inhibit the 26S proteasome by binding to the 20S subunit, preventing substrate entry. The causes increased PrP^{Sc} aggregation and accumulation of poly-ubiquitinated proteins in the cytoplasm.

A series of studies using Tg37 hemizygous mice infected with Rocky Mountain Laboratory (RML) prions identified aberrant UPR signaling as a pathogenic mechanism in prion disease. These mice overexpress wild type murine PrP around 3-fold normal levels and follow a well-documented disease progression after prion inoculation. Loss of synapse regeneration is evident at 7 weeks post inoculation (wpi), and PERK activation and phosphorylation of eIF2α at 9 wpi precedes neuronal loss, which is apparent at 10 wpi [122]. Lentivirally expressing GADD34 reduces eIF2α phosphorylation, restores translation and prevents neurodegeneration. In contrast, administration of salubrinal (an inhibitor of eIF2α dephosphorylation) accelerates disease in these mice. Interestingly, targeting the PERK pathway in this model confers neuroprotection without affecting levels of PrP^C or PrP^{Sc}. In a subsequent study, prion-infected Tg37 hemizygous mice were treated with GSK2606414 [121] a potent and bioavailable PERK inhibitor [128]. This compound effectively prevented neurodegeneration, even when dosed after the emergence of early neurological disease indicators. Again, this protection was independent of PrP^{Sc} levels.

Another compound, ISRIB (integrated stress response inhibitor), which acts downstream of eIF2α phosphorylation, has been used to successfully delay neurodegeneration in prion-infected mice [119]. ISRIB binds to eIF2B, a guanine exchange factor essential for supplying the energy needed for the initiation of translation, stabilizing it in its dimeric form [129]. This allows eIF2B to act as a guanine exchange factor even in the presence of eIF2α-P, which normally inhibits this action. The result is a partial restoration of protein synthesis in ER-stressed conditions. Administration of ISRIB to prion mice resulted in an increase in survival and marked neuroprotection of hippocampal neurons, with the treated mice also performing better in behavioural tests [119]. These studies validate PERK and the PERK-eIF2α pathway as therapeutic targets, with the restoration of translation proposed to be the neuroprotective process.

A further study aimed at repurposing drugs for use in neurodegeneration identified two molecules, trazodone and dibenzoylmethane (DBM), that partially restore protein synthesis rates during ER stress [120]. Both were neuroprotective in prion disease and a model of fronto-temporal dementia [120]. Trazodone is a serotonin agonist and reuptake inhibitor that can be rapidly repurposed, and DBM is a curcumin analogue under investigation as an anti-cancer agent.

The results of these studies suggest that the decline in protein synthesis rates mediated by PERK pathway signaling is a main contributing factor to prion disease associated neurodegeneration (Figure 4). Neuroprotection was observed despite the continued replication of PrP^Sc, further demonstrating the lack of direct toxicity of PrP^Sc and instead implicating downstream processes as mediating the toxicity in prion disease.

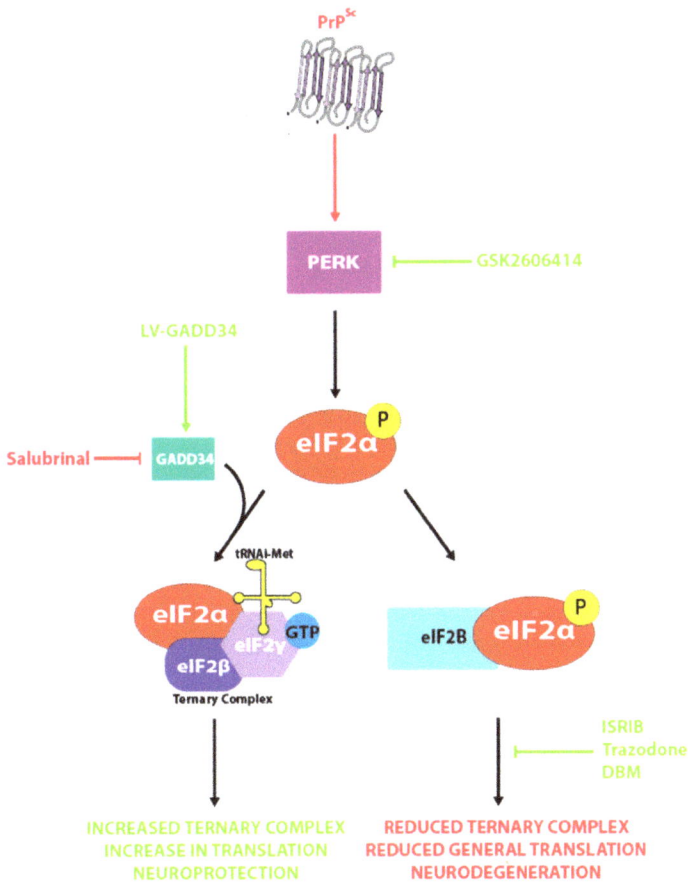

Figure 4. The role of the unfolded protein response (UPR) in prion neurodegeneration. Aggregates of PrP^Sc activate PERK signaling, leading to a reduction in protein synthesis rates mediated by the phosphorylation of eIF2α. This starves neurons of essential proteins, leading to neurodegeneration. Restoring translation rates via lentiviral expression of GADD34 or treatment with a variety of small molecule inhibitors increases translation and is substantially neuroprotective.

3.5. Oxidative Stress

Increased oxidative stress and prion disease have been linked in a plethora of previous studies. However, it is unclear if prion disease progression causes increased oxidative stress or if prion-associated pathology results from oxidative stress.

Conversion of PrP^C to PrP^{Sc} as an effect of oxidative stress has been reported in various in vitro studies [130–132]. It has been postulated that prion misfolding in disease, particularly those of a sporadic nature, could be triggered by oxidative stress. It has also been suggested that PrP^C plays a role in cellular resilience to oxidative stress via copper metabolism associated with PrP^C and superoxide dismutase (SOD) activity [133,134]. Loss of the antioxidant functions of PrP^C after conversion to PrP^{Sc} could hence contribute to pathogenesis of prion diseases. Studies performed in scrapie-infected mice and hamsters show an increase in markers of oxidative stress such as heme oxidase 1 [135] during disease progression. Alterations in free radical metabolism and increased oxidative stress can cause mitochondrial dysfunction in the brains of scrapie-infected animals, suggesting that this mitochondrial dysfunction is a contributing factor to prion disease progression [136,137].

3.6. Synaptic Dysfunction

One of the starkest phenotypes of prion disease is the gradual but continuous loss of synapses as the disease progresses. In mice, the number of synapses progressively decreases across a number of brain regions, including the hippocampus, with both pre- and post-synaptic processes affected [138–142]. This loss of synapses is correlated with a number of behavioural phenotypes [122,143,144], and precedes neuronal loss, suggesting that it is the loss of synapses that is causing the behavioral phenotypes. In human disease, there is also evidence for large quantities of synaptic loss and neurological deficits without extensive neuronal loss [145], but examples are limited due to the necessity of waiting until a post-mortem for investigation, where neuronal loss leaves a much more catastrophic signature than synaptic dysfunction.

PrP can modulate the activity of N-methyl-d-aspartate (NMDA) receptors, an important ion channel involved in long-term potentiation, but also to the detriment of neurons during excitotoxicity. In prion disease, an increase in activity through NMDA receptors is observed facilitating neuronal death, which can be attenuated by expressing PrP^C [146–148]. There is also evidence for PrP^{Sc} directly forming pores in the plasma membrane, disrupting the electrochemical balance of the neuron [149]. Here, soluble oligomers of misfolded PrP insert into the plasma membrane forming ion channel like structures that allow the passage of ions, causing homeostatic disruption [150].

3.7. Microglia

Microglial accumulation is observed in a range of neurodegenerative diseases, including prion disease. Microglia are the macrophages of the CNS and release cytokines in response to a wide variety of stressors. There is some evidence that microglia can actually be involved in the dissemination of PrP^{Sc} throughout the brain. Baker and colleagues demonstrated that purified microglia from CJD-infected mice show similar infectivity to crude homogenate, despite having 50x less PrP^{Sc} [151]. Microglia have been shown to be required for the toxicity of a human PrP fragment in vitro, while release of reactive oxygen species (ROS) from activated microglia is suggested as the cause of neuronal apoptosis [152,153].

However, one study [154] suggests a protective role of Cx3cl1/Cx3cr1 signaling in prion disease. The chemokine Cx3cl1 is expressed by neurons and its receptor Cxcr1 is solely expressed by microglia. Prion incubation times in Cx3cr1 null mice were significantly reduced, with no observed changes in microglial activation or chemokine/cytokine expression.

Brain slices with microglia ablation aggravated prion-induced neurotoxicity, and IL34$^{-/-}$ mice (which present with reduced microglia) amplified deposition and accelerated prion disease [155]. However, contrasting work suggested microglia are not efficient at clearing PrP^{Sc} [156].

4. Discussion

The prion diseases display a unique pathological and biochemical profile characterized by spongiosis, astrocytosis and neuronal death. These disorders are accompanied by a unique set of biological features, and remain among some of the most puzzling and inscrutable diseases known. Uniquely for a transmissible disease, the infectious agent is a protein encoded by the host's own genome. The protein only hypothesis of prion disease is now widely accepted, but many unanswered questions still remain. At the center of the prion phenomenon is the conversion of PrP^C to PrP^{Sc}, usually followed by the widespread aggregation of PrP^{Sc} throughout the brain. One of the first major milestones of prion disease research was the discovery that disease was not caused by the loss of function of PrP^C [56]. Subsequent research has uncovered a number of proposed cellular roles for PrP^C, including some neuroprotective processes [2], so the loss of function of PrP^C cannot be completely ruled out as a contributor to pathology, especially as levels drop during the disease course [157]. Another interesting finding, and one that has yet to be fully explored, is that PrP^C is required for the toxicity of prion disease to manifest [57,58]. Uncovering why will surely be a major step in the search for a cure for prion disease.

If the loss of function of PrP^C is not the main pathological cause, then the aggregation of PrP^{Sc} becomes the most likely suspect for the mediator of toxicity. Unfortunately, prion disease is rarely so simple. Aggregates of PrP^{Sc} correlate poorly with disease progression, subclinical cases of prion disease with large amounts of PrP^{Sc} have been discovered, and prion disease without aggregates of PrP^{Sc} has also been observed [68,69]. In addition, the extra-neuronal replication of PrP^{Sc} by glial cells is not toxic to neurons if they no longer express PrP^C [58]. Even the infectivity of PrP^{Sc} is not fully understood, as only a tiny proportion of PrP^{Sc} appears to be infectious, and protease resistance, usually the main marker of the presence of PrP^{Sc}, is poorly correlated with infectivity [71,75,76].

At the heart of prion disease is the structural change from α-helical PrP^C to β-sheet rich PrP^{Sc} [17]. This, along with the puzzling prion strain phenomenon, suggests that determining the structure of PrP^{Sc} and its various misfolded states will greatly illuminate the molecular mechanisms of PrP^{Sc} associated toxicity. Unfortunately, this has proven extremely difficult due to the insolubility of PrP^{Sc}, hampering our understanding. However, growing evidence is suggesting that it is oligomers of PrP^{Sc} that are most toxic, compared to the larger fibrils of PrP^{Sc} or single monomers [158]. Why PrP^C is more prone to misfolding than other proteins, and exactly how this happens, especially in sporadic cases where no template of PrP^{Sc} is present, is an extremely important topic of research. It has been suggested that there are cofactors that may contribute to the misfolding [159], but so far no definitive explanation has been found. Another suggestion is that an intermediate between PrP^C and PrP^{Sc} may be the most toxic species, but as even stable PrP^{Sc} is proving difficult to determine the structure for, more temporary species are likely to be even harder to uncover.

Despite the evidence that argues against a direct toxic role of PrP^{Sc}, a number of detrimental processes are associated with the misfolded protein. Among the best characterized are inhibition of the proteasome and over-activation of the UPR, but a number of other processes including synaptic disruptions, the initiation of apoptosis and induction of oxidative stress are also observed. A consensus on the molecular underpinnings of PrP^{Sc} associated toxicity is far from clear, and the heterogeneity of PrP^{Sc} may mean some of these processes are strain/structure dependent. It is also possible that the mere presence of an aggregating misfolded protein, rather than the intrinsic properties of PrP^{Sc}, explain some of the associated toxicity. Similar damaging pathways are activated in other neurodegenerative diseases such as Alzheimer's and Parkinson's, where the identity of the misfolding proteins are different, but many similarities remain [88,160].

Unfortunately, treatment options for the prion diseases are extremely limited, but several strategies have been suggested. Reducing levels of PrP^C will remove the substrate of the conversion to PrP^{Sc} [161,162], but the increasing number of beneficial cellular roles attributed to PrP^C reduces the attractiveness of this approach. Another is to prevent the conversion of PrP^C to PrP^{Sc} by either stabilizing PrP^C or inhibiting the conversion process [163]. This has been the most popular approach, as it is amenable to small molecule or therapeutic antibody intervention, while preserving the function of PrP^C [164,165]. A third method is to accelerate clearance of PrP^{Sc} from the brain [166]. A new approach is to target the downstream effects of prion replication, by inhibiting processes such as the UPR. This is surprisingly effective, as the underlying prion conversion and aggregation is unaffected, but extensive neuroprotection is still observed [119–121]. The effectiveness of UPR inhibitors again illustrates the complicated relationship between PrP^{Sc} and neurotoxicity; the elucidation of these underlying molecular mechanisms will undoubtedly improve our ability to treat these devastating diseases.

Conflicts of Interest: The authors declare no conflict of interest.

References

1. Rodriguez, J.A.; Jiang, L.; Eisenberg, D.S. Toward the atomic structure of PrP^{Sc}. *Cold Spring Harb. Perspect. Biol.* **2017**, *9*, a031336. [CrossRef] [PubMed]
2. Castle, A.R.; Gill, A.C. Physiological functions of the cellular prion protein. *Front. Mol. Biosci.* **2017**, *4*, 19. [CrossRef] [PubMed]
3. Stahl, N.; Borchelt, D.R.; Hsiao, K.; Prusiner, S.B. Scrapie prion protein contains a phosphatidylinositol glycolipid. *Cell* **1987**, *51*, 229–240. [CrossRef]
4. Peralta, O.A.; Huckle, W.R.; Eyestone, W.H. Developmental expression of the cellular prion protein (prp(c)) in bovine embryos. *Mol. Reprod. Dev.* **2012**, *79*, 488–498. [CrossRef] [PubMed]
5. Bamborough, P.; Wille, H.; Telling, G.C.; Yehiely, F.; Prusiner, S.B.; Cohen, F.E. Prion protein structure and scrapie replication: Theoretical, spectroscopic, and genetic investigations. *Cold Spring Harb. Symp. Quant. Biol.* **1996**, *61*, 495–509. [PubMed]
6. Wulf, M.A.; Senatore, A.; Aguzzi, A. The biological function of the cellular prion protein: An update. *BMC Biol.* **2017**, *15*, 34. [CrossRef] [PubMed]
7. Bremer, J.; Baumann, F.; Tiberi, C.; Wessig, C.; Fischer, H.; Schwarz, P.; Steele, A.D.; Toyka, K.V.; Nave, K.A.; Weis, J.; et al. Axonal prion protein is required for peripheral myelin maintenance. *Nat. Neurosci.* **2010**, *13*, 310–318. [CrossRef] [PubMed]
8. Brown, D.R.; Qin, K.; Herms, J.W.; Madlung, A.; Manson, J.; Strome, R.; Fraser, P.E.; Kruck, T.; von Bohlen, A.; Schulz-Schaeffer, W.; et al. The cellular prion protein binds copper in vivo. *Nature* **1997**, *390*, 684–687. [CrossRef] [PubMed]
9. Kuwahara, C.; Takeuchi, A.M.; Nishimura, T.; Haraguchi, K.; Kubosaki, A.; Matsumoto, Y.; Saeki, K.; Matsumoto, Y.; Yokoyama, T.; Itohara, S.; et al. Prions prevent neuronal cell-line death. *Nature* **1999**, *400*, 225–226. [CrossRef] [PubMed]
10. Bounhar, Y.; Zhang, Y.; Goodyer, C.G.; LeBlanc, A. Prion protein protects human neurons against bax-mediated apoptosis. *J. Biol. Chem.* **2001**, *276*, 39145–39149. [CrossRef] [PubMed]
11. Zeng, F.; Watt, N.T.; Walmsley, A.R.; Hooper, N.M. Tethering the n-terminus of the prion protein compromises the cellular response to oxidative stress. *J. Neurochem.* **2003**, *84*, 480–490. [CrossRef] [PubMed]
12. Mallucci, G.R.; Ratte, S.; Asante, E.A.; Linehan, J.; Gowland, I.; Jefferys, J.G.; Collinge, J. Post-natal knockout of prion protein alters hippocampal ca1 properties, but does not result in neurodegeneration. *EMBO J.* **2002**, *21*, 202–210. [CrossRef] [PubMed]
13. Nishida, N.; Katamine, S.; Shigematsu, K.; Nakatani, A.; Sakamoto, N.; Hasegawa, S.; Nakaoke, R.; Atarashi, R.; Kataoka, Y.; Miyamoto, T. Prion protein is necessary for latent learning and long-term memory retention. *Cell. Mol. Neurobiol.* **1997**, *17*, 537–545. [CrossRef] [PubMed]
14. Sanchez-Alavez, M.; Conti, B.; Moroncini, G.; Criado, J.R. Contributions of neuronal prion protein on sleep recovery and stress response following sleep deprivation. *Brain Res.* **2007**, *1158*, 71–80. [CrossRef] [PubMed]

15. Lopes, M.H.; Hajj, G.N.; Muras, A.G.; Mancini, G.L.; Castro, R.M.; Ribeiro, K.C.; Brentani, R.R.; Linden, R.; Martins, V.R. Interaction of cellular prion and stress-inducible protein 1 promotes neuritogenesis and neuroprotection by distinct signaling pathways. *J. Neurosci.* **2005**, *25*, 11330–11339. [CrossRef] [PubMed]

16. Loubet, D.; Dakowski, C.; Pietri, M.; Pradines, E.; Bernard, S.; Callebert, J.; Ardila-Osorio, H.; Mouillet-Richard, S.; Launay, J.M.; Kellermann, O.; et al. Neuritogenesis: The prion protein controls beta1 integrin signaling activity. *FASEB J.* **2012**, *26*, 678–690. [CrossRef] [PubMed]

17. Pan, K.M.; Baldwin, M.; Nguyen, J.; Gasset, M.; Serban, A.; Groth, D.; Mehlhorn, I.; Huang, Z.; Fletterick, R.J.; Cohen, F.E.; et al. Conversion of alpha-helices into beta-sheets features in the formation of the scrapie prion proteins. *Proc. Natl. Acad. Sci. USA* **1993**, *90*, 10962–10966. [CrossRef] [PubMed]

18. Silva, C.J.; Vazquez-Fernandez, E.; Onisko, B.; Requena, J.R. Proteinase k and the structure of PrP^{Sc}: The good, the bad and the ugly. *Virus Res.* **2015**, *207*, 120–126. [CrossRef] [PubMed]

19. Laurent, M. Prion diseases and the 'protein only' hypothesis: A theoretical dynamic study. *Biochem. J.* **1996**, *318*, 35–39. [CrossRef] [PubMed]

20. Castilla, J.; Saa, P.; Hetz, C.; Soto, C. In vitro generation of infectious scrapie prions. *Cell* **2005**, *121*, 195–206. [CrossRef] [PubMed]

21. Liao, Y.C.; Lebo, R.V.; Clawson, G.A.; Smuckler, E.A. Human prion protein cdna: Molecular cloning, chromosomal mapping, and biological implications. *Science* **1986**, *233*, 364–367. [CrossRef] [PubMed]

22. Kretzschmar, H.A.; Stowring, L.E.; Westaway, D.; Stubblebine, W.H.; Prusiner, S.B.; Dearmond, S.J. Molecular cloning of a human prion protein cdna. *DNA* **1986**, *5*, 315–324. [CrossRef] [PubMed]

23. Collinge, J. Prion diseases of humans and animals: Their causes and molecular basis. *Annu. Rev. Neurosci.* **2001**, *24*, 519–550. [CrossRef] [PubMed]

24. Prusiner, S.B. Molecular biology of prion diseases. *Science* **1991**, *252*, 1515–1522. [CrossRef] [PubMed]

25. Palmer, M.S.; Dryden, A.J.; Hughes, J.T.; Collinge, J. Homozygous prion protein genotype predisposes to sporadic creutzfeldt-jakob disease. *Nature* **1991**, *352*, 340–342. [CrossRef] [PubMed]

26. Hizume, M.; Kobayashi, A.; Teruya, K.; Ohashi, H.; Ironside, J.W.; Mohri, S.; Kitamoto, T. Human prion protein (prp) 219k is converted to PrP^{Sc} but shows heterozygous inhibition in variant creutzfeldt-jakob disease infection. *J. Biol. Chem.* **2009**, *284*, 3603–3609. [CrossRef] [PubMed]

27. Colby, D.W.; Prusiner, S.B. Prions. *Cold Spring Harb. Perspect. Biol.* **2011**, *3*, a006833. [CrossRef] [PubMed]

28. Kovacs, G.G.; Trabattoni, G.; Hainfellner, J.A.; Ironside, J.W.; Knight, R.S.; Budka, H. Mutations of the prion protein gene phenotypic spectrum. *J. Neurol.* **2002**, *249*, 1567–1582. [PubMed]

29. Imran, M.; Mahmood, S. An overview of human prion diseases. *Virol. J.* **2011**, *8*, 559. [CrossRef] [PubMed]

30. Gambetti, P.; Kong, Q.; Zou, W.; Parchi, P.; Chen, S.G. Sporadic and familial cjd: Classification and characterisation. *Br. Med. Bull.* **2003**, *66*, 213–239. [CrossRef] [PubMed]

31. Gajdusek, D.C.; Zigas, V. Degenerative disease of the central nervous system in new guinea: The endemic occurrence of kuru in the native population. *N. Engl. J. Med.* **1957**, *257*, 974–978. [CrossRef] [PubMed]

32. Gibbs, C.J., Jr.; Joy, A.; Heffner, R.; Franko, M.; Miyazaki, M.; Asher, D.M.; Parisi, J.E.; Brown, P.W.; Gajdusek, D.C. Clinical and pathological features and laboratory confirmation of creutzfeldt-jakob disease in a recipient of pituitary-derived human growth hormone. *N. Engl. J. Med.* **1985**, *313*, 734–738. [CrossRef] [PubMed]

33. Britton, T.C.; Al-Sarraj, S.; Shaw, C.; Campbell, T.; Collinge, J. Sporadic creutzfeldt-jakob disease in a 16-year-old in the UK. *Lancet* **1995**, *346*, 1155. [CrossRef]

34. Bateman, D.; Hilton, D.; Love, S.; Zeidler, M.; Beck, J.; Collinge, J. Sporadic creutzfeldt-jakob disease in a 18-year-old in the UK. *Lancet* **1995**, *346*, 1155–1156. [CrossRef]

35. Will, R.G.; Ironside, J.W.; Zeidler, M.; Cousens, S.N.; Estibeiro, K.; Alperovitch, A.; Poser, S.; Pocchiari, M.; Hofman, A.; Smith, P.G. A new variant of creutzfeldt-jakob disease in the uk. *Lancet* **1996**, *347*, 921–925. [CrossRef]

36. Collinge, J.; Sidle, K.C.; Meads, J.; Ironside, J.; Hill, A.F. Molecular analysis of prion strain variation and the aetiology of 'new variant' cjd. *Nature* **1996**, *383*, 685–690. [CrossRef] [PubMed]

37. Hill, A.F.; Desbruslais, M.; Joiner, S.; Sidle, K.C.; Gowland, I.; Collinge, J.; Doey, L.J.; Lantos, P. The same prion strain causes vcjd and bse. *Nature* **1997**, *389*, 448–450. [CrossRef] [PubMed]

38. Nathanson, N.; Wilesmith, J.; Griot, C. Bovine spongiform encephalopathy (bse): Causes and consequences of a common source epidemic. *Am. J. Epidemiol.* **1997**, *145*, 959–969. [CrossRef] [PubMed]

39. Miller, M.W.; Williams, E.S.; McCarty, C.W.; Spraker, T.R.; Kreeger, T.J.; Larsen, C.T.; Thorne, E.T. Epizootiology of chronic wasting disease in free-ranging cervids in colorado and wyoming. *J. Wildl. Dis.* **2000**, *36*, 676–690. [CrossRef] [PubMed]

40. Haley, N.J.; Hoover, E.A. Chronic wasting disease of cervids: Current knowledge and future perspectives. *Annu. Rev. Anim. Biosci.* **2015**, *3*, 305–325. [CrossRef] [PubMed]

41. Benestad, S.L.; Mitchell, G.; Simmons, M.; Ytrehus, B.; Vikoren, T. First case of chronic wasting disease in europe in a norwegian free-ranging reindeer. *Vet. Res.* **2016**, *47*, 88. [CrossRef] [PubMed]

42. Jackson, W.S. Selective vulnerability to neurodegenerative disease: The curious case of prion protein. *Dis. Model. Mech.* **2014**, *7*, 21–29. [CrossRef] [PubMed]

43. Collins, S.; McLean, C.A.; Masters, C.L. Gerstmann-straussler-scheinker syndrome, fatal familial insomnia, and kuru: A review of these less common human transmissible spongiform encephalopathies. *J. Clin. Neurosci.* **2001**, *8*, 387–397. [CrossRef] [PubMed]

44. Montagna, P.; Gambetti, P.; Cortelli, P.; Lugaresi, E. Familial and sporadic fatal insomnia. *Lancet Neurol.* **2003**, *2*, 167–176. [CrossRef]

45. Bueler, H.; Fischer, M.; Lang, Y.; Bluethmann, H.; Lipp, H.P.; DeArmond, S.J.; Prusiner, S.B.; Aguet, M.; Weissmann, C. Normal development and behaviour of mice lacking the neuronal cell-surface prp protein. *Nature* **1992**, *356*, 577–582. [CrossRef] [PubMed]

46. Manson, J.C.; Clarke, A.R.; Hooper, M.L.; Aitchison, L.; McConnell, I.; Hope, J. 129/ola mice carrying a null mutation in prp that abolishes mrna production are developmentally normal. *Mol. Neurobiol.* **1994**, *8*, 121–127. [CrossRef] [PubMed]

47. Yu, G.; Chen, J.; Yu, H.; Liu, S.; Chen, J.; Xu, X.; Sha, H.; Zhang, X.; Wu, G.; Xu, S.; et al. Functional disruption of the prion protein gene in cloned goats. *J. Gen. Virol.* **2006**, *87*, 1019–1027. [CrossRef] [PubMed]

48. Richt, J.A.; Kasinathan, P.; Hamir, A.N.; Castilla, J.; Sathiyaseelan, T.; Vargas, F.; Sathiyaseelan, J.; Wu, H.; Matsushita, H.; Koster, J.; et al. Production of cattle lacking prion protein. *Nat. Biotechnol.* **2007**, *25*, 132–138. [CrossRef] [PubMed]

49. Steele, A.D.; Lindquist, S.; Aguzzi, A. The prion protein knockout mouse: A phenotype under challenge. *Prion* **2007**, *1*, 83–93. [CrossRef] [PubMed]

50. McLennan, N.F.; Brennan, P.M.; McNeill, A.; Davies, I.; Fotheringham, A.; Rennison, K.A.; Ritchie, D.; Brannan, F.; Head, M.W.; Ironside, J.W.; et al. Prion protein accumulation and neuroprotection in hypoxic brain damage. *Am. J. Pathol.* **2004**, *165*, 227–235. [CrossRef]

51. Sakaguchi, S.; Katamine, S.; Nishida, N.; Moriuchi, R.; Shigematsu, K.; Sugimoto, T.; Nakatani, A.; Kataoka, Y.; Houtani, T.; Shirabe, S.; et al. Loss of cerebellar purkinje cells in aged mice homozygous for a disrupted prp gene. *Nature* **1996**, *380*, 528–531. [CrossRef] [PubMed]

52. Moore, R.C.; Lee, I.Y.; Silverman, G.L.; Harrison, P.M.; Strome, R.; Heinrich, C.; Karunaratne, A.; Pasternak, S.H.; Chishti, M.A.; Liang, Y.; et al. Ataxia in prion protein (prp)-deficient mice is associated with upregulation of the novel prp-like protein doppel. *J. Mol. Biol.* **1999**, *292*, 797–817. [CrossRef] [PubMed]

53. Katamine, S.; Nishida, N.; Sugimoto, T.; Noda, T.; Sakaguchi, S.; Shigematsu, K.; Kataoka, Y.; Nakatani, A.; Hasegawa, S.; Moriuchi, R.; et al. Impaired motor coordination in mice lacking prion protein. *Cell. Mol. Neurobiol.* **1998**, *18*, 731–742. [CrossRef] [PubMed]

54. Yamaguchi, N.; Sakaguchi, S.; Shigematsu, K.; Okimura, N.; Katamine, S. Doppel-induced purkinje cell death is stoichiometrically abrogated by prion protein. *Biochem. Biophys. Res. Commun.* **2004**, *319*, 1247–1252. [CrossRef] [PubMed]

55. Peoc'h, K.; Volland, H.; De Gassart, A.; Beaudry, P.; Sazdovitch, V.; Sorgato, M.C.; Creminon, C.; Laplanche, J.L.; Lehmann, S. Prion-like protein doppel expression is not modified in scrapie-infected cells and in the brains of patients with creutzfeldt-jakob disease. *FEBS Lett.* **2003**, *536*, 61–65. [CrossRef]

56. Bueler, H.; Aguzzi, A.; Sailer, A.; Greiner, R.A.; Autenried, P.; Aguet, M.; Weissmann, C. Mice devoid of prp are resistant to scrapie. *Cell* **1993**, *73*, 1339–1347. [CrossRef]

57. Brandner, S.; Isenmann, S.; Raeber, A.; Fischer, M.; Sailer, A.; Kobayashi, Y.; Marino, S.; Weissmann, C.; Aguzzi, A. Normal host prion protein necessary for scrapie-induced neurotoxicity. *Nature* **1996**, *379*, 339–343. [CrossRef] [PubMed]

58. Mallucci, G.; Dickinson, A.; Linehan, J.; Klohn, P.C.; Brandner, S.; Collinge, J. Depleting neuronal prp in prion infection prevents disease and reverses spongiosis. *Science* **2003**, *302*, 871–874. [CrossRef] [PubMed]

59. Chesebro, B.; Trifilo, M.; Race, R.; Meade-White, K.; Teng, C.; LaCasse, R.; Raymond, L.; Favara, C.; Baron, G.; Priola, S.; et al. Anchorless prion protein results in infectious amyloid disease without clinical scrapie. *Science* **2005**, *308*, 1435–1439. [CrossRef] [PubMed]

60. Hill, A.F.; Collinge, J. Subclinical prion infection in humans and animals. *Br. Med. Bull.* **2003**, *66*, 161–170. [CrossRef] [PubMed]

61. Lasmezas, C.I.; Deslys, J.P.; Robain, O.; Jaegly, A.; Beringue, V.; Peyrin, J.M.; Fournier, J.G.; Hauw, J.J.; Rossier, J.; Dormont, D. Transmission of the bse agent to mice in the absence of detectable abnormal prion protein. *Science* **1997**, *275*, 402–405. [CrossRef] [PubMed]

62. Manuelidis, L.; Fritch, W.; Xi, Y.G. Evolution of a strain of cjd that induces bse-like plaques. *Science* **1997**, *277*, 94–98. [CrossRef] [PubMed]

63. Nitrini, R.; Rosemberg, S.; Passos-Bueno, M.R.; da Silva, L.S.; Iughetti, P.; Papadopoulos, M.; Carrilho, P.M.; Caramelli, P.; Albrecht, S.; Zatz, M.; et al. Familial spongiform encephalopathy associated with a novel prion protein gene mutation. *Ann. Neurol.* **1997**, *42*, 138–146. [CrossRef] [PubMed]

64. Collins, S.; Boyd, A.; Fletcher, A.; Byron, K.; Harper, C.; McLean, C.A.; Masters, C.L. Novel prion protein gene mutation in an octogenarian with creutzfeldt-jakob disease. *Arch. Neurol.* **2000**, *57*, 1058–1063. [CrossRef] [PubMed]

65. Grasbon-Frodl, E.; Lorenz, H.; Mann, U.; Nitsch, R.M.; Windl, O.; Kretzschmar, H.A. Loss of glycosylation associated with the t183a mutation in human prion disease. *Acta Neuropathol.* **2004**, *108*, 476–484. [CrossRef] [PubMed]

66. Hegde, R.S.; Mastrianni, J.A.; Scott, M.R.; DeFea, K.A.; Tremblay, P.; Torchia, M.; DeArmond, S.J.; Prusiner, S.B.; Lingappa, V.R. A transmembrane form of the prion protein in neurodegenerative disease. *Science* **1998**, *279*, 827–834. [CrossRef] [PubMed]

67. Hill, A.F.; Joiner, S.; Linehan, J.; Desbruslais, M.; Lantos, P.L.; Collinge, J. Species-barrier-independent prion replication in apparently resistant species. *Proc. Natl. Acad. Sci. USA* **2000**, *97*, 10248–10253. [CrossRef] [PubMed]

68. Race, R.; Raines, A.; Raymond, G.J.; Caughey, B.; Chesebro, B. Long-term subclinical carrier state precedes scrapie replication and adaptation in a resistant species: Analogies to bovine spongiform encephalopathy and variant creutzfeldt-jakob disease in humans. *J. Virol.* **2001**, *75*, 10106–10112. [CrossRef] [PubMed]

69. Hill, A.F.; Collinge, J. Subclinical prion infection. *Trends Microbiol.* **2003**, *11*, 578–584. [CrossRef] [PubMed]

70. Bolton, D.C.; Bendheim, P.E. Purification of scrapie agents: How far have we come? *Curr. Top. Microbiol. Immunol.* **1991**, *172*, 39–55. [PubMed]

71. Silveira, J.R.; Raymond, G.J.; Hughson, A.G.; Race, R.E.; Sim, V.L.; Hayes, S.F.; Caughey, B. The most infectious prion protein particles. *Nature* **2005**, *437*, 257–261. [CrossRef] [PubMed]

72. Tzaban, S.; Friedlander, G.; Schonberger, O.; Horonchik, L.; Yedidia, Y.; Shaked, G.; Gabizon, R.; Taraboulos, A. Protease-sensitive scrapie prion protein in aggregates of heterogeneous sizes. *Biochemistry* **2002**, *41*, 12868–12875. [PubMed]

73. Pastrana, M.A.; Sajnani, G.; Onisko, B.; Castilla, J.; Morales, R.; Soto, C.; Requena, J.R. Isolation and characterization of a proteinase k-sensitive PrP^{Sc} fraction. *Biochemistry* **2006**, *45*, 15710–15717. [CrossRef] [PubMed]

74. Sajnani, G.; Silva, C.J.; Ramos, A.; Pastrana, M.A.; Onisko, B.C.; Erickson, M.L.; Antaki, E.M.; Dynin, I.; Vazquez-Fernandez, E.; Sigurdson, C.J.; et al. Pk-sensitive prp is infectious and shares basic structural features with pk-resistant prp. *PLoS Pathog.* **2012**, *8*, e1002547. [CrossRef] [PubMed]

75. Bueler, H.; Raeber, A.; Sailer, A.; Fischer, M.; Aguzzi, A.; Weissmann, C. High prion and PrP^{Sc} levels but delayed onset of disease in scrapie-inoculated mice heterozygous for a disrupted prp gene. *Mol. Med.* **1994**, *1*, 19–30. [PubMed]

76. Biasini, E.; Seegulam, M.E.; Patti, B.N.; Solforosi, L.; Medrano, A.Z.; Christensen, H.M.; Senatore, A.; Chiesa, R.; Williamson, R.A.; Harris, D.A. Non-infectious aggregates of the prion protein react with several PrP^{Sc}-directed antibodies. *J. Neurochem.* **2008**, *105*, 2190–2204. [CrossRef] [PubMed]

77. Riesner, D.; Kellings, K.; Post, K.; Wille, H.; Serban, H.; Groth, D.; Baldwin, M.A.; Prusiner, S.B. Disruption of prion rods generates 10-nm spherical particles having high alpha-helical content and lacking scrapie infectivity. *J. Virol.* **1996**, *70*, 1714–1722. [PubMed]

78. Morales, R.; Abid, K.; Soto, C. The prion strain phenomenon: Molecular basis and unprecedented features. *Biochim. Biophys. Acta* **2007**, *1772*, 681–691. [CrossRef] [PubMed]

79. Safar, J.; Wille, H.; Itri, V.; Groth, D.; Serban, H.; Torchia, M.; Cohen, F.E.; Prusiner, S.B. Eight prion strains have prp(sc) molecules with different conformations. *Nat. Med.* **1998**, *4*, 1157–1165. [CrossRef] [PubMed]

80. Meyer, R.K.; McKinley, M.P.; Bowman, K.A.; Braunfeld, M.B.; Barry, R.A.; Prusiner, S.B. Separation and properties of cellular and scrapie prion proteins. *Proc. Natl. Acad. Sci. USA* **1986**, *83*, 2310–2314. [CrossRef] [PubMed]

81. Korth, C.; Stierli, B.; Streit, P.; Moser, M.; Schaller, O.; Fischer, R.; Schulz-Schaeffer, W.; Kretzschmar, H.; Raeber, A.; Braun, U.; et al. Prion (PrPSc)-specific epitope defined by a monoclonal antibody. *Nature* **1997**, *390*, 74–77. [CrossRef] [PubMed]

82. Eisenberg, D.; Jucker, M. The amyloid state of proteins in human diseases. *Cell* **2012**, *148*, 1188–1203. [CrossRef] [PubMed]

83. Sandberg, M.K.; Al-Doujaily, H.; Sharps, B.; Clarke, A.R.; Collinge, J. Prion propagation and toxicity in vivo occur in two distinct mechanistic phases. *Nature* **2011**, *470*, 540–542. [CrossRef] [PubMed]

84. Singh, J.; Udgaonkar, J.B. Molecular mechanism of the misfolding and oligomerization of the prion protein: Current understanding and its implications. *Biochemistry* **2015**, *54*, 4431–4442. [CrossRef] [PubMed]

85. Kazlauskaite, J.; Young, A.; Gardner, C.E.; Macpherson, J.V.; Venien-Bryan, C.; Pinheiro, T.J. An unusual soluble beta-turn-rich conformation of prion is involved in fibril formation and toxic to neuronal cells. *Biochem. Biophys. Res. Commun.* **2005**, *328*, 292–305. [CrossRef] [PubMed]

86. Novitskaya, V.; Bocharova, O.V.; Bronstein, I.; Baskakov, I.V. Amyloid fibrils of mammalian prion protein are highly toxic to cultured cells and primary neurons. *J. Biol. Chem.* **2006**, *281*, 13828–13836. [CrossRef] [PubMed]

87. Simoneau, S.; Rezaei, H.; Sales, N.; Kaiser-Schulz, G.; Lefebvre-Roque, M.; Vidal, C.; Fournier, J.G.; Comte, J.; Wopfner, F.; Grosclaude, J.; et al. In vitro and in vivo neurotoxicity of prion protein oligomers. *PLoS Pathog.* **2007**, *3*, e125. [CrossRef] [PubMed]

88. Ugalde, C.L.; Finkelstein, D.I.; Lawson, V.A.; Hill, A.F. Pathogenic mechanisms of prion protein, amyloid-beta and alpha-synuclein misfolding: The prion concept and neurotoxicity of protein oligomers. *J. Neurochem.* **2016**, *139*, 162–180. [CrossRef] [PubMed]

89. Resenberger, U.K.; Harmeier, A.; Woerner, A.C.; Goodman, J.L.; Muller, V.; Krishnan, R.; Vabulas, R.M.; Kretzschmar, H.A.; Lindquist, S.; Hartl, F.U.; et al. The cellular prion protein mediates neurotoxic signalling of beta-sheet-rich conformers independent of prion replication. *EMBO J.* **2011**, *30*, 2057–2070. [CrossRef] [PubMed]

90. Boellaard, J.W.; Schlote, W.; Tateishi, J. Neuronal autophagy in experimental creutzfeldt-jakob's disease. *Acta Neuropathol.* **1989**, *78*, 410–418. [CrossRef] [PubMed]

91. Boellaard, J.W.; Kao, M.; Schlote, W.; Diringer, H. Neuronal autophagy in experimental scrapie. *Acta Neuropathol.* **1991**, *82*, 225–228. [CrossRef] [PubMed]

92. Liberski, P.P.; Yanagihara, R.; Gibbs, C.J., Jr.; Gajdusek, D.C. Neuronal autophagic vacuoles in experimental scrapie and creutzfeldt-jakob disease. *Acta Neuropathol.* **1992**, *83*, 134–139. [CrossRef] [PubMed]

93. Sikorska, B.; Liberski, P.P.; Giraud, P.; Kopp, N.; Brown, P. Autophagy is a part of ultrastructural synaptic pathology in creutzfeldt-jakob disease: A brain biopsy study. *Int. J. Biochem. Cell Biol.* **2004**, *36*, 2563–2573. [CrossRef] [PubMed]

94. Xu, Y.; Tian, C.; Wang, S.B.; Xie, W.L.; Guo, Y.; Zhang, J.; Shi, Q.; Chen, C.; Dong, X.P. Activation of the macroautophagic system in scrapie-infected experimental animals and human genetic prion diseases. *Autophagy* **2012**, *8*, 1604–1620. [CrossRef] [PubMed]

95. Heiseke, A.; Aguib, Y.; Riemer, C.; Baier, M.; Schatzl, H.M. Lithium induces clearance of protease resistant prion protein in prion-infected cells by induction of autophagy. *J. Neurochem.* **2009**, *109*, 25–34. [CrossRef] [PubMed]

96. Karapetyan, Y.E.; Sferrazza, G.F.; Zhou, M.; Ottenberg, G.; Spicer, T.; Chase, P.; Fallahi, M.; Hodder, P.; Weissmann, C.; Lasmezas, C.I. Unique drug screening approach for prion diseases identifies tacrolimus and astemizole as antiprion agents. *Proc. Natl. Acad. Sci. USA* **2013**, *110*, 7044–7049. [CrossRef] [PubMed]

97. Nakagaki, T.; Satoh, K.; Ishibashi, D.; Fuse, T.; Sano, K.; Kamatari, Y.O.; Kuwata, K.; Shigematsu, K.; Iwamaru, Y.; Takenouchi, T.; et al. Fk506 reduces abnormal prion protein through the activation of autolysosomal degradation and prolongs survival in prion-infected mice. *Autophagy* **2013**, *9*, 1386–1394. [CrossRef] [PubMed]

98. Cortes, C.J.; Qin, K.; Cook, J.; Solanki, A.; Mastrianni, J.A. Rapamycin delays disease onset and prevents prp plaque deposition in a mouse model of gerstmann-straussler-scheinker disease. *J. Neurosci.* **2012**, *32*, 12396–12405. [CrossRef] [PubMed]

99. Elmore, S. Apoptosis: A review of programmed cell death. *Toxicol. Pathol.* **2007**, *35*, 495–516. [CrossRef] [PubMed]

100. Giese, A.; Groschup, M.H.; Hess, B.; Kretzschmar, H.A. Neuronal cell death in scrapie-infected mice is due to apoptosis. *Brain Pathol.* **1995**, *5*, 213–221. [CrossRef] [PubMed]

101. Jesionek-Kupnicka, D.; Buczynski, J.; Kordek, R.; Liberski, P.P. Neuronal loss and apoptosis in experimental creutzfeldt-jakob disease in mice. *Folia Neuropathol.* **1999**, *37*, 283–286. [PubMed]

102. Gray, F.; Chretien, F.; Adle-Biassette, H.; Dorandeu, A.; Ereau, T.; Delisle, M.B.; Kopp, N.; Ironside, J.W.; Vital, C. Neuronal apoptosis in creutzfeldt-jakob disease. *J. Neuropathol. Exp. Neurol.* **1999**, *58*, 321–328. [CrossRef] [PubMed]

103. Kovacs, G.G.; Budka, H. Distribution of apoptosis-related proteins in sporadic creutzfeldt-jakob disease. *Brain Res.* **2010**, *1323*, 192–199. [CrossRef] [PubMed]

104. Drew, S.C.; Haigh, C.L.; Klemm, H.M.; Masters, C.L.; Collins, S.J.; Barnham, K.J.; Lawson, V.A. Optical imaging detects apoptosis in the brain and peripheral organs of prion-infected mice. *J. Neuropathol. Exp. Neurol.* **2011**, *70*, 143–150. [CrossRef] [PubMed]

105. Coulpier, M.; Messiaen, S.; Hamel, R.; de Marco, M.F.; Lilin, T.; Eloit, M. Bax deletion does not protect neurons from bse-induced death. *Neurobiol. Dis.* **2006**, *23*, 603–611. [CrossRef] [PubMed]

106. Steele, A.D.; King, O.D.; Jackson, W.S.; Hetz, C.A.; Borkowski, A.W.; Thielen, P.; Wollmann, R.; Lindquist, S. Diminishing apoptosis by deletion of bax or overexpression of bcl-2 does not protect against infectious prion toxicity in vivo. *J. Neurosci.* **2007**, *27*, 13022–13027. [CrossRef] [PubMed]

107. Steele, A.D.; Hetz, C.; Yi, C.H.; Jackson, W.S.; Borkowski, A.W.; Yuan, J.; Wollmann, R.H.; Lindquist, S. Prion pathogenesis is independent of caspase-12. *Prion* **2007**, *1*, 243–247. [CrossRef] [PubMed]

108. Glickman, M.H.; Ciechanover, A. The ubiquitin-proteasome proteolytic pathway: Destruction for the sake of construction. *Physiol. Rev.* **2002**, *82*, 373–428. [CrossRef] [PubMed]

109. McKinnon, C.; Goold, R.; Andre, R.; Devoy, A.; Ortega, Z.; Moonga, J.; Linehan, J.M.; Brandner, S.; Lucas, J.J.; Collinge, J.; et al. Prion-mediated neurodegeneration is associated with early impairment of the ubiquitin-proteasome system. *Acta Neuropathol.* **2016**, *131*, 411–425. [CrossRef] [PubMed]

110. Andre, R.; Tabrizi, S.J. Misfolded prp and a novel mechanism of proteasome inhibition. *Prion* **2012**, *6*, 32–36. [CrossRef] [PubMed]

111. Deriziotis, P.; Andre, R.; Smith, D.M.; Goold, R.; Kinghorn, K.J.; Kristiansen, M.; Nathan, J.A.; Rosenzweig, R.; Krutauz, D.; Glickman, M.H.; et al. Misfolded prp impairs the ups by interaction with the 20s proteasome and inhibition of substrate entry. *EMBO J.* **2011**, *30*, 3065–3077. [CrossRef] [PubMed]

112. Kristiansen, M.; Deriziotis, P.; Dimcheff, D.E.; Jackson, G.S.; Ovaa, H.; Naumann, H.; Clarke, A.R.; van Leeuwen, F.W.; Menendez-Benito, V.; Dantuma, N.P.; et al. Disease-associated prion protein oligomers inhibit the 26s proteasome. *Mol. Cell* **2007**, *26*, 175–188. [CrossRef] [PubMed]

113. Lin, Z.; Zhao, D.; Yang, L. Interaction between misfolded prp and the ubiquitin-proteasome system in prion-mediated neurodegeneration. *Acta Biochim. Biophys. Sin.* **2013**, *45*, 477–484. [CrossRef] [PubMed]

114. Dron, M.; Dandoy-Dron, F.; Farooq Salamat, M.K.; Laude, H. Proteasome inhibitors promote the sequestration of PrP^{Sc} into aggresomes within the cytosol of prion-infected cad neuronal cells. *J. Gen. Virol.* **2009**, *90*, 2050–2060. [CrossRef] [PubMed]

115. Homma, T.; Ishibashi, D.; Nakagaki, T.; Satoh, K.; Sano, K.; Atarashi, R.; Nishida, N. Increased expression of p62/sqstm1 in prion diseases and its association with pathogenic prion protein. *Sci. Rep.* **2014**, *4*, 4504. [CrossRef] [PubMed]

116. Hetz, C.; Russelakis-Carneiro, M.; Maundrell, K.; Castilla, J.; Soto, C. Caspase-12 and endoplasmic reticulum stress mediate neurotoxicity of pathological prion protein. *EMBO J.* **2003**, *22*, 5435–5445. [CrossRef] [PubMed]

117. Torres, M.; Matamala, J.M.; Duran-Aniotz, C.; Cornejo, V.H.; Foley, A.; Hetz, C. Er stress signaling and neurodegeneration: At the intersection between alzheimer's disease and prion-related disorders. *Virus Res.* **2015**, *207*, 69–75. [CrossRef] [PubMed]

118. Halliday, M.; Radford, H.; Mallucci, G.R. Prions: Generation and spread versus neurotoxicity. *J. Biol. Chem.* **2014**, *289*, 19862–19868. [CrossRef] [PubMed]

119. Halliday, M.; Radford, H.; Sekine, Y.; Moreno, J.; Verity, N.; le Quesne, J.; Ortori, C.A.; Barrett, D.A.; Fromont, C.; Fischer, P.M.; et al. Partial restoration of protein synthesis rates by the small molecule isrib prevents neurodegeneration without pancreatic toxicity. *Cell Death Dis.* **2015**, *6*, e1672. [CrossRef] [PubMed]

120. Halliday, M.; Radford, H.; Zents, K.A.M.; Molloy, C.; Moreno, J.A.; Verity, N.C.; Smith, E.; Ortori, C.A.; Barrett, D.A.; Bushell, M.; et al. Repurposed drugs targeting eif2alpha-p-mediated translational repression prevent neurodegeneration in mice. *Brain* **2017**, *140*, 1768–1783. [CrossRef] [PubMed]

121. Moreno, J.A.; Halliday, M.; Molloy, C.; Radford, H.; Verity, N.; Axten, J.M.; Ortori, C.A.; Willis, A.E.; Fischer, P.M.; Barrett, D.A.; et al. Oral treatment targeting the unfolded protein response prevents neurodegeneration and clinical disease in prion-infected mice. *Sci. Transl. Med.* **2013**, *5*, 206ra138. [CrossRef] [PubMed]

122. Moreno, J.A.; Radford, H.; Peretti, D.; Steinert, J.R.; Verity, N.; Martin, M.G.; Halliday, M.; Morgan, J.; Dinsdale, D.; Ortori, C.A.; et al. Sustained translational repression by eif2alpha-p mediates prion neurodegeneration. *Nature* **2012**, *485*, 507–511. [PubMed]

123. Hetz, C. The unfolded protein response: Controlling cell fate decisions under er stress and beyond. *Nat. Rev. Mol. Cell. Biol.* **2012**, *13*, 89–102. [CrossRef] [PubMed]

124. Schroder, M. The unfolded protein response. *Mol. Biotechnol.* **2006**, *34*, 279–290. [CrossRef]

125. Schroder, M.; Kaufman, R.J. The mammalian unfolded protein response. *Annu. Rev. Biochem.* **2005**, *74*, 739–789. [CrossRef] [PubMed]

126. Walter, P.; Ron, D. The unfolded protein response: From stress pathway to homeostatic regulation. *Science* **2011**, *334*, 1081–1086. [CrossRef] [PubMed]

127. Novoa, I.; Zeng, H.; Harding, H.P.; Ron, D. Feedback inhibition of the unfolded protein response by gadd34-mediated dephosphorylation of eif2alpha. *J. Cell Biol.* **2001**, *153*, 1011–1022. [CrossRef] [PubMed]

128. Axten, J.M.; Medina, J.R.; Feng, Y.; Shu, A.; Romeril, S.P.; Grant, S.W.; Li, W.H.; Heerding, D.A.; Minthorn, E.; Mencken, T.; et al. Discovery of 7-methyl-5-(1-{[3-(trifluoromethyl)phenyl]acetyl}-2,3-dihydro-1h-indol-5-yl)-7h-p yrrolo[2,3-d]pyrimidin-4-amine (gsk2606414), a potent and selective first-in-class inhibitor of protein kinase r (pkr)-like endoplasmic reticulum kinase (perk). *J. Med. Chem.* **2012**, *55*, 7193–7207. [CrossRef] [PubMed]

129. Sekine, Y.; Zyryanova, A.; Crespillo-Casado, A.; Fischer, P.M.; Harding, H.P.; Ron, D. Stress responses. Mutations in a translation initiation factor identify the target of a memory-enhancing compound. *Science* **2015**, *348*, 1027–1030. [CrossRef] [PubMed]

130. Doronina, V.A.; Staniforth, G.L.; Speldewinde, S.H.; Tuite, M.F.; Grant, C.M. Oxidative stress conditions increase the frequency of de novo formation of the yeast [psi+] prion. *Mol. Microbiol.* **2015**, *96*, 163–174. [CrossRef] [PubMed]

131. Redecke, L.; Binder, S.; Elmallah, M.I.; Broadbent, R.; Tilkorn, C.; Schulz, B.; May, P.; Goos, A.; Eich, A.; Rubhausen, M.; et al. Uv-light-induced conversion and aggregation of prion proteins. *Free Radic. Biol. Med.* **2009**, *46*, 1353–1361. [CrossRef] [PubMed]

132. Yuan, F.; Yang, L.; Zhang, Z.; Wu, W.; Zhou, X.; Yin, X.; Zhao, D. Cellular prion protein (PrP^C) of the neuron cell transformed to a pk-resistant protein under oxidative stress, comprising main mitochondrial damage in prion diseases. *J. Mol. Neurosci.* **2013**, *51*, 219–224. [CrossRef] [PubMed]

133. Brown, D.R.; Besinger, A. Prion protein expression and superoxide dismutase activity. *Biochem. J.* **1998**, *334 Pt 2*, 423–429. [CrossRef] [PubMed]

134. Sakudo, A.; Lee, D.C.; Li, S.; Nakamura, T.; Matsumoto, Y.; Saeki, K.; Itohara, S.; Ikuta, K.; Onodera, T. Prp cooperates with sti1 to regulate sod activity in prp-deficient neuronal cell line. *Biochem. Biophys. Res. Commun.* **2005**, *328*, 14–19. [CrossRef] [PubMed]

135. Choi, Y.G.; Kim, J.I.; Lee, H.P.; Jin, J.K.; Choi, E.K.; Carp, R.I.; Kim, Y.S. Induction of heme oxygenase-1 in the brains of scrapie-infected mice. *Neurosci. Lett.* **2000**, *289*, 173–176. [CrossRef]

136. Choi, S.I.; Ju, W.K.; Choi, E.K.; Kim, J.; Lea, H.Z.; Carp, R.I.; Wisniewski, H.M.; Kim, Y.S. Mitochondrial dysfunction induced by oxidative stress in the brains of hamsters infected with the 263 k scrapie agent. *Acta Neuropathol.* **1998**, *96*, 279–286. [CrossRef] [PubMed]

137. Lee, D.W.; Sohn, H.O.; Lim, H.B.; Lee, Y.G.; Kim, Y.S.; Carp, R.I.; Wisniewski, H.M. Alteration of free radical metabolism in the brain of mice infected with scrapie agent. *Free Radic. Res.* **1999**, *30*, 499–507. [CrossRef] [PubMed]

138. Belichenko, P.V.; Brown, D.; Jeffrey, M.; Fraser, J.R. Dendritic and synaptic alterations of hippocampal pyramidal neurones in scrapie-infected mice. *Neuropathol. Appl. Neurobiol.* **2000**, *26*, 143–149. [CrossRef] [PubMed]

139. Jeffrey, M.; Halliday, W.G.; Bell, J.; Johnston, A.R.; MacLeod, N.K.; Ingham, C.; Sayers, A.R.; Brown, D.A.; Fraser, J.R. Synapse loss associated with abnormal prp precedes neuronal degeneration in the scrapie-infected murine hippocampus. *Neuropathol. Appl. Neurobiol.* **2000**, *26*, 41–54. [CrossRef] [PubMed]

140. Brown, D.; Belichenko, P.; Sales, J.; Jeffrey, M.; Fraser, J.R. Early loss of dendritic spines in murine scrapie revealed by confocal analysis. *Neuroreport* **2001**, *12*, 179–183. [CrossRef] [PubMed]

141. Gray, B.C.; Siskova, Z.; Perry, V.H.; O'Connor, V. Selective presynaptic degeneration in the synaptopathy associated with me7-induced hippocampal pathology. *Neurobiol. Dis.* **2009**, *35*, 63–74. [CrossRef] [PubMed]

142. Mallucci, G.R.; White, M.D.; Farmer, M.; Dickinson, A.; Khatun, H.; Powell, A.D.; Brandner, S.; Jefferys, J.G.; Collinge, J. Targeting cellular prion protein reverses early cognitive deficits and neurophysiological dysfunction in prion-infected mice. *Neuron* **2007**, *53*, 325–335. [CrossRef] [PubMed]

143. Cunningham, C.; Deacon, R.; Wells, H.; Boche, D.; Waters, S.; Diniz, C.P.; Scott, H.; Rawlins, J.N.; Perry, V.H. Synaptic changes characterize early behavioural signs in the me7 model of murine prion disease. *Eur. J. Neurosci.* **2003**, *17*, 2147–2155. [CrossRef] [PubMed]

144. Chiti, Z.; Knutsen, O.M.; Betmouni, S.; Greene, J.R. An integrated, temporal study of the behavioural, electrophysiological and neuropathological consequences of murine prion disease. *Neurobiol. Dis.* **2006**, *22*, 363–373. [CrossRef] [PubMed]

145. Clinton, J.; Forsyth, C.; Royston, M.C.; Roberts, G.W. Synaptic degeneration is the primary neuropathological feature in prion disease: A preliminary study. *Neuroreport* **1993**, *4*, 65–68. [CrossRef] [PubMed]

146. Sassoon, J.; Daniels, M.; Brown, D.R. Astrocytic regulation of nmda receptor subunit composition modulates the toxicity of prion peptide prp106-126. *Mol. Cell. Neurosci.* **2004**, *25*, 181–191. [CrossRef] [PubMed]

147. Ratte, S.; Prescott, S.A.; Collinge, J.; Jefferys, J.G. Hippocampal bursts caused by changes in nmda receptor-dependent excitation in a mouse model of variant cjd. *Neurobiol. Dis.* **2008**, *32*, 96–104. [CrossRef] [PubMed]

148. Khosravani, H.; Zhang, Y.; Tsutsui, S.; Hameed, S.; Altier, C.; Hamid, J.; Chen, L.; Villemaire, M.; Ali, Z.; Jirik, F.R.; et al. Prion protein attenuates excitotoxicity by inhibiting nmda receptors. *J. Cell Biol.* **2008**, *181*, 551–565. [CrossRef] [PubMed]

149. Lashuel, H.A.; Hartley, D.; Petre, B.M.; Walz, T.; Lansbury, P.T., Jr. Neurodegenerative disease: Amyloid pores from pathogenic mutations. *Nature* **2002**, *418*, 291. [CrossRef] [PubMed]

150. Quist, A.; Doudevski, I.; Lin, H.; Azimova, R.; Ng, D.; Frangione, B.; Kagan, B.; Ghiso, J.; Lal, R. Amyloid ion channels: A common structural link for protein-misfolding disease. *Proc. Natl. Acad. Sci. USA* **2005**, *102*, 10427–10432. [CrossRef] [PubMed]

151. Baker, C.A.; Martin, D.; Manuelidis, L. Microglia from creutzfeldt-jakob disease-infected brains are infectious and show specific mrna activation profiles. *J. Virol.* **2002**, *76*, 10905–10913. [CrossRef] [PubMed]

152. Giese, A.; Brown, D.R.; Groschup, M.H.; Feldmann, C.; Haist, I.; Kretzschmar, H.A. Role of microglia in neuronal cell death in prion disease. *Brain Pathol.* **1998**, *8*, 449–457. [CrossRef] [PubMed]

153. Brown, D.R.; Schmidt, B.; Kretzschmar, H.A. Role of microglia and host prion protein in neurotoxicity of a prion protein fragment. *Nature* **1996**, *380*, 345–347. [CrossRef] [PubMed]

154. Grizenkova, J.; Akhtar, S.; Brandner, S.; Collinge, J.; Lloyd, S.E. Microglial cx3cr1 knockout reduces prion disease incubation time in mice. *BMC Neurosci.* **2014**, *15*, 44. [CrossRef] [PubMed]

155. Zhu, C.; Herrmann, U.S.; Falsig, J.; Abakumova, I.; Nuvolone, M.; Schwarz, P.; Frauenknecht, K.; Rushing, E.J.; Aguzzi, A. A neuroprotective role for microglia in prion diseases. *J. Exp. Med.* **2016**, *213*, 1047–1059. [CrossRef] [PubMed]

156. Hughes, M.M.; Field, R.H.; Perry, V.H.; Murray, C.L.; Cunningham, C. Microglia in the degenerating brain are capable of phagocytosis of beads and of apoptotic cells, but do not efficiently remove PrP^{Sc}, even upon lps stimulation. *Glia* **2010**, *58*, 2017–2030. [CrossRef] [PubMed]

157. Mays, C.E.; Kim, C.; Haldiman, T.; van der Merwe, J.; Lau, A.; Yang, J.; Grams, J.; Di Bari, M.A.; Nonno, R.; Telling, G.C.; et al. Prion disease tempo determined by host-dependent substrate reduction. *J. Clin. Investig.* **2014**, *124*, 847–858. [CrossRef] [PubMed]

158. Huang, P.; Lian, F.; Wen, Y.; Guo, C.; Lin, D. Prion protein oligomer and its neurotoxicity. *Acta Biochim. Biophys. Sin.* **2013**, *45*, 442–451. [CrossRef] [PubMed]

159. Kupfer, L.; Hinrichs, W.; Groschup, M.H. Prion protein misfolding. *Curr. Mol. Med.* **2009**, *9*, 826–835. [CrossRef] [PubMed]

160. Verma, M.; Vats, A.; Taneja, V. Toxic species in amyloid disorders: Oligomers or mature fibrils. *Ann. Indian Acad. Neurol.* **2015**, *18*, 138–145. [PubMed]

161. Safar, J.G.; DeArmond, S.J.; Kociuba, K.; Deering, C.; Didorenko, S.; Bouzamondo-Bernstein, E.; Prusiner, S.B.; Tremblay, P. Prion clearance in bigenic mice. *J. Gen. Virol.* **2005**, *86*, 2913–2923. [CrossRef] [PubMed]

162. White, M.D.; Farmer, M.; Mirabile, I.; Brandner, S.; Collinge, J.; Mallucci, G.R. Single treatment with rnai against prion protein rescues early neuronal dysfunction and prolongs survival in mice with prion disease. *Proc. Natl. Acad. Sci. USA* **2008**, *105*, 10238–10243. [CrossRef] [PubMed]

163. Kawasaki, Y.; Kawagoe, K.; Chen, C.J.; Teruya, K.; Sakasegawa, Y.; Doh-ura, K. Orally administered amyloidophilic compound is effective in prolonging the incubation periods of animals cerebrally infected with prion diseases in a prion strain-dependent manner. *J. Virol.* **2007**, *81*, 12889–12898. [CrossRef] [PubMed]

164. Trevitt, C.R.; Collinge, J. A systematic review of prion therapeutics in experimental models. *Brain* **2006**, *129*, 2241–2265. [CrossRef] [PubMed]

165. Sim, V.L.; Caughey, B. Recent advances in prion chemotherapeutics. *Infect. Disord. Drug Targets* **2009**, *9*, 81–91. [CrossRef] [PubMed]

166. Supattapone, S.; Nguyen, H.O.; Cohen, F.E.; Prusiner, S.B.; Scott, M.R. Elimination of prions by branched polyamines and implications for therapeutics. *Proc. Natl. Acad. Sci. USA* **1999**, *96*, 14529–14534. [CrossRef] [PubMed]

Review

Pharmacological Agents Targeting the Cellular Prion Protein

Maria Letizia Barreca [1,*], **Nunzio Iraci** [1], **Silvia Biggi** [2], **Violetta Cecchetti** [1] and **Emiliano Biasini** [2,3,*]

[1] Department of Pharmaceutical Sciences, University of Perugia, 06123 Perugia, Italy; nunzio.iraci@gmail.com (N.I.); violetta.cecchetti@unipg.it (V.C.)
[2] Dulbecco Telethon Laboratory of Prions and Amyloids, Centre for Integrative Biology (CIBIO), University of Trento, 38123 Trento, Italy; silvia.biggi@unitn.it
[3] Department of Neuroscience, IRCCS-Istituto di Ricerche Farmacologiche Mario Negri, 20156 Milan, Italy
* Correspondence: maria.barreca@unipg.it (M.L.B.); emiliano.biasini@unitn.it (E.B.)

Received: 8 February 2018; Accepted: 2 March 2018; Published: 7 March 2018

Abstract: Prion diseases are associated with the conversion of the cellular prion protein (PrP^C), a glycoprotein expressed at the surface of a wide variety of cell types, into a misfolded conformer (the scrapie form of PrP, or PrP^{Sc}) that accumulates in brain tissues of affected individuals. PrP^{Sc} is a self-catalytic protein assembly capable of recruiting native conformers of PrP^C, and causing their rearrangement into new PrP^{Sc} molecules. Several previous attempts to identify therapeutic agents against prion diseases have targeted PrP^{Sc}, and a number of compounds have shown potent anti-prion effects in experimental models. Unfortunately, so far, none of these molecules has successfully been translated into effective therapies for prion diseases. Moreover, mounting evidence suggests that PrP^{Sc} might be a difficult pharmacological target because of its poorly defined structure, heterogeneous composition, and ability to generate different structural conformers (known as prion strains) that can elude pharmacological intervention. In the last decade, a less intuitive strategy to overcome all these problems has emerged: targeting PrP^C, the common substrate of any prion strain replication. This alternative approach possesses several technical and theoretical advantages, including the possibility of providing therapeutic effects also for other neurodegenerative disorders, based on recent observations indicating a role for PrP^C in delivering neurotoxic signals of different misfolded proteins. Here, we provide an overview of compounds claimed to exert anti-prion effects by directly binding to PrP^C, discussing pharmacological properties and therapeutic potentials of each chemical class.

Keywords: cellular prion protein; prion diseases; PrP ligands; pharmacological chaperones

---

1. Introduction

With few exceptions, proteins evolved their biological function in parallel with the ability to remain soluble under physiological conditions. However, in several pathological situations, specific proteins lose their native fold and acquire a different tertiary and quaternary conformation, clustering into aberrant aggregates. This phenomenon, known as protein misfolding, lays at the root of a wide variety of human diseases, such as neurodegenerative disorders, in which protein aggregation occurs in the brain [1]. Examples include common disorders such as Alzheimer's and Parkinson's diseases, or rarer disorders such as amyotrophic lateral sclerosis and prion diseases. Despite the fact that the pathological protein component is different in each neurodegenerative disorder, compelling evidence coming from genetic, biophysical and biochemical studies indicate that misfolded proteins are toxic to neurons. In fact, they often expose regions that are normally buried in the native state, leading to aggregation and aberrant interaction with cellular components such as membranes, proteins, or other macromolecules. These events may negatively affect neuronal homeostasis, for example, by blocking axonal transport, damaging synaptic endings or sequestering essential proteins, ultimately leading to cell death [2]. Possible strategies for tackling protein aggregation include breaking-up aggregates, increasing their degradation, or blocking their formation by stabilizing the native conformation of the monomeric protein precursors. While the first two have largely been explored in the past, the latter is a relatively new concept, and may possibly provide theoretical and technical advantages. For example, although detailed information about the structure of protein aggregates is rarely available, the three-dimensional organization of the monomeric precursors is often well characterized. A particularly meaningful example is represented by prion diseases. These disorders have the peculiarity of manifesting in a sporadic, inherited or transmissible fashion, and are associated with the conformational conversion of the cellular prion protein (PrP^C), a glycoprotein of uncertain function anchored to the outer surface of the plasma membrane, into a misfolded isoform (called PrP^{Sc}) that accumulates in the central nervous system of affected organisms [3]. PrP^{Sc} is a proteinaceous infectious particle (prion), capable of multiplying by directly recruiting native conformers of PrP^C, and causing their conformational rearrangement into new PrP^{Sc} molecules [4].

The vast majority of experimental strategies aimed at identifying therapeutics for human prion diseases has so far targeted PrP^{Sc}, the most direct, pathologically-relevant form of PrP [5]. However, the structure of PrP^{Sc} is poorly defined, and this form is also likely to be heterogeneous in composition and conformation. In fact, one of the most puzzling aspects of prion diseases is the phenomenon of prion strains [6]. It is believed that distinct conformations of PrP^{Sc} may explain the unusually wide spectrum of biochemical, neuropathological and clinical features that characterize prion diseases [7]. Prion strains are of particular relevance for the treatment of prion diseases, as their appearance may cause the acquisition of drug resistance to therapeutic treatments [8,9]. Indeed, a number of previously discovered anti-prion compounds have been shown to act in a strain-specific fashion, a property that severely limits their therapeutic potentials [10–12].

A possible, perhaps less intuitive strategy to overcome these limitations could be to target PrP^C, the common substrate of any prion strain replication. The structure of PrP^C is known at atomic level resolution, thanks to multiple previous reports employing nuclear magnetic resonance (NMR) or X-ray crystallography [13–15]. This provides a convenient ground to carry out rational drug design campaigns. Moreover, from a theoretical standpoint, a molecule binding to PrP^C with sufficiently high-affinity might in principle stabilize its folding by reducing the Gibbs free energy. Consequently, the activation energy (ΔG) required for the unfolding process will increase proportionally, with the result that the rate of formation of any PrP^{Sc} strain will be kinetically and thermodynamically disfavored. Small molecules acting with such mechanisms are known as pharmacological chaperones. Interestingly, two or more ligands with independent binding sites on PrP^C could synergize to completely block the formation of any unfolded form, since the relationship between ΔG and the stability constant of a folded polypeptide chain is exponential. In light of these conclusions, PrP^C appears as a convenient molecular target for tackling prion propagation [16]. Is this protein also the right pharmacological target for preventing prion diseases? It is widely agreed that PrP^C plays a crucial role in the pathogenesis of prion diseases not only by virtue of its ability to serve as substrate for generation of PrP^{Sc}. In fact, it has been reported that genetically depleting neuronal PrP^C in mice with established prion infection reverses neuronal loss and progression of clinical signs, despite the continuous production of infectious PrP^{Sc} by surrounding astrocytes [17]. Similarly, the absence of endogenous PrP^C renders host brain tissue resistant to the toxic effects of PrP^{Sc} emanating from implanted graft tissue [18]. These data indicate that other toxic species, rather than fully aggregated PrP^{Sc}, are responsible for the pathology of prion diseases. This conclusion is consistent with a number of previous reports underscoring the distinction between prion infectivity and prion toxicity [19–22]. In particular, recent experiments indicate that accumulation of infectivity and neurodegeneration proceed in distinct chronological and mechanistic phases [23]. While infectivity accumulates relatively rapidly, and requires only a minimum expression of PrP^C, neurodegeneration takes much longer and is directly dependent on the amount of PrP^C expressed in the brain. Taken together, these lines of evidence suggest that an unknown PrP conformer, either "on" or "off" pathway to PrP^{Sc}, could be the pathological form in prion diseases. These data provide a possible explanation for the evidence that, with few exceptions [12,24], none of the anti-prion compounds identified so far has shown a substantial effect in vivo. In fact, these molecules could disfavor PrP^{Sc} accumulation without hampering the neurotoxicity originating from other toxic conformers. Conversely, stabilizing the folded state of PrP^C has the potential to block not only PrP^{Sc} formation and propagation, but also the appearance of any putative toxic conformer. Another potential advantage of targeting PrP^C arises from recent observations indicating that PrP^C may exert a toxicity-transducing activity upon binding to PrP^{Sc}, as well as to various disease-associated, misfolded oligomeric assemblies, such as those formed by the amyloid β (Aβ) peptide, or by the protein alpha-synuclein, linked to Alzheimer's and Parkinson's diseases, respectively [25–29]. Importantly, mice depleted for PrP expression develop normally, with subtle phenotypic changes appearing only later in life, thus suggesting that pharmacological decrease of PrP^C function could produce little side effects. This conclusion is also supported by the recent identification of loss-of-function PrP alleles in healthy subjects [30]. Overall, these data support the potential value of targeting PrP^C, as this approach may provide therapeutic benefits not only for prion diseases, but possibly also for other neurodegenerative disorders. In this manuscript, we review the main chemical classes reported to act against prion replication in a PrP^C-directed fashion, focusing our discussion on molecules for which binding constant (K$_D$), structural information and anti-prion half-maximal effective concentration (EC$_{50}$) have experimentally been determined (Figure 1).

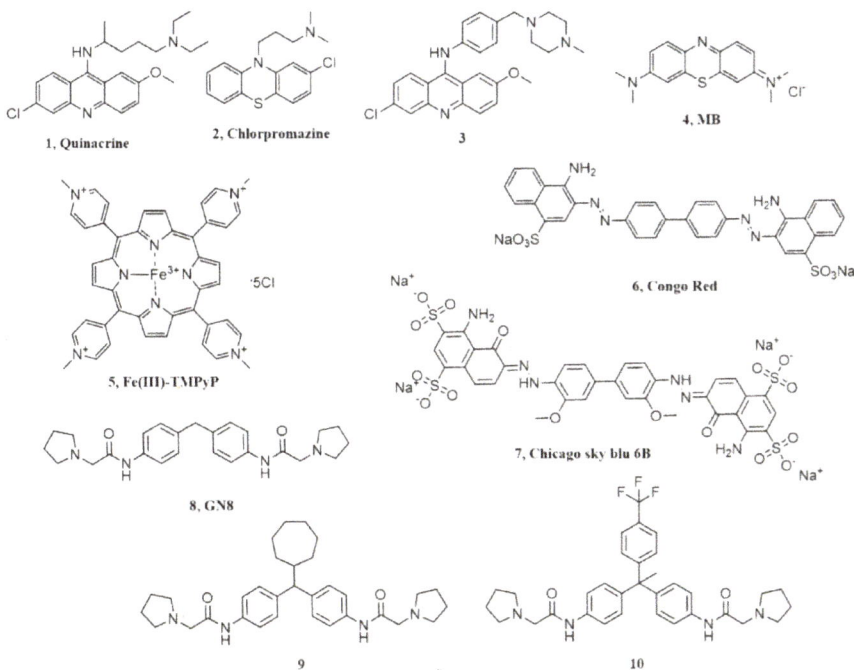

Figure 1. *Cont.*

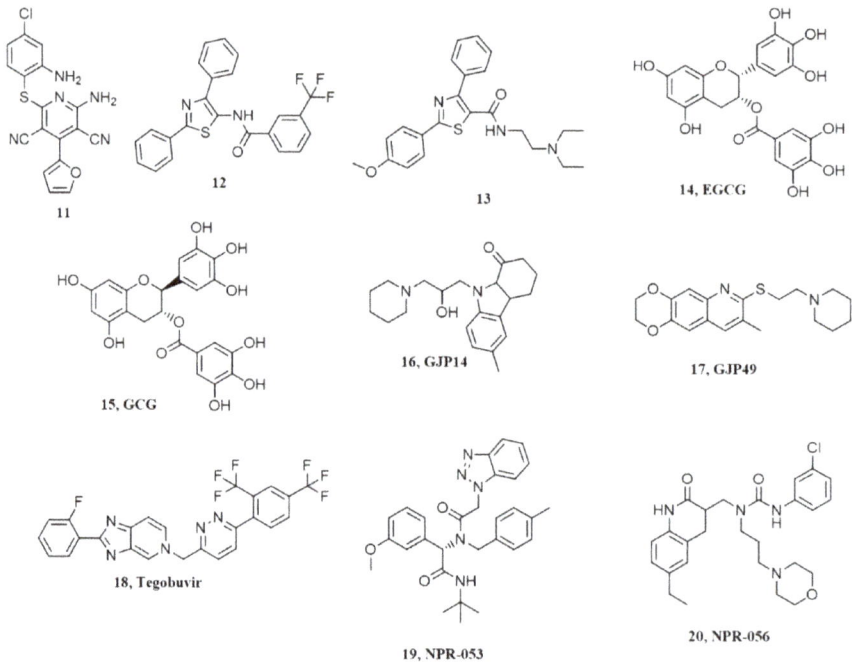

Figure 1. Chemical structures of the different compounds claimed to directly bind PrP^C.

2. Acridine and Phenothiazine Derivatives

Tricyclic derivatives of acridines (compound **1** in Figure 1, quinacrine) and phenothiazines like chlorpromazine (compound **2** in Figure 1) were initially reported to be promising candidates for the treatment of prion diseases [31,32]. Indeed, these drugs have already been used in humans for many years, and are known to cross the blood–brain barrier, thus giving hope to their repurposing for prion diseases. The antimalarial agent quinacrine and the antipsychotic drug chlorpromazine showed inhibition of PrP^{Sc} formation in prion-infected N2a cells, with EC$_{50}$ values of ~0.3 μM and ~3 μM, respectively. The acridine derivative quinacrine deserves particular attention, as it showed better potency in cell cultures, and was tested in human trials for prion diseases (more extensively than chlorpromazine, which was tested only in combination with the antimalarial agent). Quinacrine enantiomers showed stereoselectivity against prions, with the (S)-quinacrine exhibiting superior activity in eradicating PrP^{Sc} from cells [33]. Unfortunately, despite the promising in vitro profile, no beneficial effects were observed in vivo, using prion-infected rodent models of prion disease [34,35]. In addition to animal models, the activity and safety of quinacrine was assessed in clinical trials in human Creutzfeldt–Jakob disease (CJD) patients, but no effects were observed either on survival at the two-month time point or on the clinical course of the disease [36,37]. Pharmacokinetic studies unveiled that free quinacrine concentration in the brain reached only ~1 μM, which is a lower value than the cellular EC$_{50}$ observed in vitro [11,38]. These results highlighted the difficulty of translating results obtained by in vitro or cell-based methods to the clinical context. The lack of clinical efficacy of quinacrine against CJD was mainly attributed to metabolic instability, scarce accumulation of the drug into the brain due to active efflux by P-glycoprotein (P-gp) and the formation of drug resistant prion strains [11]. Original studies suggested that the anti-prion activity of quinacrine was directly connected to its ability to modify the lysosomal environment, causing improved clearance of PrP^{Sc} [31]. However, later studies reported that quinacrine binds to the globular domain of human recombinant PrP (residues 121–230), as observed by NMR spectroscopy. Tyr225, Tyr226, and Gln227 of helix 3 (H3) were identified as key residues in such ligand–protein interaction (region 1 in Figure 2) [39]. Of note, these experiments were conducted at very high concentrations, and the obtained dissociation constant of quinacrine (K_D = 4.6 mM) was about four orders of magnitude higher than its cellular EC_{50} value (required to clear PrP^{Sc} from prion-infected cells in vitro). Similar data (K_D ~1 mM) were obtained in another study where quinacrine binding to recombinant human PrP was analyzed by surface plasmon resonance (SPR) [40], although other SPR studies reported the ability of quinacrine to bind human recombinant PrP with a K_D of 15 μM [41]. In another report, dynamic light scattering studies and circular dichroism (CD) measurements suggested that quinacrine binding induces a conformational change in PrP, disfavouring PrP^{Sc} formation [42]. It is worth noting that the potential of quinacrine as a prion inhibitor has stimulated great interest in the 9-aminoacridine family as therapeutic candidates for prion diseases, and intensive research efforts have been spent on the synthesis, biological evaluation and structure–activity relationship (SAR) studies of quinacrine derivatives [43–46]. In particular, the nature of the aliphatic side-chain on 9-amino group of the tricyclic scaffold was found to be one key feature for enhancing binding affinity to PrP, PAMPA permeability and inhibition of PrP^{Sc} accumulation. As an example, a quinacrine derivative (compound 3 in Figure 1) showed improved anti-prion activity, as compared to the parent compound, across different prion-infected murine cell models (ScN2a, N167, F3). In addition, this compound exhibited stronger binding affinity by SPR, and seemed to be a weaker substrate for P-gp [46]. However, more recent SPR- and NMR-based studies have highlighted a non-specific binding interaction of quinacrine to PrP^C, reiterating the original observation that its mode of action involves PrP-independent mechanisms [47,48].

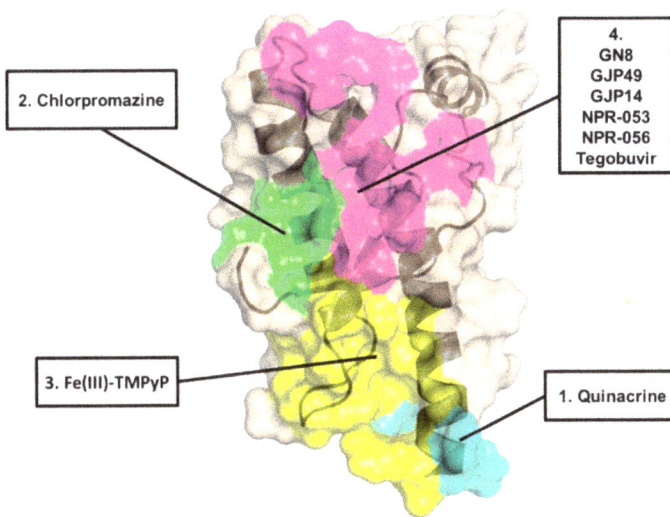

Figure 2. Visualization of the proposed binding regions for the different PrP^C ligands (indicated).

Similarly to quinacrine, the direct binding of phenothiazine derivative chlorpromazine to PrP^C was originally investigated by NMR [39] and SPR [41], showing a weaker interaction with recombinant PrP, as compared to quinacrine. A subsequent study based on NMR and X-ray crystallography (PDB ID 4MA8) reported a precise binding site of phenothiazines on PrP^C, located in a hydrophobic pocket formed by helix-2 (H2) and the two anti-parallel β-sheets (S1 and S2; region 2 in Figure 2) [49]. The data also indicated that an unexpected intramolecular reorganization of the N-terminal, unstructured tail of PrP^C around the C-terminal domain, through the formation of a hydrophobic anchor, directly suggesting a mechanism by which phenothiazines may act as pharmacological chaperone of PrP^C. Unfortunately, the study did not provide an affinity value for the binding of phenothiazines to PrP^C.

Such value was instead precisely defined in the following report, employing SPR and dynamic mass redistribution (DMR) [50]. The results confirmed original observations indicating a weak interaction of chlorpromazine to PrP^C, with an estimated K_D higher than 400 μM, compatible with data collected in the original study [49], which employed millimolar concentrations of chlorpromazine to carry out NMR and X-ray crystallography experiments. A K_D value in the high micromolar concentration range is incompatible with the reported anti-prion effects of chlorpromazine in cells, indicating that its mode of action is independent from direct PrP binding [50]. Moreover, chlorpromazine also failed to inhibit prion replication in vitro (by the protein misfolding cyclic amplification reaction, PMCA), as instead it would be expected for a pharmacological chaperone of PrP^C. Interestingly, the same study reported compelling evidence indicating that the mechanism of action underlying the anti-prion effect of chlorpromazine is related to the previously known ability of the compound to inhibit clathrin-mediated endocytosis, leading to decreased levels of PrP^C at the cell surface. Consistent with this conclusion, two inhibitors of dynamins, proteins involved in the regulation of the scission of membrane vesicles, and recently reported to be targeted by chlorpromazine [51], mimicked PrP^C-relocalizing effects, and blocked the replication of two different prion strains in cell cultures [50]. An additional recent work provided evidence for a chlorpromazine-induced redistribution of PrP^{Sc} from the endocytic-recycling pathway to the lysosomal compartment, an effect that could be the direct consequence of the relocalization of PrP^C from the cell surface [52].

Methylene Blue (MB, compound **4** in Figure 1), a phenothiazine derivative, has been shown to affect the kinetics of PrP oligomerization by binding to a surface cleft on PrP^C [53]. Using size exclusion chromatography, static light scattering, differential scanning calorimetry and transmission electron microscopy, the authors studied the influence of methylene blue on the oligomerization and fibrillation of human, ovine and murine recombinant PrP, observing a decrease in oligomerization kinetics and overall levels. NMR experiments mapped MB binding sites in a surface cleft delimited by residues belonging to S1-H1 and H2-H3 loops, and H1, H2 and H3 helices (residues Asn146, Asn156, Tyr160, Lys188, Thr191, Val192, Thr194, Thr195, Gln215). Of note, MB has been investigated as potential therapeutic agent in other proteinopathies [54–57], which is consistent with the number of potential applications that have been tested for this compound, likely reflecting its ability to engage non-specific interactions with a broad range of proteins.

3. Cyclic Tetrapyrroles

Cyclic tetrapyrroles, planar aromatic ring systems coordinating metal ions and bearing pendants of different chemical nature, were originally found to be effective in prion-infected cells, and later claimed to act by directly binding to PrP^C [58,59]. In particular, by employing isothermal titration calorimetry (ITC), the cationic tetrapyrrole Fe(III)-TMPyP (compound **5** in Figure 1) was shown to bind human recombinant PrP in the C-terminal, globular domain (K_D = 4.52 μM), which was consistent with its cellular EC$_{50}$ of 1.6 μM in cells (as tested in rocky mountain laboratory, RML-infected PK1 cells) and the range of concentrations (1–11 μM) active in the protein-misfolding cyclic amplification (PMCA) reaction [48]. NMR studies allowed the identification of the binding site of Fe(III)-TMPyP on human PrP, with key interacting residues clustered at the C terminus of H3 and in the loop between residues 160 and 180 (region 3 in Figure 2). Importantly, Fe(III)-TMPyP, or highly similar porphyrins, also showed the ability to inhibit the cytotoxic activity of a mutant PrP carrying a deletion in the central region (Δ105–125), abrogated the PrP^C-mediated synaptotoxic effects of Aβ oligomers in primary hippocampal neurons, and significantly prolonged survival time in prion-infected mice [60,61]. Unfortunately, the therapeutic potentials of porphyrins like Fe(III)-TMPyP is dampened by their poor pharmacokinetic properties, such as possible non-specific interactions with plasma proteins, and unlikelihood to cross the blood–brain barrier [62]. However, as assayed by in vitro and cell-based tests, these compounds appear as the most effective pharmacological chaperones of PrP^C, and have already been employed to gain insights into the physiological activity of PrP^C, and its functional connection to neurodegenerative pathways. Performing extensive pharmacokinetic profiling of this class of molecules, coupled to chemical optimization efforts and/or innovative ways of delivery to the central nervous system, could provide effective therapeutic strategies for prion diseases, and possibly other neurodegenerative disorders linked to the toxicity-transducing activity of PrP^C.

4. Diazo Dyes

The diazo dye Congo red (compound **6** in Figure 1) was found to possess anti-prion activity in cells and in vivo, using scrapie-infected golden Syrian hamsters [63–67]. In particular, Congo red prevented the formation and accumulation of PrP^{Sc} in neuroblastoma cells with an EC$_{50}$ of about 0.015 µM. The binding of Congo red to human recombinant PrP was investigated by SPR, and showed a K$_D$ value of 1.6 µM [40]. However, other studies reported that, in physiological conditions, the molecule binds non-specifically to PrP^C as an aggregated polyanion [47]. Congo Red itself has a number of shortfalls, such as non-specific interactions with various macromolecules, self-polymerization, toxicity and poor permeability through BBB. For this reason, several Congo red derivatives were designed and synthesized to improve the pharmacological profile of the compound, and a number of analogues showed anti-prion effects at nanomolar concentrations, even though no information about their possible interaction with PrP^C was reported [68–70].

5. Chicago Sky Blue 6B

This molecule emerged from a screen of 1200 approved drugs and pharmacological tool compounds (Prestwick Chemical Library) based on a fluorescence polarization (FP) assay, and aimed at identifying compounds capable of inhibiting the binding of Aβ oligomers to PrP^C [71]. Chicago Sky Blue 6B (compound **7** in Figure 1) was identified as the best-ranked candidate, with EC$_{50}$ values of 0.41 µM and 19.7 µM in FP and ELISA assays, respectively. ITC experiments confirmed that Chicago Sky Blue 6B is able to interact with human recombinant PrP, with a K$_D$ value of 0.55 µM. Importantly, the compound did not bind a PrP construct containing only residues 119–231, indicating that its binding site lies within the N-terminal, unstructured tail of the protein. Since Aβ oligomers are known to bind PrP in the same region, the data suggested that Chicago Sky Blue 6B may act by a mechanism of direct competition. Of note, Chicago Sky Blue 6B also showed anti-prion effects in RML-infected N2a cells, with EC$_{50}$ values in the low micromolar range, and in absence of evident cytotoxicity. At the time this manuscript was prepared, no other studies have employed Chicago Sky Blue 6B in the context of prion diseases.

6. Diphenylmethane Derivatives

A compound known as GN8 (compound **8** in Figure 1) emerged from an in silico, dynamics-based drug screen of ~320,000 compounds aimed at directly identifying pharmacological chaperones for PrP^C [72]. In vitro validation studies estimated the affinity of GN8 for recombinant, mouse PrP in the low micromolar range (KD ~5 µM). Heteronuclear NMR and molecular modeling mapped the PrP binding region of GN8 at the C-terminal domain, particularly involving residues N159 and E196 (region 4 in Figure 2). Furthermore, the authors employed CD in a thermal-denaturation assay to confirm that the binding of GN8 stabilizes the PrP^C conformation significantly ($\Delta\Delta H$ = 6.7 kcal/mol). Biological validation showed that GN8 efficiently inhibits prion replication in cells, with an estimated EC$_{50}$ of ~1.35 µM. Importantly, GN8 was also found to prolong the survival of prion-infected mice, thus confirming the effective anti-prion activity of this molecule. Subsequent studies focused on the synthesis and evaluation of anti-prion effects for a series of GN8 analogues with the main objective of generating a SAR profile [73]. Two derivatives (compounds **9** and **10** in Figure 1) were found to be approximately three times more potent than the parent compound, with EC$_{50}$ values around 0.5 µM, in absence of detectable toxicity. CD-coupled thermal-denaturation assays indicated that one of these molecules significantly stabilized recombinant PrP, with a degree of stabilization by this ligand approximately doubled, as compared with that of GN8 ($\Delta\Delta H$ = 14.2 kcal/mol). Binding was also confirmed by SPR. According to these data, GN8 and its derivatives appear as promising pharmacological chaperones of PrP^C. However, it is worth noting that two subsequent studies failed to confirm binding of GN8 to mouse or human recombinant PrP, using a battery of biophysical techniques [48,60]. Such experimental discrepancy is currently unresolved.

7. Pyridine Dicarbonitriles

Four pyridine dicarbonitrile analogues, originally identified as anti-prion compounds in prion-infected cells [74], were later tested for their direct interaction with PrP^C using SPR [41]. One derivative (compound **11** in Figure 1) showed anti-prion activity (EC_{50} values ~20 µM) and detectable binding to recombinant PrP. This observation justified the following efforts to generate small libraries of pyridine dicarbonitrile derivatives, which were then tested by SPR for binding to PrP^C, and in cellular assays to evaluate anti-prion activity [75,76]. Unexpectedly, no direct correlation was observed between binding to PrP^C and anti-prion efficacy, with the most potent anti-prion pyridine dicarbonitrile showing either weak or no binding to PrP^C. Collectively, these data suggested that pyridine dicarbonitrile likely inhibit prion replication in a PrP^C-independent fashion.

8. Diarylthiazoles

The same team originally involved in the study on the dicarbonitrile derivatives also reported the synthesis and screening of 2,4-diarylthiazole-based compounds as potential anti-prion agents [77,78]. The authors stated that original 2,4-diarylthiazole scaffold was identified as a PrP ligand through a virtual screening campaign, although details of such screening were not described. SPR was then employed to test the binding of several derivatives to mouse or human recombinant PrP. Only one compound (compound **12** in Figure 1) showed a high-affinity interaction with PrP. All the molecules were also tested in prion-infected SMB cells, but once again no correlation was found between PrP binding and anti-prion activity in cells. A second series of reverse amide 2,4-diarylthiazole-based anti-prion compounds was later reported in a following study. The molecules were first tested for prion inhibition in SMB cells and then evaluated for binding to recombinant PrP, as assayed by SPR. Among the compounds active in cells, one derivative (compound **13** in Figure 1) (EC_{50} = 4 µM) also showed affinity for PrP, although a careful evaluation of SPR data suggested the possibility of a non-specific interaction. Overall, these studies highlighted a general lack of correlation between anti-prion activity and PrP binding for 2,4-diarylthiazole-based compounds, suggesting that other PrP-independent modes of action account for the anti-prion effects of this chemical class.

9. Natural Polyphenols

In search of small molecules able to interfere with prion propagation, another study screened a collection of natural compounds with proven activity against amyloid formation in vitro [79,80]. The major polyphenols component of green tea, i.e., epigallocatechin gallate (EGCG, compound **14** in Figure 1) and its stereoisomer gallocatechin gallate (GCG, compound **15** in Figure 1), showed anti-prion activity in prion-infected N2a cells. The direct interaction of EGCG with recombinant PrP (residues 90–232) was experimentally tested by ITC, showing a strong affinity (K_D = 0.13 µM) and a remarkable stabilization effect (ΔH of −43 KJ). Further experiments on the effect of EGCG binding revealed an unexpected destabilization effect of the compound on the native conformation of PrP^C, inducing its rapid transition into detergent-insoluble species, which were rapidly degraded intracellularly. The authors also observed that the anti-prion activity depended on the gallate side chain and the three hydroxyl groups of the trihydroxyphenyl side chain. Unfortunately, a subsequent study characterized the binding properties of EGCG to PrP^C by SPR and NMR, concluding that the compound binds to the protein in a non-specific fashion [47]. These results dampened the enthusiasm for the treatment of prion diseases with EGCG-like polyphenols.

10. Miscellanea

Several structurally diverse compounds identified by virtual screening campaigns on the proposed binding pocket for GN8 have been claimed to be specific PrP^C ligands, capable of acting as chemical chaperones. In 2009, a virtual screening study led to the selection of 205 commercially available compounds to be evaluated for their effects on the PrP^C conversion process [81]. Ex vivo-experiments identified 24 non-cytotoxic molecules that significantly inhibited prion replication in GT-FK cells, at a concentration of 10 μM. To further elucidate their mechanism of action, the authors measured the binding affinity for recombinant PrP by SPR, and then compared anti-prion activity in cells with affinity values. Eleven compounds were classified as PrP-directed anti-prion compounds; for example, for a molecule named GJP14 (compound **16** in Figure 1), the authors reported an EC$_{50}$ = 8.54 μM [82]. Compounds GJP14 and GJ49 (compound **17** in Figure 1) were further characterized for their binding properties by SPR and NMR (for example, for GJ49 K$_D$ = 50.8 μM), showing a ligand-binding pocket in the C-terminal, globular domain of PrP^C (region 4 in Figure 2) [47].

A related study performed a 3D pharmacophore-based virtual screen of an in-house chemical library, and selected 37 potential anti-prion compounds to be assessed by cell-based and SPR-based assays [83]. The results identified a molecule named BMD42-29 (a benoxazole derivative whose structure was not disclosed) as the best hit among the screened molecules, with an EC$_{50}$ value against prion replication in cells in the low micromolar range (<5 μM). Of note, in prion-infected N2a cells, the compound did not produce a marked reduction in total PrP levels. SPR experiments revealed that BMD42-29 had strong binding affinity to PrP^C (K$_D$ = 21.5 μM), with kinetic rates characterized by rapid association and slow dissociation constants. The predicted binding mode of BMD42-29 was located in the same pocket of GN8, and was characterized by two hydrogen bonds with Asn159 and Glu196, and hydrophobic interactions with Leu130 and Arg156. The author concluded that BMD42-29 may act by stabilizing PrP^C, thus inhibiting its pathological conformational change to PrP^{Sc}. In 2016, another group built a platform called "NAGARA", aimed at unifying docking simulation, molecular dynamics and quantum chemistry to perform large-scale screening of commercially available compounds [84]. One hundred hits predicted in silico to bind PrP^C were subjected to cell-based validation to evaluate anti-prion effects. Tegobuvir (previously known as an anti-hepatitis C agent, compound **18** in Figure 1) emerged as one of the most promising candidates, with an estimated EC$_{50}$ of 1.7 μM, as assayed in immortalized neuronal mouse cells persistently infected with the human Fukuoka-1 prion strain. The molecule also showed detectable binding to PrP^C in the low micromolar range (K$_D$ = 19 μM, region 4 in Figure 2). In the same year, by coupling docking simulations of a large virtual library (~200K compounds) and binding interaction analyses, another group reported the identification of 96 novel small molecules capable of binding PrP^C in the same pocket of GN8 [85]. The ability of the in silico-predicted hits to target PrP^C was evaluated by SPR and thermal shift assay (TSA), whereas their anti-prion effects were estimated using persistently infected cells and animal models of prion diseases. Compounds NPR-053 (compound **19** in Figure 1) and NPR-056 (compound **20** in Figure 1) emerged as the most promising candidates, in light of their ability to reduce PrP^{Sc} levels in cultured cells, with EC$_{50}$ values of 7.68 μM and 3.72 μM, respectively. Both SPR and TSA provided evidence for a direct binding of both compounds to PrP^C (region 4 in Figure 2), with NPR-053 inducing the strongest stabilization effect on PrP^C native folding (ΔT$_m$ = 2.69 °C). All of these compounds represent promising candidate pharmacological chaperones for PrP^C, although further experimental validation is needed before considering them as promising therapeutic agents for prion diseases.

Table 1. Summary of main chemical scaffolds reported to exert anti-prion effects by directly targeting PrP^C.

Chemical Scaffold	Compound (Figure 1)	K_D *	EC_{50} **	Effect In Vivo ***	Conclusions
Acridine derivatives	1	~1mM	~0.3 μM	Not significant	Primary effects are PrP-independent
Phenothiazine derivatives	2	>400 μM	~3 μM	Not significant	Likely acting by inducing PrP^C re-localization from the cell surface
Tetrapyrroles	5	4.52 μM	1.6 μM	Prolongation of survival time in prion-infected mice	Low specificity and possible poor pharmacokinetics
Diazo dyes	6	1.6 μM	0.015 μM	Not available	Low specificity
Chicago sky blue 6B	7	0.55 μM	Low μM	Not available	Need confirmation
Diphenylmethane derivatives	8	5 μM	1.35 μM	Prolongation of survival time in prion-infected mice	PrP^C binding not reproduced in some study
Pyridine Dicarbonitriles	11	~20μM	18.6 μM	Not available	No correlation between anti-prion activity and binding to PrP^C
Diarylthiazoles	13	3.8 μM	4 μM	Not available	No correlation between anti-prion activity and binding to PrP^C
Natural polyphenols	14	0.13 μM	-	Not available	Possible non-specific interaction with PrP^C
Miscellanea	17	50.8 μM	Not available	Not available	Need confirmation
	20	19 μM	3.72 μM	Not available	Need confirmation

* Reported affinity for PrP^C; ** Anti-prion activity measured in cell cultures; *** Tested in prion-infected rodent models and/or human patients.

11. Conclusions

Mounting evidence indicates that the accumulation of PrP^{Sc} alone could not account for the wide spectrum of neurotoxic events occurring in prion diseases. Instead, an unexpected role for PrP^C as toxicity–transducer receptor for PrP^{Sc} and other disease-associated misfolded oligomeric assemblies, such as Aβ and alpha-synuclein, has raised great interest for targeting this protein pharmacologically. In this manuscript, we reviewed previous efforts to identify PrP^C-directed compounds, taking into account limitations and reproducibility of each experimental attempt. A number of chemical scaffolds, identified by combining computational methods with biochemical, biophysical and cell-based assays, have been claimed to exert anti-prion effects by targeting PrP^C (Table 1). Some of these molecules, such as the cationic tetrapyrrole Fe(III)-TMPyP, provide a proof-of-principle for targeting PrP^C pharmacologically. Others, such as chlorpromazine, reveal unexpected mechanisms to counteract prion replication by lowering cell surface PrP^C. However, the vast majority of compounds show inconsistencies between affinity for PrP^C and biologically-active concentrations, low binding specificity, and/or lack of reproducibility. At the moment, none of these molecules appear as immediate candidates for clinical testing in the near future. Moreover, the great deal of negative data eventually provide further support to the notion that the vast majority of anti-prion molecules identified so far exert their activity through unknown targets, or by altering the homeostasis of PrP^C, rather than binding the protein directly. What could be the reason for such a lack of success in identifying small ligands of PrP^C? We believe the answer to this question may lie in a few, non-mutually exclusive possibilities. First, the screening techniques employed so far (e.g., in silico approaches coupled to biophysical assays) could have been inadequate for effectively identifying PrP^C-directed molecules. Moreover, most of the approaches reviewed in this manuscript relied on recombinant PrP for testing the binding of small molecules, while physiological, post-translational modifications of the protein (sugar and lipid moieties) may heavily influence ligand binding. It is also possible that a single PrP^C ligand will never be truly effective in preventing prion replication, since its stabilization effect on PrP^C folding could be counteracted by the strong affinity of PrP^{Sc} for its substrate. In this scenario, testing the combination of two or three ligands binding PrP^C in distinct pockets may produce the expected anti-prion effects. Ultimately, it is also possible that PrP^C simply lies among the proteins that can be classified as "undraggable". We like to believe that the latter conclusion will soon be refuted by direct experimental evidence.

Acknowledgments: The study was supported by a Young Investigator Award from the Italian Ministry of Health (GR-2010-2312769), and a grant from the CJD Foundation. EB is an Assistant Telethon Scientist at the Dulbecco Telethon Institute (TCP14009, Fondazione Telethon, Italy).

Author Contributions: M.L.B. and E.B. conceived the main aspects of the review, M.L.B., N.I., S.B., V.C. and E.B. wrote the paper.

Conflicts of Interest: The authors declare no conflict of interest. M.L.B. and E.B. are co-founders of Sibylla Biotech (www.sibyllabiotech.it), a startup company focused on developing new therapeutics for neurodegenerative disorders, including prion diseases.

References

1. Chiti, F.; Dobson, C.M. Protein misfolding, functional amyloid, and human disease. *Annu. Rev. Biochem.* **2006**, *75*, 333–366. [CrossRef] [PubMed]
2. Bucciantini, M.; Giannoni, E.; Chiti, F.; Baroni, F.; Formigli, L.; Zurdo, J.; Taddei, N.; Ramponi, G.; Dobson, C.M.; Stefani, M. Inherent toxicity of aggregates implies a common mechanism for protein misfolding diseases. *Nature* **2002**, *416*, 507–511. [CrossRef] [PubMed]
3. Prusiner, S.B. Biology and genetics of prions causing neurodegeneration. *Annu. Rev. Genet.* **2013**, *47*, 601–623. [CrossRef] [PubMed]
4. Prusiner, S.B. Novel proteinaceous infectious particles cause scrapie. *Science* **1982**, *216*, 136–144. [CrossRef] [PubMed]

5. Giles, K.; Olson, S.H.; Prusiner, S.B. Developing therapeutics for prp prion diseases. *Cold Spring Harbor Perspect. Med.* **2017**, *7*, a023747. [CrossRef] [PubMed]

6. Baskakov, I.V. The many shades of prion strain adaptation. *Prion* **2014**, *8*, 27836. [CrossRef]

7. Espinosa, J.C.; Nonno, R.; Di Bari, M.; Aguilar-Calvo, P.; Pirisinu, L.; Fernandez-Borges, N.; Vanni, I.; Vaccari, G.; Marin-Moreno, A.; Frassanito, P.; et al. PrPc governs susceptibility to prion strains in bank vole, while other host factors modulate strain features. *J. Virol.* **2016**, *90*, 10660–10669. [CrossRef] [PubMed]

8. Collinge, J. Medicine. Prion strain mutation and selection. *Science* **2010**, *328*, 1111–1112. [CrossRef] [PubMed]

9. Li, J.; Browning, S.; Mahal, S.P.; Oelschlegel, A.M.; Weissmann, C. Darwinian evolution of prions in cell culture. *Science* **2010**, *327*, 869–872. [CrossRef] [PubMed]

10. Sim, V.L. Prion disease: Chemotherapeutic strategies. *Infect. Disord. Drug Targets* **2012**, *12*, 144–160. [CrossRef] [PubMed]

11. Ghaemmaghami, S.; Ahn, M.; Lessard, P.; Giles, K.; Legname, G.; DeArmond, S.J.; Prusiner, S.B. Continuous quinacrine treatment results in the formation of drug-resistant prions. *PLoS Pathog.* **2009**, *5*, e1000673. [CrossRef] [PubMed]

12. Giles, K.; Berry, D.B.; Condello, C.; Hawley, R.C.; Gallardo-Godoy, A.; Bryant, C.; Oehler, A.; Elepano, M.; Bhardwaj, S.; Patel, S.; et al. Different 2-aminothiazole therapeutics produce distinct patterns of scrapie prion neuropathology in mouse brains. *J. Pharmacol. Exp. Ther.* **2015**, *355*, 2–12. [CrossRef] [PubMed]

13. Riek, R.; Hornemann, S.; Wider, G.; Billeter, M.; Glockshuber, R.; Wüthrich, K. NMR structure of the mouse prion protein domain PrP(121–231). *Nature* **1996**, *382*, 180–182. [CrossRef] [PubMed]

14. Zahn, R.; Liu, A.; Luhrs, T.; Riek, R.; von Schroetter, C.; Lopez Garcia, F.; Billeter, M.; Calzolai, L.; Wider, G.; Wüthrich, K. NMR solution structure of the human prion protein. *Proc. Natl. Acad. Sci. USA* **2000**, *97*, 145–150. [CrossRef] [PubMed]

15. Antonyuk, S.V.; Trevitt, C.R.; Strange, R.W.; Jackson, G.S.; Sangar, D.; Batchelor, M.; Cooper, S.; Fraser, C.; Jones, S.; Georgiou, T.; et al. Crystal structure of human prion protein bound to a therapeutic antibody. *Proc. Natl. Acad. Sci. USA* **2009**, *106*, 2554–2558. [CrossRef] [PubMed]

16. Nicoll, A.J.; Collinge, J. Preventing prion pathogenicity by targeting the cellular prion protein. *Infect. Disord. Drug Targets* **2009**, *9*, 48–57. [CrossRef] [PubMed]

17. Mallucci, G.R.; White, M.D.; Farmer, M.; Dickinson, A.; Khatun, H.; Powell, A.D.; Brandner, S.; Jefferys, J.G.; Collinge, J. Targeting cellular prion protein reverses early cognitive deficits and neurophysiological dysfunction in prion-infected mice. *Neuron* **2007**, *53*, 325–335. [CrossRef] [PubMed]

18. Brandner, S.; Isenmann, S.; Raeber, A.; Fischer, M.; Sailer, A.; Kobayashi, Y.; Marino, S.; Weissmann, C.; Aguzzi, A. Normal host prion protein necessary for scrapie-induced neurotoxicity. *Nature* **1996**, *379*, 339–343. [CrossRef] [PubMed]

19. Biasini, E.; Medrano, A.Z.; Thellung, S.; Chiesa, R.; Harris, D.A. Multiple biochemical similarities between infectious and non-infectious aggregates of a prion protein carrying an octapeptide insertion. *J. Neurochem.* **2008**, *104*, 1293–1308. [CrossRef] [PubMed]

20. Biasini, E.; Seegulam, M.E.; Patti, B.N.; Solforosi, L.; Medrano, A.Z.; Christensen, H.M.; Senatore, A.; Chiesa, R.; Williamson, R.A.; Harris, D.A. Non-infectious aggregates of the prion protein react with several PrPsc-directed antibodies. *J. Neurochem.* **2008**, *105*, 2190–2204. [CrossRef] [PubMed]

21. Chiesa, R.; Harris, D.A. Prion diseases: What is the neurotoxic molecule? *Neurobiol. Dis.* **2001**, *8*, 743–763. [CrossRef] [PubMed]

22. Aguzzi, A.; Falsig, J. Prion propagation, toxicity and degradation. *Nat. Neurosci.* **2012**, *15*, 936–939. [CrossRef] [PubMed]

23. Sandberg, M.K.; Al-Doujaily, H.; Sharps, B.; Clarke, A.R.; Collinge, J. Prion propagation and toxicity in vivo occur in two distinct mechanistic phases. *Nature* **2011**, *470*, 540–542. [CrossRef] [PubMed]

24. Wagner, J.; Ryazanov, S.; Leonov, A.; Levin, J.; Shi, S.; Schmidt, F.; Prix, C.; Pan-Montojo, F.; Bertsch, U.; Mitteregger-Kretzschmar, G.; et al. ANLE138B: A novel oligomer modulator for disease-modifying therapy of neurodegenerative diseases such as prion and Parkinson's disease. *Acta Neuropathol.* **2013**, *125*, 795–813. [CrossRef] [PubMed]

25. Elezgarai, S.R.; Biasini, E. Common therapeutic strategies for prion and alzheimer's diseases. *Biol. Chem.* **2016**, *397*, 1115–1124. [CrossRef] [PubMed]

26. Lauren, J.; Gimbel, D.A.; Nygaard, H.B.; Gilbert, J.W.; Strittmatter, S.M. Cellular prion protein mediates impairment of synaptic plasticity by amyloid-beta oligomers. *Nature* **2009**, *457*, 1128–1132. [CrossRef] [PubMed]

27. Biasini, E.; Harris, D.A. Targeting the cellular prion protein to treat neurodegeneration. *Future Med. Chem.* **2012**, *4*, 1655–1658. [CrossRef] [PubMed]

28. Biasini, E.; Turnbaugh, J.A.; Unterberger, U.; Harris, D.A. Prion protein at the crossroads of physiology and disease. *Trends Neurosci.* **2012**, *35*, 92–103. [CrossRef] [PubMed]

29. Ferreira, D.G.; Temido-Ferreira, M.; Miranda, H.V.; Batalha, V.L.; Coelho, J.E.; Szego, E.M.; Marques-Morgado, I.; Vaz, S.H.; Rhee, J.S.; Schmitz, M.; et al. Alpha-synuclein interacts with PrP(c) to induce cognitive impairment through mGluR5 and NMDAR2B. *Nat. Neurosci.* **2017**, *20*, 1569–1579. [CrossRef] [PubMed]

30. Minikel, E.V.; Vallabh, S.M.; Lek, M.; Estrada, K.; Samocha, K.E.; Sathirapongsasuti, J.F.; McLean, C.Y.; Tung, J.Y.; Yu, L.P.; Gambetti, P.; et al. Quantifying prion disease penetrance using large population control cohorts. *Sci. Transl. Med.* **2016**, *20*, 322–323. [CrossRef] [PubMed]

31. Doh-Ura, K.; Iwaki, T.; Caughey, B. Lysosomotropic agents and cysteine protease inhibitors inhibit scrapie-associated prion protein accumulation. *J. Virol.* **2000**, *74*, 4894–4897. [CrossRef] [PubMed]

32. Korth, C.; May, B.C.; Cohen, F.E.; Prusiner, S.B. Acridine and phenothiazine derivatives as pharmacotherapeutics for prion disease. *Proc. Natl. Acad. Sci. USA* **2001**, *98*, 9836–9841. [CrossRef] [PubMed]

33. Ryou, C.; Legname, G.; Peretz, D.; Craig, J.C.; Baldwin, M.A.; Prusiner, S.B. Differential inhibition of prion propagation by enantiomers of quinacrine. *Lab. Investig.* **2003**, *83*, 837–843. [CrossRef] [PubMed]

34. Collins, S.J.; Lewis, V.; Brazier, M.; Hill, A.F.; Fletcher, A.; Masters, C.L. Quinacrine does not prolong survival in a murine Creutzfeldt-JaKob disease model. *Ann. Neurol.* **2002**, *52*, 503–506. [CrossRef] [PubMed]

35. Barret, A.; Tagliavini, F.; Forloni, G.; Bate, C.; Salmona, M.; Colombo, L.; De Luigi, A.; Limido, L.; Suardi, S.; Rossi, G.; et al. Evaluation of quinacrine treatment for prion diseases. *J. Virol.* **2003**, *77*, 8462–8469. [CrossRef] [PubMed]

36. Collinge, J.; Gorham, M.; Hudson, F.; Kennedy, A.; Keogh, G.; Pal, S.; Rossor, M.; Rudge, P.; Siddique, D.; Spyer, M.; et al. Safety and efficacy of quinacrine in human prion disease (prion-1 study): A patient-preference trial. *Lancet Neurol.* **2009**, *8*, 334–344. [CrossRef]

37. Geschwind, M.D.; Kuo, A.L.; Wong, K.S.; Haman, A.; Devereux, G.; Raudabaugh, B.J.; Johnson, D.Y.; Torres-Chae, C.C.; Finley, R.; Garcia, P.; et al. Quinacrine treatment trial for sporadic Creutzfeldt-JaKob disease. *Neurology* **2013**, *81*, 2015–2023. [CrossRef] [PubMed]

38. Ahn, M.; Ghaemmaghami, S.; Huang, Y.; Phuan, P.W.; May, B.C.; Giles, K.; DeArmond, S.J.; Prusiner, S.B. Pharmacokinetics of quinacrine efflux from mouse brain via the P-glycoprotein efflux transporter. *PLoS ONE* **2012**, *7*, e39112. [CrossRef] [PubMed]

39. Vogtherr, M.; Grimme, S.; Elshorst, B.; Jacobs, D.M.; Fiebig, K.; Griesinger, C.; Zahn, R. Antimalarial drug quinacrine binds to C-terminal helix of cellular prion protein. *J. Med. Chem.* **2003**, *46*, 3563–3564. [CrossRef] [PubMed]

40. Kawatake, S.; Nishimura, Y.; Sakaguchi, S.; Iwaki, T.; Doh-ura, K. Surface plasmon resonance analysis for the screening of anti-prion compounds. *Biol. Pharm. Bull.* **2006**, *29*, 927–932. [CrossRef] [PubMed]

41. Touil, F.; Pratt, S.; Mutter, R.; Chen, B. Screening a library of potential prion therapeutics against cellular prion proteins and insights into their mode of biological activities by surface plasmon resonance. *J. Pharm. Biomed. Anal.* **2006**, *40*, 822–832. [CrossRef] [PubMed]

42. Georgieva, D.; Schwark, D.; von Bergen, M.; Redecke, L.; Genov, N.; Betzel, C. Interactions of recombinant prions with compounds of therapeutical significance. *Biochem. Biophys. Res. Commun.* **2006**, *344*, 463–470. [CrossRef] [PubMed]

43. Cope, H.; Mutter, R.; Heal, W.; Pascoe, C.; Brown, P.; Pratt, S.; Chen, B. Synthesis and SAR study of acridine, 2-methylquinoline and 2-phenylquinazoline analogues as anti-prion agents. *Eur. J. Med. Chem.* **2006**, *41*, 1124–1143. [CrossRef] [PubMed]

44. Huang, Y.; Okochi, H.; May, B.C.; Legname, G.; Prusiner, S.B.; Benet, L.Z.; Guglielmo, B.J.; Lin, E.T. Quinacrine is mainly metabolized to mono-desethyl quinacrine by CYP3A4/5 and its brain accumulation is limited by P-glycoprotein. *Drug Metab. Dispos.* **2006**, *34*, 1136–1144. [CrossRef] [PubMed]

45. Nguyen, T.; Sakasegawa, Y.; Doh-Ura, K.; Go, M.L. Anti-prion activities and drug-like potential of functionalized quinacrine analogs with basic phenyl residues at the 9-amino position. *Eur. J. Med. Chem.* **2011**, *46*, 2917–2929. [CrossRef] [PubMed]

46. Nguyen, T.H.; Lee, C.Y.; Teruya, K.; Ong, W.Y.; Doh-ura, K.; Go, M.L. Antiprion activity of functionalized 9-aminoacridines related to quinacrine. *Bioorg. Med. Chem.* **2008**, *16*, 6737–6746. [CrossRef] [PubMed]

47. Kamatari, Y.O.; Hayano, Y.; Yamaguchi, K.; Hosokawa-Muto, J.; Kuwata, K. Characterizing antiprion compounds based on their binding properties to prion proteins: Implications as medical chaperones. *Protein Sci.* **2013**, *22*, 22–34. [CrossRef] [PubMed]

48. Nicoll, A.J.; Trevitt, C.R.; Tattum, M.H.; Risse, E.; Quarterman, E.; Ibarra, A.A.; Wright, C.; Jackson, G.S.; Sessions, R.B.; Farrow, M.; et al. Pharmacological chaperone for the structured domain of human prion protein. *Proc. Natl. Acad. Sci. USA* **2010**, *107*, 17610–17615. [CrossRef] [PubMed]

49. Baral, P.K.; Swayampakula, M.; Rout, M.K.; Kav, N.N.; Spyracopoulos, L.; Aguzzi, A.; James, M.N. Structural basis of prion inhibition by phenothiazine compounds. *Structure* **2014**, *22*, 291–303. [CrossRef] [PubMed]

50. Stincardini, C.; Massignan, T.; Biggi, S.; Elezgarai, S.R.; Sangiovanni, V.; Vanni, I.; Pancher, M.; Adami, V.; Moreno, J.; Stravalaci, M.; et al. An antipsychotic drug exerts anti-prion effects by altering the localization of the cellular prion protein. *PLoS ONE* **2017**, *12*, e0182589. [CrossRef] [PubMed]

51. Daniel, J.A.; Chau, N.; Abdel-Hamid, M.K.; Hu, L.; von Kleist, L.; Whiting, A.; Krishnan, S.; Maamary, P.; Joseph, S.R.; Simpson, F.; et al. Phenothiazine-derived antipsychotic drugs inhibit dynamin and clathrin-mediated endocytosis. *Traffic* **2015**, *16*, 635–654. [CrossRef] [PubMed]

52. Yamasaki, T.; Suzuki, A.; Hasebe, R.; Horiuchi, M. Comparison of the anti-prion mechanism of four different anti-prion compounds, anti-PrP monoclonal antibody 44B1, pentosan polysulfate, chlorpromazine, and u18666a, in prion-infected mouse neuroblastoma cells. *PLoS ONE* **2014**, *9*, e106516. [CrossRef] [PubMed]

53. Cavaliere, P.; Torrent, J.; Prigent, S.; Granata, V.; Pauwels, K.; Pastore, A.; Rezaei, H.; Zagari, A. Binding of methylene blue to a surface cleft inhibits the oligomerization and fibrillization of prion protein. *Biochim. Biophys. Acta* **2013**, *1832*, 20–28. [CrossRef] [PubMed]

54. Mori, T.; Koyama, N.; Segawa, T.; Maeda, M.; Maruyama, N.; Kinoshita, N.; Hou, H.; Tan, J.; Town, T. Methylene blue modulates beta-secretase, reverses cerebral amyloidosis, and improves cognition in transgenic mice. *J. Biol. Chem.* **2014**, *289*, 30303–30317. [CrossRef] [PubMed]

55. Sontag, E.M.; Lotz, G.P.; Agrawal, N.; Tran, A.; Aron, R.; Yang, G.; Necula, M.; Lau, A.; Finkbeiner, S.; Glabe, C.; et al. Methylene blue modulates huntingtin aggregation intermediates and is protective in huntington's disease models. *J. Neurosci.* **2012**, *32*, 11109–11119. [CrossRef] [PubMed]

56. Wischik, C.M.; Edwards, P.C.; Lai, R.Y.; Roth, M.; Harrington, C.R. Selective inhibition of alzheimer disease-like tau aggregation by phenothiazines. *Proc. Natl. Acad. Sci. USA* **1996**, *93*, 11213–11218. [CrossRef] [PubMed]

57. Yamashita, M.; Nonaka, T.; Arai, T.; Kametani, F.; Buchman, V.L.; Ninkina, N.; Bachurin, S.O.; Akiyama, H.; Goedert, M.; Hasegawa, M. Methylene blue and dimebon inhibit aggregation of TDP-43 in cellular models. *FEBS Lett.* **2009**, *583*, 2419–2424. [CrossRef] [PubMed]

58. Caughey, W.S.; Raymond, L.D.; Horiuchi, M.; Caughey, B. Inhibition of protease-resistant prion protein formation by porphyrins and phthalocyanines. *Proc. Natl. Acad. Sci. USA* **1998**, *95*, 12117–12122. [CrossRef] [PubMed]

59. Priola, S.A.; Raines, A.; Caughey, W.S. Porphyrin and phthalocyanine antiscrapie compounds. *Science* **2000**, *287*, 1503–1506. [CrossRef] [PubMed]

60. Massignan, T.; Cimini, S.; Stincardini, C.; Cerovic, M.; Vanni, I.; Elezgarai, S.R.; Moreno, J.; Stravalaci, M.; Negro, A.; Sangiovanni, V.; et al. A cationic tetrapyrrole inhibits toxic activities of the cellular prion protein. *Sci. Rep.* **2016**, *6*, 23180. [CrossRef] [PubMed]

61. Kocisko, D.A.; Caughey, W.S.; Race, R.E.; Roper, G.; Caughey, B.; Morrey, J.D. A porphyrin increases survival time of mice after intracerebral prion infection. *Antimicrob. Agents Chemother.* **2006**, *50*, 759–761. [CrossRef] [PubMed]

62. Rajora, M.A.; Lou, J.W.H.; Zheng, G. Advancing porphyrin's biomedical utility via supramolecular chemistry. *Chem. Soc. Rev.* **2017**, *46*, 6433–6469. [CrossRef] [PubMed]

63. Caspi, S.; Halimi, M.; Yanai, A.; Sasson, S.B.; Taraboulos, A.; Gabizon, R. The anti-prion activity of congo red. Putative mechanism. *J. Biol. Chem.* **1998**, *273*, 3484–3489. [CrossRef] [PubMed]

64. Milhavet, O.; Mange, A.; Casanova, D.; Lehmann, S. Effect of congo red on wild-type and mutated prion proteins in cultured cells. *J. Neurochem.* **2000**, *74*, 222–230. [CrossRef] [PubMed]

65. Caughey, B.; Brown, K.; Raymond, G.J.; Katzenstein, G.E.; Thresher, W. Binding of the protease-sensitive form of prp (prion protein) to sulfated glycosaminoglycan and congo red [corrected]. *J. Virol.* **1994**, *68*, 2135–2141. [PubMed]

66. Caughey, B.; Ernst, D.; Race, R.E. Congo red inhibition of scrapie agent replication. *J. Virol.* **1993**, *67*, 6270–6272. [PubMed]

67. Ingrosso, L.; Ladogana, A.; Pocchiari, M. Congo red prolongs the incubation period in scrapie-infected hamsters. *J. Virol.* **1995**, *69*, 506–508. [PubMed]

68. Rudyk, H.; Knaggs, M.H.; Vasiljevic, S.; Hope, J.; Birkett, C.; Gilbert, I.H. Synthesis and evaluation of analogues of congo red as potential compounds against transmissible spongiform encephalopathies. *Eur. J. Med. Chem.* **2003**, *38*, 567–579. [CrossRef]

69. Rudyk, H.; Vasiljevic, S.; Hennion, R.M.; Birkett, C.R.; Hope, J.; Gilbert, I.H. Screening congo red and its analogues for their ability to prevent the formation of PrP-res in scrapie-infected cells. *J. Gen. Virol.* **2000**, *81*, 1155–1164. [CrossRef] [PubMed]

70. Sellarajah, S.; Lekishvili, T.; Bowring, C.; Thompsett, A.R.; Rudyk, H.; Birkett, C.R.; Brown, D.R.; Gilbert, I.H. Synthesis of analogues of congo red and evaluation of their anti-prion activity. *J. Med. Chem.* **2004**, *47*, 5515–5534. [CrossRef] [PubMed]

71. Risse, E.; Nicoll, A.J.; Taylor, W.A.; Wright, D.; Badoni, M.; Yang, X.; Farrow, M.A.; Collinge, J. Identification of a compound that disrupts binding of amyloid-beta to the prion protein using a novel fluorescence-based assay. *J. Biol. Chem.* **2015**, *290*, 17020–17028. [CrossRef] [PubMed]

72. Kuwata, K.; Nishida, N.; Matsumoto, T.; Kamatari, Y.O.; Hosokawa-Muto, J.; Kodama, K.; Nakamura, H.K.; Kimura, K.; Kawasaki, M.; Takakura, Y.; et al. Hot spots in prion protein for pathogenic conversion. *Proc. Natl. Acad. Sci. USA* **2007**, *104*, 11921–11926. [CrossRef] [PubMed]

73. Kimura, T.; Hosokawa-Muto, J.; Kamatari, Y.O.; Kuwata, K. Synthesis of GN8 derivatives and evaluation of their antiprion activity in TSE-infected cells. *Bioorg. Med. Chem. Lett.* **2011**, *21*, 1502–1507. [CrossRef] [PubMed]

74. Perrier, V.; Wallace, A.C.; Kaneko, K.; Safar, J.; Prusiner, S.B.; Cohen, F.E. Mimicking dominant negative inhibition of prion replication through structure-based drug design. *Proc. Natl. Acad. Sci. USA* **2000**, *97*, 6073–6078. [CrossRef] [PubMed]

75. Guo, K.; Mutter, R.; Heal, W.; Reddy, T.R.; Cope, H.; Pratt, S.; Thompson, M.J.; Chen, B. Synthesis and evaluation of a focused library of pyridine dicarbonitriles against prion disease. *Eur. J. Med. Chem.* **2008**, *43*, 93–106. [CrossRef] [PubMed]

76. Reddy, T.R.; Mutter, R.; Heal, W.; Guo, K.; Gillet, V.J.; Pratt, S.; Chen, B. Library design, synthesis, and screening: Pyridine dicarbonitriles as potential prion disease therapeutics. *J. Med. Chem.* **2006**, *49*, 607–615. [CrossRef] [PubMed]

77. Heal, W.; Thompson, M.J.; Mutter, R.; Cope, H.; Louth, J.C.; Chen, B. Library synthesis and screening: 2,4-diphenylthiazoles and 2,4-diphenyloxazoles as potential novel prion disease therapeutics. *J. Med. Chem.* **2007**, *50*, 1347–1353. [CrossRef] [PubMed]

78. Thompson, M.J.; Louth, J.C.; Greenwood, G.K.; Sorrell, F.J.; Knight, S.G.; Adams, N.B.; Chen, B. Improved 2,4-diarylthiazole-based antiprion agents: Switching the sense of the amide group at C5 leads to an increase in potency. *ChemMedChem* **2010**, *5*, 1476–1488. [CrossRef] [PubMed]

79. Porat, Y.; Abramowitz, A.; Gazit, E. Inhibition of amyloid fibril formation by polyphenols: Structural similarity and aromatic interactions as a common inhibition mechanism. *Chem. Biol. Drug Des.* **2006**, *67*, 27–37. [CrossRef] [PubMed]

80. Rambold, A.S.; Miesbauer, M.; Olschewski, D.; Seidel, R.; Riemer, C.; Smale, L.; Brumm, L.; Levy, M.; Gazit, E.; Oesterhelt, D.; et al. Green tea extracts interfere with the stress-protective activity of PrP and the formation of PrP. *J. Neurochem.* **2008**, *107*, 218–229. [CrossRef] [PubMed]

81. Hosokawa-Muto, J.; Kamatari, Y.O.; Nakamura, H.K.; Kuwata, K. Variety of antiprion compounds discovered through an in silico screen based on cellular-form prion protein structure: Correlation between antiprion activity and binding affinity. *Antimicrob. Agents Chemother.* **2009**, *53*, 765–771. [CrossRef] [PubMed]

82. Kimura, T.; Hosokawa-Muto, J.; Asami, K.; Murai, T.; Kuwata, K. Synthesis of 9-substituted 2,3,4,9-tetrahydro-1H-carbazole derivatives and evaluation of their anti-prion activity in tse-infected cells. *Eur. J. Med. Chem.* **2011**, *46*, 5675–5679. [CrossRef] [PubMed]

83. Hyeon, J.W.; Choi, J.; Kim, S.Y.; Govindaraj, R.G.; Jam Hwang, K.; Lee, Y.S.; An, S.S.; Lee, M.K.; Joung, J.Y.; No, K.T.; et al. Discovery of novel anti-prion compounds using in silico and in vitro approaches. *Sci. Rep.* **2015**, *5*, 14944. [CrossRef] [PubMed]

84. Ma, B.; Yamaguchi, K.; Fukuoka, M.; Kuwata, K. Logical design of anti-prion agents using nagara. *Biochem. Biophys. Res. Commun.* **2016**, *469*, 930–935. [CrossRef] [PubMed]

85. Ishibashi, D.; Nakagaki, T.; Ishikawa, T.; Atarashi, R.; Watanabe, K.; Cruz, F.A.; Hamada, T.; Nishida, N. Structure-based drug discovery for prion disease using a novel binding simulation. *EBioMedicine* **2016**, *9*, 238–249. [CrossRef] [PubMed]

pathogens

MDPI

Review

Comparing the Folds of Prions and Other Pathogenic Amyloids

José Miguel Flores-Fernández [†], Vineet Rathod [†] and Holger Wille *

Department of Biochemistry & Centre for Prions and Protein Folding Diseases, University of Alberta, 204 Brain and Aging Research Building, Edmonton, AB T6G 2M8, Canada; floresfe@ualberta.ca (J.M.F.-F.); vrathod@ualberta.ca (V.R.)
* Correspondence: wille@ualberta.ca; Tel.: +1-780-248-1712
† These authors contributed equally to this work.

Received: 16 April 2018; Accepted: 2 May 2018; Published: 4 May 2018

Abstract: Pathogenic amyloids are the main feature of several neurodegenerative disorders, such as Creutzfeldt–Jakob disease, Alzheimer's disease, and Parkinson's disease. High resolution structures of tau paired helical filaments (PHFs), amyloid-β(1-42) (Aβ(1-42)) fibrils, and α-synuclein fibrils were recently reported using cryo-electron microscopy. A high-resolution structure for the infectious prion protein, PrP^{Sc}, is not yet available due to its insolubility and its propensity to aggregate, but cryo-electron microscopy, X-ray fiber diffraction, and other approaches have defined the overall architecture of PrP^{Sc} as a 4-rung β-solenoid. Thus, the structure of PrP^{Sc} must have a high similarity to that of the fungal prion HET-s, which is part of the fungal heterokaryon incompatibility system and contains a 2-rung β-solenoid. This review compares the structures of tau PHFs, Aβ(1-42), and α-synuclein fibrils, where the β-strands of each molecule stack on top of each other in a parallel in-register arrangement, with the β-solenoid folds of HET-s and PrP^{Sc}.

Keywords: prion structure; β-solenoid; PHF-tau structure; Aβ(1-42) fibril structure; α-synuclein amyloid structure; parallel in-register β-structure

---

1. Introduction

Pathogenic amyloids are the hallmark of many neurodegenerative diseases, such as Creutzfeldt-Jakob disease (CJD) [1], Alzheimer's disease (AD) [2], and Parkinson's disease (PD) [3]. Most proteins are subjected to covalent or non-covalent post-translational modifications, including N-linked glycosylation, disulfide bond formation, and protein folding, during which the protein obtains its native conformation and its functional state. Proper folding allows for the protein to reach and maintain a stable state and to exert its biological function(s). However, the amyloid state results from a process by which monomeric proteins or poorly folded peptides self-assemble into fibrillar aggregates, termed "amyloid". Amyloid fibrils have a common architecture, consisting of a "cross-β structure", which is independent of the native fold of the protein [4]. The misfolded proteins for CJD, AD, and PD are PrP^{Sc}, amyloid-β and microtubule-associated protein tau (tau), and α-synuclein, respectively. These proteins misfold and aggregate into a very stable β-sheet architecture, and eventually form amyloid fibrils. The molecular mechanisms how the altered conformations of these disease-associated proteins lead to slow, progressive neurodegeneration remain elusive. Recent studies on the spread of toxic forms of the tau protein and of truncated Aβ, which results from the proteolytic cleavage of the amyloid precursor protein, indicated that they share characteristics with PrP^{Sc} [5]. To date, there is no effective treatment for the prion diseases and most other pathogenic amyloids [6], thus making it necessary to understand the structure, formation, and aggregation processes of these pathogenic proteins to devise therapeutic strategies.

The infectious prion protein, termed PrP^{Sc}, has the ability to convert native PrP^C into a copy of itself, adopting a non-native conformation that has the propensity to self-assemble into amyloid fibrils [7]. The main features that distinguish PrP^{Sc} from PrP^C are its high content in β-structure [8], its partial resistance to proteases [9], its insolubility, and its propensity to aggregate into amyloid fibrils and other quaternary structures [10], which accumulate over time, resulting in brain cell and tissue damage [11]. Based on its structural properties, PrP^{Sc} differs in a number of epitopes from those that are recognized by antibodies targeting PrP^C [12]. Over the years, many structural models for PrP^{Sc} have been proposed [11]; however, the most prominent ones are the β-solenoid [13], the β-spiral [14], and the parallel in-register β-sheet models [15]. The insoluble nature and the propensity of PrP^{Sc} to aggregate make it difficult to determine its structure. Hence, the high-resolution structure of PrP^{Sc} is still unknown, although efforts have been made to gain insights by combining techniques, such as X-ray fiber diffraction [16], electron microscopy (EM) [17], and limited proteolysis using proteinase K [18]. Together, these data indicated that PrP^{Sc} is a β-solenoid protein, consisting of 4-stacked β-rungs [7]. However, it is still unknown whether the β-solenoid twists in a left- or right-handed sense.

Recent insights into the structure of PrP^{Sc} have revealed similarities and differences with the structures of non-pathogenic proteins adopting a β-solenoid fold. Moreover, other classes of pathogenic amyloids, such as Aβ(1-42) fibrils, tau PHFs, and α-synuclein fibrils, which have been recently solved by solid-state NMR spectroscopy (ssNMR) and cryo-EM [19–23], provide novel insights into the folds of self-propagating amyloids that cause neurodegenerative diseases. While these last three amyloids do not fall under the criteria of a β-solenoid, they are similar enough to warrant a side-by-side comparison. Thus, this review provides a comparison between the folds of PrP^{Sc} and those of other pathogenic amyloids.

2. β-Solenoid Amyloids: PrP^{Sc} and HET-s

2.1. The Structure of PrP^{Sc}

The current knowledge on the structure of PrP^{Sc} has been summarized in another review of this special issue [7]. Therefore, we provide only a short overview on what is known about the structure of PrP^{Sc}, its proteolytically truncated variant (PrP27-30), and related molecular species.

Fourier-transform infrared (FTIR) spectroscopy provided the first experimental evidence that the N-terminally truncated PrP27-30 contains predominantly β-structure [8,24]. Electron crystallography analyses on 2D crystals of PrP27-30 and an engineered variant of only 106 residues, PrP^{Sc}106, suggested the presence of a β-solenoid fold as a key feature of the infectious conformer [13,17]. Subsequently, X-ray fiber diffraction determined the molecular height of PrP27-30 in amyloid fibrils to be 19.2 Å, corresponding to the height of 4 β-strands (19.2 Å = 4 × 4.8 Å) [16]. In addition, the diffraction data confirmed that the core of PrP27-30 adopts a β-solenoid fold, consisting of 4-stacked β-rungs (Figure 1A). The repeating unit size of 19.2 Å was also found in the diffraction patterns that were obtained from PrP^{Sc}106 amyloid fibrils [25]. Recently, cryo-EM and subsequent three-dimensional (3D) reconstructions demonstrated that PrP27-30 amyloid fibrils can be formed by two intertwined protofilaments. Furthermore, the cryo-EM analysis corroborated that the structure of PrP27-30 consists of a 4-rung β-solenoid [26].

PrP^{Sc} has a β-sheet core that is assumed to be water-inaccessible with individual β-strands that are connected by short turns and loops [27], although it is still unknown which residues are in each β-strand and which ones are facing outward or inward with respect to the β-solenoid core [7,28]. A proposed model, based on mass spectrometry of proteinase K-resistant fragments obtained from PrP27-30, suggests which amino acids may be located in β-strands or connecting loops [18,28]. The structure of PrP^{Sc} has a high degree of stability, which provides resistance against denaturation and decontamination, but not all of the factors contributing to this stability are fully understood. It is clear that the highly-ordered hydrogen bonds that run up and down the β-sheets are essential for the structure and stability of the PrP^{Sc} β-solenoid fold [27]. However, van-der-Waals forces, hydrophobic and electrostatic interactions, as well as aromatic side-chain stacking also contribute to its stability, as it has been demonstrated in other amyloids [29] and β-solenoid proteins [30].

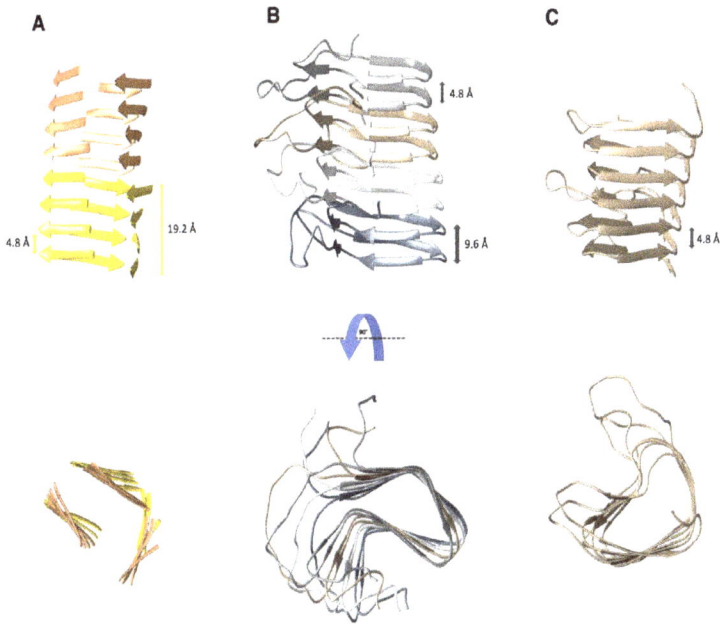

Figure 1. Three examples of proteins that adopt a β-solenoid fold. The characteristic distance between individual β-rungs is 4.8 ± 0.2 Å. (**A**) Representative model of the 4-rung β-solenoid architecture of PrP^{Sc} based on X-ray fiber diffraction and cryo-EM [16,26]. Characteristic distances of the 4-rung β-solenoid spacing are labeled. Each cartoon color represents a single PrP^{Sc} monomer. (**B**) Structure of an amyloid fibril formed by the prion domain (residues 218–289) of the fungal prion HET-s. Each monomer adopted a 2-rung left-handed β-solenoid fold and is shown in a different color. PDB access code: 2rnm [31]. (**C**) Structure of the right-handed β-solenoid protein pectate lyase C from *Erwinia chrysanthemi*. Its N- and C-terminal caps were removed for representation of the β-solenoid structure (residues 118–285). PDB access code: 2pec [32].

2.2. The Structure of the HET-s Prion Domain

HET-s is a known functional prion of the filamentous fungus *Podospora anserina* and it is involved in regulating heterokaryon incompatibility among different mating types [33,34]. The prion domain of HET-s (HET-s(218-289)) is able to form amyloid fibrils and it becomes protease resistant in the process [35]. Its structure, which has been solved by ssNMR, consists of a left-handed, 2-rung β-solenoid (Figure 1B) [31,36]. The β-solenoid rungs are connected by a flexible loop of 15 residues [37]. There are eight β-strands per molecule, and each β-solenoid rung has four β-strands connected by short loops. The first two β-strands are connected by a short, 2-residue β-arc, changing the orientation of the peptide backbone by ~90°; the second and third β-strands are connected by a 3-residue β-arc, changing the orientation by ~150°, and, lastly, the third and fourth β-strands are connected by a single glycine residue [31].

HET-s(218-289) has a triangular hydrophobic core formed by the first three β-strands of each β-solenoid rung, which also includes two buried polar amino acids (T233 and S273) and two asparagine ladders consisting of N226/N262 and N243/N279 [31]. The fourth β-strand is pointing away from the β-solenoid core, forming part of the loop that connects the first and the second rungs of the β-solenoid structure. The β-strands of the second β-solenoid rung stack on top of the β-strands of the first rung, forming four intramolecular β-sheets. These β-sheets can connect with the β-strands of the next HET-s(218-289) molecule through intermolecular hydrogen bonds/β-sheet contacts [31]. HET-s(218-289) contains polar and charged amino acids that are exposed on the surface of the β-solenoid structure, where three salt bridges are formed between the residues K229, E234 and R236 from the first β-rung and the residues E265, K270, and E272 from the second β-rung [36]. The HET-s(1-227) N-terminal domain consists of nine α-helices and two short β-strands [38,39]. Overall, HET-s(218-289) has an amino acid composition that is very different to the yeast prions which are Q/N rich [40].

In addition to the characterization via ssNMR, HET-s(218-289) has also been analyzed by cryo-EM [41] and X-ray fiber diffraction [42,43]. In the latter studies, meridional reflections at ~4.8 Å and ~9.6 Å confirmed that HET-s(218-289) adopts a 2-rung β-solenoid structure with clear similarities to the structure of PrP^{Sc} [16,42].

3. The β-solenoid Fold of Non-Pathogenic Proteins

β-solenoid proteins are characterized by a polypeptide chain that folds into more or less regular "solenoidal windings" (Figure 1B,C), while the canonical β-helical proteins follow a more stringent helical geometry [44]. β-solenoid proteins contain between three to well above 100 β-rungs [45]. Each β-rung contains two to four β-strands and they are connected by tight turns, β-arcs (two to six residues), or longer loops (Figure 1C). Overall, the β-rungs have a length between 12 and 30 amino acids. A β-rung corresponds to a complete turn of the amino acid backbone to where the next β-rung begins with an axial rise of 4.8 ± 0.2 Å [44,46]. The β-rungs that form the β-solenoid structure are connected by hydrogen bonds to the β-rungs above and below, forming a hydrophobic core with solvent-exposed side-chains on the surface [44,47]. A distinctive feature of β-solenoid proteins is the stacking of identical residues on the same position in subsequent β-rungs [44,48]. Such "ladders" are usually comprised of polar residues, with asparagine being the most commonly followed by serine and threonine, but aromatic residues can also form separate ladders that are stabilized by aromatic stacking [44,47,49,50].

β-solenoid proteins can be classified into right- or left-handed polymers, depending on the direction in which the polypeptide chain winds around the axis [30]. In addition, β-solenoid proteins can display a twist, which is determined by an angular offset between individual β-rungs [30,49]. The shape of a β-solenoid cross-section is defined by the β-arcs connecting each β-strand, with the most frequent shapes being generally triangular, rectangular, or oval [44]. The short β-arcs that connect individual β-strands are mainly formed by non-polar and uncharged polar residues [46], but longer loops can also connect subsequent β-rungs, while retaining the overall shape of the β-solenoid. Lastly, the N- and C-termini of β-solenoid proteins generally have polymerization-inhibiting caps, which contain polar and charged amino acids and protect the hydrophobic core from solvent exposure [51,52].

4. Amyloid Folds of Other Pathogenic Proteins

4.1. Short Amyloid Peptides

Short, amyloidogenic peptides can be studied in great detail due to their small size, propensity to form highly regular amyloid structures, and straightforward availability through synthetic routes. Thus, short peptides of the prion protein have been analyzed as microcrystals that were obtained from amyloid fibril preparations [53,54], while the structure of the more complex, full-length PrP^{Sc} remains unsolved [7].

These peptide structures have a minimal cross-β-sheet structure, and the structural motif was named a "steric zipper", based on the interdigitation of the side chains from neighboring β-sheets. In total, eight classes of steric zippers were described for these short amyloid peptides [55]. The steric zipper classes are defined by the orientation and the stacking of the β-strands that make up individual protofilaments—running either in parallel or antiparallel orientation. Moreover, adjacent β-sheets may pack face-to-face or face-to-back, and they also could be oriented up-up or up-down [54]. Examples of peptides that are derived from human (SNQNNF) or elk prion proteins (NNQNTF) adopted different steric zipper conformations [54,56], indicating the structural variability that is accessible to such short peptides. It was speculated that these sequence stretches would adopt similar conformations in PrP^Sc, but the X-ray fiber diffraction data contained no indication of such tight stacking of β-sheets in PrP^Sc/PrP27-30 [16,25]. Similarly, the tau peptide VQIVYK [54] does not adopt the same steric zipper arrangement in the PHF-tau structure that was recently elucidated by cryo-EM [20] (see below).

4.2. The Structure of PHF-Tau

The microtubule-associated protein tau is a natively unfolded protein consisting of up to 441 amino acids. It plays a key role in maintaining the elongated morphology of neurons by stabilizing microtubules in the axon [57,58]. The tau protein contains three or four sequence repeats (R1, R2, R3, and R4) as part of the microtubule-binding domain, with the presence of the repeat R2 being determined by alternative splicing of the mRNA. Tau inclusions are observed in the brains of diseased patients, either as paired helical filaments (PHFs) or straight filaments (SFs), with similar protofilament structures [5,59,60].

Studies involving X-ray fiber diffraction, circular dichroism (CD) spectroscopy, and FTIR spectroscopy on PHFs and SFs revealed a common cross-β architecture with stacks of β-strands being arranged perpendicular to the fiber axis [61–63]. A recent, high-resolution cryo-EM structure of PHFs and SFs at a resolution of 3.4 Å showed that R3 and R4 (residues V306–F378) form the core of intertwined protofilaments that are based on a C-shaped subunit (Figure 2A) [20]. The protofilament cores in PHFs and SFs are similar in structure and contain 8 β-strands that form the C-shaped subunit by packing in an overall antiparallel arrangement [20]. The organization of these β-strands is highly ordered through parallel in-register stacking of tau molecules, allowing for identical amino acids to stack on top of one another and forming hydrogen bond networks along the fiber axis with a helical rise of 4.8 Å [20,64]. The first β-strand of the PHF core contains the hexapeptide $_{306}$VQIVYK$_{311}$, which aligns through face-to-face packing with the residues $_{373}$THKLTF$_{378}$ of the eighth β-strand [20]. The interaction is based on hydrophobic contacts that are similar to a steric zipper, but distinct from the homotypic VQIVYK steric zipper that was predicted earlier by analyzing short, amyloidogenic peptides only [54].

An interesting feature that was observed in the structure of PHFs is a β-helix-like structure spanning residues 337–353 in R3, which resembles a single β-solenoid rung of the HET-s prion domain [20,31]. Thus, the core of PHFs contains two structural motifs parallel in-register β-sheets and β-helices. Overall, the side chains of the amino acids facing the exterior of the PHFs help to increase the stability by forming hydrogen bonds between charged residues, asparagines, and glutamines, while the β-helix-like structure and the interior of the PHF protofilaments are stabilized by hydrophobic clustering and aliphatic and aromatic stacking [20,55].

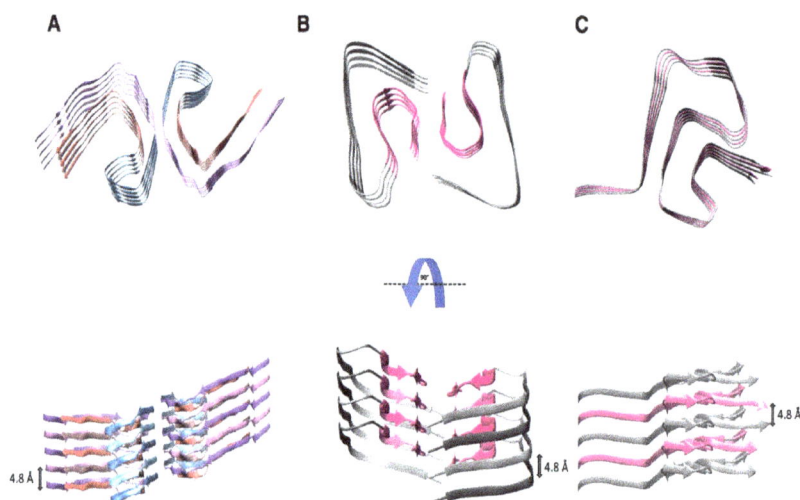

Figure 2. Structures of pathogenic amyloid fibrils (**A**) paired helical filament (PHF) of tau; (**B**) Aß(1-42); and, (**C**) α-synuclein. The fibril cores of these proteins are arranged as parallel in-register ß-sheet structures, which are stabilized by hydrophobic interactions, salt bridges, and hydrogen bonds up and down the fibril axis. The axial distance between the stacked protein molecules is 4.8 ± 0.2 Å. (**A**) Top and the side views of a high-resolution structure of PHFs obtained by cryo-EM [20]. Five successive layers of the tau protein along the fibril axis revealed that the fibril core is composed of two C-shaped subunits (residues 306–378). PDB access code: 5o3l [20]. (**B**) Top and side views of a high-resolution structure of an Aß(1-42) fibril produced by cryo-EM [19]. The LS-shaped cross-sections of each protofilament reveal a staggered stacking of molecules along the fibril axis. PDB access code: 5oqv [19]. (**C**) Top and side views of a high-resolution structure of α-synuclein determined by ssNMR and X-ray fiber diffraction [21]. The fibril core of α -synuclein contains a Greek key motif based on a parallel in-register ß-sheet topology (residues 42–96). PDB access code: 2n0a [21].

4.3. The Structure of Aβ(1-42)

Aβ is a peptide that is released through proteolytic cleavage of the amyloid precursor protein (APP), which is expressed in many tissues, including neurons [5,65]. It has been shown that the Aβ C-terminal variants Aβ(1-40) and Aβ(1-42) are prone to aggregation, forming toxic oligomers and amyloid fibrils. Aβ(1-42) is the predominant Aβ molecule that is found in neuritic plaques of Alzheimer's disease patients [65,66], but it is unclear how Aβ aggregates damage the brain and give rise to neuronal dysfunction and cellular toxicity [67].

Among others, studies using X-ray fiber diffraction, EM, and ssNMR have shown that Aβ fibrils consist of two protofilaments that intertwine along the fibril axis, suggesting a two-fold symmetry [67,68]. The fibrils were observed to have protofilaments with a diameter of ~5 nm and a ~110 nm cross over distance with a left-handed twist [65,69,70]. ssNMR indicated that each Aβ molecule contains four β-strands between amino acids 15 to 42, resulting in a highly ordered structure at the C-terminus of the peptide [55]. Within each Aβ molecule, the four β-strands stack on top of each other, forming an S-shaped structure (Figure 2B). The individual β-strands were stabilized by hydrogen bonds to the molecules above and below, forming contiguous β-sheets that run parallel to the fibril axis. Moreover, the C-terminus of each S-shaped Aβ molecule formed the fibril core with residues 23 to 26 interacting in a steric zipper-like fashion [55,67].

More recently, a high-resolution cryo-EM structure revealed that Aβ(1-42) fibrils do not possess a two-fold symmetry, but instead adopt a screw symmetry with a rise of ~4.8 Å. The staggered arrangement, which produced the screw symmetry and cannot be detected via ssNMR, resulted in each molecule of Aβ(1-42) interacting with six other molecules in the minimal fibril unit. Moreover, as a consequence of the screw symmetry, the Aβ(1-42) fibrils possess distinct end surfaces on the two protofilaments, referred to as "groove" and "ridge", based on residues 27–33, which either formed a protrusion or indentation at the protofilament end [19]. In addition, the Aβ(1-42) fibril structure was stabilized by salt bridges between residues D1 and K28, D7 and R5, as well as E11 and H6 / H13. The hydrophobic residues were buried in the protofilament core, while the polar side chains were facing the outside of the protofilaments [19].

4.4. The Structure α-Synuclein

Relatively little is known about the structure and the function of the cytoplasmic protein, α-synuclein, which is mainly found in the presynaptic terminals of neuronal cells. The native form consists of three domains namely: the N-terminal lipid-binding α-helix (residues 1–60), the amyloid binding central domain (NAC; residues 61–95), and the C-terminal acidic tail (residues 96–140) [21,71,72]. α-synuclein is a natively unfolded protein in aqueous solution and an α-helical protein in association with phospholipids. Monomers of α-synuclein tend to aggregate into oligomers through an unknown mechanism, which then further assemble forming long amyloid fibrils. These fibrils are seen in the Lewy bodies and Lewy neurites from patients with Parkinson's disease (PD) [73]. A 3D structure of α-synuclein amyloid fibrils has been determined using ssNMR and was further validated by EM and X-ray fiber diffraction [21,74]. The ssNMR analysis showed that the core of the α-synuclein fibril consists of residues 44 to 97, forming seven β-strands that adopt a Greek-key topology [21,75]. The resulting β-sheets form a parallel in register cross-β structure that is stabilized by hydrogen bonds along the fibril axis.

Recently, a high-resolution structure of the full-length α-synuclein amyloid fibril was deciphered using cryo-EM at 3.4 Å resolution, which revealed two intertwined protofilaments that were composed of a staggered arrangement of subunits [23]. The fibrils adopt a 2_1 symmetry similar to the one seen with PHF and Aβ(1-42) fibrils [19,20,23]. The fibril core consisted of residues 42 to 95 (as shown in Figure 2C) and contained 8 β-strands forming a parallel in-register β-sheet structure with a spacing of ~4.8 Å between the β-sheets [23]. The side chains in the core were tightly packed through hydrophobic and aromatic interactions, forming a hydrophobic pocket containing residues I88, A91, and F94 [21,76]. The observed Greek-key topology was supported through strong hydrophobic interactions between residues V77, V82, A89, and A90, which formed β-strands [21]. Moreover, the β-strands in each molecule of α-synuclein were supported by a glutamine ladder along the fibril axis and turns that mainly consisted of small hydrophobic amino acids, such as alanine and glycine.

5. Similarities and Differences between Common Amyloid Folds

Many different techniques have been used to elucidate the structures of amyloid fibrils from neurodegenerative diseases as well as functional amyloids [77]. These structures have been studied using recombinant proteins, short peptides, as well as full-length proteins (see above). Substantial progress has been made using the classic X-ray fiber diffraction approach, as well as more recent techniques, such as ssNMR and cryo-EM [78]. A spate of structural studies on amyloids containing Aβ(1-42), α-synuclein, the tau protein (as PHFs and SFs), HET-s, and PrP^{Sc} revealed two predominant folds: the parallel in-register β-structure and the β-solenoid structure (compare Table 1).

Table 1. Comparison of structural features in pathogenic and non-pathogenic amyloids.

	Protein/Protein Aggregate	β-Solenoid	Parallel in-Register	Steric Zippers		Salt Bridges[2]		Hydro-Phobic Core	Symmetry of Proto-Filaments
				Homo-Steric[1]	Hetero-Steric[2]	Intra-Molecular	Inter-Molecular		
Pathogenic proteins	PHF-tau	(partial)	+	1	6	5	2	+	2_1 screw
	Aβ(1-42)	–	+	1	2	0	2	+	2_1 screw
	α-synuclein	–	+	1	2	2	2	+	2_1 screw
	PrP^{Sc}	+	–	0	0	unknown	unknown	+	unknown
Non-pathogenic proteins	HET-s	+	–	0	0	3	0	+	N. A.
	Pectate lyase C	+	–	0	0	15	N. A.	+	N. A.

[1] observed at the protofilament interface; [2] based on two protofilaments.

5.1. Similarities between PrP^{Sc}, HET-s, and β-solenoid Proteins

Interestingly, the two β-solenoid structures among the amyloids are both linked to prions and the autocatalytic conversion from an innocuous precursor into the prion state. Thus, it is informative to compare these two structures (to the degree that they are known) and those of other, non-pathogenic β-solenoid proteins. The HET-s(218-289) prion domain contains 71 residues, and its structure forms a 2-rung β-solenoid [31], which means that for a similar 4-rung β-solenoid structure, 142 residues would be needed. Intriguingly, this matches perfectly with the ~143 residues in the proteinase K-resistant core of PrP^{Sc} (PrP27-30), which was revealed to contain a 4-rung β-solenoid structure (Figure 1A,B) [7,16,26]. Many β-solenoids proteins contain asparagine ladders as a prominent surface feature, which contributes to the stability of the fold. This feature can also be found in the HET-s(218-289) 2-rung β-solenoid structure with N226/N262 and N243/N279 stacking on top of one another [31]. Other amino acids that are known to form stabilizing ladders on the surface of β-solenoids proteins are threonine and tyrosine, while glycine is prominently found in the turns that are connecting β-strands due to its small size [44,46,47]. The high proportion of these residues in sequence of PrP27-30 (Thr (7.7%), Tyr (7.7%) and Asn (6.3%)) suggests that such ladders could also stabilize the 4-rung β-solenoid fold of PrP^{Sc}. In addition, PrP27-30 contains even more Gly (9.9%), which could serve to make tight turns between adjacent β-strands [28]. Moreover, β-solenoid proteins usually have a hydrophobic core that stabilizes its structure, while the side chains of solvent exposed residues tend to be hydrophilic [44,46,47]. This arrangement is also found in the structure of HET-s(218-289) [31], and are assumed to apply for the structure of PrP^{Sc} as well [7,11,28].

Lastly, most β-solenoid proteins have a cap at the N- or C-termini (or both), which prevents the propagation of the β-solenoid fold into other regions of the protein [51]. This cap also prevents β-solenoid proteins from aggregating, since the removal of the cap has been observed to result in insoluble and/or amyloidogenic proteins [79]. Since PrP^{Sc} has not been subjected to evolutionary selection it does not appear to contain a capping structure, which would prevent aggregation and the conversion of other protein molecules (e.g., PrP^C) [7]. It could be argued that the N-terminal, α-helix rich domain of full-length HET-s could be considered a cap for the β-solenoid structure of the prion domain [39].

5.2. Comparisons of Aβ(1-42), α-Synuclein, PHF-tau and β-Solenoid Amyloids

Recent high-resolution structures for Aβ(1-42) and α-synuclein amyloid fibrils, as well as tau PHFs that were obtained via cryo-EM, demonstrated that all these fibrils are based on a parallel in-register β-sheet architecture [19,20,23]. This means that each successive layer in the fibril consists of another protein molecule that stacks on top of the preceding one without substantial translation or rotation. However, in all three pathogenic fibrils a staggered conformation of the subunits was observed in the structures determined by cryo-EM, which resulted in a slight tilt for each layer. In turn, this tilt imparted the fibrils with a 2_1 screw symmetry. The parallel in-register stacking of proteins renders these assemblies very sensitive to charge repulsion wherever charged amino acids are found, unless salt bridges neutralize the charge imbalance.

Similarly, the amino acid ladders that can be found in β-solenoid proteins take advantage of stabilizing interactions between identical/similar residues (see above). Furthermore, charge repulsion between charged residues also threatens the stability of β-solenoid proteins, and needs to be overcome through formation of salt bridges. Nevertheless, the stacking of subsequent β-solenoid rungs allows for sequence variations that are not open to the parallel in-register β-sheet architecture. In contrast to the parallel in-register β-sheet fold, β-solenoid proteins possess a core that is encircled by a continuous peptide chain, and is mainly composed of hydrophobic residues, and which is formed by the inward facing amino acid side chains. The β-structure of the β-rungs causes the side chains of the amino acids to alternate between inward and outward facing orientations, respectively.

Thus, the structures of parallel in-register β-sheet amyloids and those of β-solenoid amyloids are governed by the same types of interactions, but they fall into separate structure classes.

Nevertheless, substantial overlap exists and both HET-s(218-289) and the mammalian prion protein (PrP) can also adopt a parallel in-register fold, as demonstrated through X-ray fiber diffraction and ssNMR [16,42,80,81].

6. Conclusions

In the last few years, high-resolution structures have been determined from Aβ(1-42) fibrils [19,22], PHF-tau filaments [20], α-synuclein fibrils [21,23], as well as HET-s(218-289) [31,36,41] using cryo-EM and ssNMR. However, a high-resolution structure of PrP^{Sc} is still lacking [7], and only lower-resolution data are available [8,13,16–18,24,26,27]. Nevertheless, we can classify the structures of these amyloid fibrils as either a parallel in-register fold (Aβ(1-42), PHF-tau, and α-synuclein) or a β-solenoid fold (HET-s(218-289) and PrP^{Sc}). While these folds differ in their principle architecture, they have many molecular details in common (compare Table 1). In addition, the pathogenic amyloids are well known for their resistance against denaturation, proteolysis, and biological clearance, which is based on the properties of the underlying amyloid fold. A detailed understanding how these molecular details influence the pathogenicity of the protein fibrils/aggregates is still missing.

Author Contributions: H.W. outlined the concept for this review; J.M.F.-F. and V.R. wrote the text; H.W. revised and edited the manuscript, and all authors approved the final version.

Acknowledgments: The authors would like to acknowledge support from the Alberta Prion Research Institute grants (APRI 201600012, 201600029, and 201700005) and the Alberta Livestock and Meat Agency (grant ALMA 2016A001R).

Conflicts of Interest: The authors declare no conflict of interest.

References

1. Gibbs, C.J.J.; Gajdusek, D.C.; Asher, D.M.; Alpers, M.P.; Beck, E.; Daniel, P.M.; Matthews, W.B. Creutzfeldt-Jakob Disease (Spongiform Encephalopathy): Transmission to Chimpanzee. *Science* **1968**, *161*, 388–389. [CrossRef] [PubMed]

2. Moller, H.J.; Graeber, M.B. The case described by Alois Alzheimer in 1911. *Eur. Arch. Psychiatry Clin. Neurosci.* **1998**, *248*, 111–122. [CrossRef] [PubMed]

3. Parkinson, J. *An Essay on the Shaking Palsy*; Whittingham & Rowland: London, UK, 1817.

4. Eisenberg, D.; Jucker, M. The amyloid state of proteins in human diseases. *Cell* **2012**, *148*, 1188–1203. [CrossRef] [PubMed]

5. Tatarnikova, O.G.; Orlov, M.A.; Bobkova, N.V. Beta-Amyloid and Tau-Protein: Structure, Interaction, and Prion-Like Properties. *Biochemistry* **2015**, *80*, 1800–1819. [CrossRef] [PubMed]

6. Giles, K.; Olson, S.H.; Prusiner, S.B. Developing Therapeutics for PrP Prion Diseases. *Cold Spring Harb. Perspect. Med.* **2017**, *7*, 1–20. [CrossRef] [PubMed]

7. Wille, H.; Requena, J.R. The Structure of PrPSc Prions. *Pathogens* **2018**, *7*, 20. [CrossRef] [PubMed]

8. Caughey, B.W.; Dong, A.; Bhat, K.S.; Ernst, D.; Hayes, S.F.; Caughey, W.S. Secondary structure analysis of the scrapie-associated protein PrP 27-30 in water by infrared spectroscopy. *Biochemistry* **1991**, *30*, 7672–7680. [CrossRef] [PubMed]

9. Prusiner, S.B.; Bolton, D.C.; Groth, D.F.; Bowman, K.A.; Cochran, S.P.; McKinley, M.P. Further purification and characterization of scrapie prions. *Biochemistry* **1982**, *21*, 6942–6950. [CrossRef] [PubMed]

10. Vázquez-Fernández, E.; Young, H.S.; Requena, J.R.; Wille, H. The Structure of Mammalian Prions and Their Aggregates. *Int. Rev. Cell Mol. Biol.* **2017**, *329*, 277–301. [PubMed]

11. Requena, J.R.; Wille, H. The Structure of the Infectious Prion Protein. *Prion* **2014**, *8*, 60–66. [CrossRef] [PubMed]

12. Horiuchi, M.; Karino, A.; Furuoka, H.; Ishiguro, N.; Kimura, K.; Shinagawa, M. Generation of Monoclonal Antibody That Distinguishes PrPSc from PrPC and Neutralizes Prion Infectivity. *Virology* **2009**, *394*, 200–207. [CrossRef] [PubMed]

13. Govaerts, C.; Wille, H.; Prusiner, S.B.; Cohen, F.E. Evidence for Assembly of Prions with Left-Handed-Helices into Trimers. *Proc. Natl. Acad. Sci. USA* **2004**, *101*, 8342–8347. [CrossRef] [PubMed]

14. DeMarco, M.L.; Daggett, V. From Conversion to Aggregation: Protofibril Formation of the Prion Protein. *Proc. Natl. Acad. Sci. USA* **2004**, *101*, 2293–2298. [CrossRef] [PubMed]

15. Cobb, N.J.; Sönnichsen, F.D.; Mchaourab, H.; Surewicz, W.K. Molecular Architecture of Human Prion Protein Amyloid: A Parallel, in-Register β-Structure. *Proc. Natl. Acad. Sci. USA* **2007**, *104*, 18946–18951. [CrossRef] [PubMed]

16. Wille, H.; Bian, W.; McDonald, M.; Kendall, A.; Colby, D.W.; Bloch, L.; Ollesch, J.; Borovinskiy, A.L.; Cohen, F.E.; Prusiner, S.B.; et al. Natural and Synthetic Prion Structure from X-Ray Fiber Diffraction. *Proc. Natl. Acad. Sci. USA* **2009**, *106*, 16990–16995. [CrossRef] [PubMed]

17. Wille, H.; Michelitsch, M.D.; Guenebaut, V.; Supattapone, S.; Serban, A.; Cohen, F.E.; Agard, D.A.; Prusiner, S.B. Structural Studies of the Scrapie Prion Protein by Electron Crystallography. *Proc. Natl. Acad. Sci. USA* **2002**, *99*, 3563–3568. [CrossRef] [PubMed]

18. Vázquez-Fernández, E.; Alonso, J.; Pastrana, M.A.; Ramos, A.; Stitz, L.; Vidal, E.; Dynin, I.; Petsch, B.; Silva, C.J.; Requena, J.R. Structural Organization of Mammalian Prions as Probed by Limited Proteolysis. *PLoS ONE* **2012**, *7*, e50111. [CrossRef] [PubMed]

19. Gremer, L.; Schölzel, D.; Schenk, C.; Reinartz, E.; Labahn, J.; Ravelli, R.B.G.; Tusche, M.; Lopez-Iglesias, C.; Hoyer, W.; Heise, H.; et al. Fibril Structure of Amyloid-β(1–42) by Cryoelectron Microscopy. *Science* **2017**, *358*, 116–119. [CrossRef] [PubMed]

20. Fitzpatrick, A.W.P.; Falcon, B.; He, S.; Murzin, A.G.; Murshudov, G.; Garringer, H.J.; Crowther, R.A.; Ghetti, B.; Goedert, M.; Scheres, S.H.W. Cryo-EM Structures of Tau Filaments from Alzheimer's Disease. *Nature* **2017**, *547*, 185–190. [CrossRef] [PubMed]

21. Tuttle, M.D.; Comellas, G.; Nieuwkoop, A.J.; Covell, D.J.; Berthold, D.A.; Kloepper, K.D.; Courtney, J.M.; Kim, J.K.; Barclay, A.M.; Kendall, A.; et al. Solid-State NMR Structure of a Pathogenic Fibril of Full-Length Human α-Synuclein. *Nat. Struct. Mol. Biol.* **2016**, *23*, 409–415. [CrossRef] [PubMed]

22. Wälti, M.A.; Ravotti, F.; Arai, H.; Glabe, C.G.; Wall, J.S.4; Böckmann, A.; Güntert, P.; Meier, B.H.; Riek, R. Atomic-resolution structure of a disease-relevant Aβ(1–42) amyloid fibril. *Proc. Natl. Acad. Sci. USA* **2016**, *113*, E4976–E4984. [CrossRef] [PubMed]

23. Guerrero-Ferreira, R.; Nicholas, M.I.; Mona, T.D.; Ringler, P.; Lauer, M.E.; Riek, R.; Britschgi, M.; Stahlberg, H. Cryo-EM structure of alpha-synuclein fibrils. *bioRxiv* **2018**, 276436. [CrossRef]

24. Pan, K.M.; Baldwin, M.; Nguyen, J.; Gasset, M.; Serban, A.; Groth, D.; Mehlhorn, I.; Huang, Z.; Fletterick, R.J.; Cohen, F.E. Conversion of α-Helices into β-Sheets Features in the Formation of the Scrapie Prion Proteins. *Proc. Natl. Acad. Sci. USA* **1993**, *90*, 10962–10966. [CrossRef] [PubMed]

25. Wan, W.; Wille, H.; Stöhr, J.; Kendall, A.; Bian, W.; McDonald, M.; Tiggelaar, S.; Watts, J.C.; Prusiner, S.B.; Stubbs, G. Structural Studies of Truncated Forms of the Prion Protein PrP. *Biophys. J.* **2015**, *108*, 1548–1554. [CrossRef] [PubMed]

26. Vázquez-Fernández, E.; Vos, M.R.; Afanasyev, P.; Cebey, L.; Sevillano, A.M.; Vidal, E.; Rosa, I.; Renault, L.; Ramos, A.; Peters, P.J.; et al. The Structural Architecture of an Infectious Mammalian Prion Using Electron Cryomicroscopy. *PLoS Pathog.* **2016**, *12*, 1–21. [CrossRef] [PubMed]

27. Smirnovas, V.; Baron, G.S.; Offerdahl, D.K.; Raymond, G.J.; Caughey, B.; Surewicz, W.K. Structural Organization of Brain-Derived Mammalian Prions Examined by Hydrogen-Deuterium Exchange. *Nat. Struct. Mol. Biol.* **2011**, *18*, 504–506. [CrossRef] [PubMed]

28. Silva, C.J.; Vázquez-Fernández, E.; Onisko, B.; Requena, J.R. Proteinase K and the Structure of PrPSc: The Good, the Bad and the Ugly. *Virus Res.* **2015**, *207*, 120–126. [CrossRef] [PubMed]

29. Tsemekhman, K.; Goldschmidt, L.; Eisenberg, D.S.; Baker, D. Cooperative Hydrogen Bonding in Amyloid Formation. *Protein Sci.* **2007**, *16*, 761–764. [CrossRef] [PubMed]

30. Kobe, B.; Kajava, A.V. When Protein Folding Is Simplified to Protein Coiling: The Continuum of Solenoid Protein Structures. *Trends Biochem. Sci.* **2000**, *25*, 509–515. [CrossRef]

31. Wasmer, C.; Lange, A.; Melckebeke, H.V.; Siemer, A.B.; Riek, R.; Meier, B.H. Amyloid Fibrils of the HET-s(218–289) Prion Form a B Solenoid with a Triangular Hydrophobic Core. *Science* **2008**, *319*, 1523–1527. [CrossRef] [PubMed]

32. Yoder, M.D.; Jurnak, F. Protein motifs. 3. The parallel beta helix and other coiled folds. *FASEB J.* **1995**, *9*, 335–342. [CrossRef] [PubMed]

33. Coustou, V.; Deleu, C.; Saupe, S.; Begueret, J. The Protein Product of the Het-S Heterokaryon Incompatibility Gene of the Fungus Podospora Anserina Behaves as a Prion Analog. *Proc. Natl. Acad. Sci. USA* **1997**, *94*, 9773–9778. [CrossRef] [PubMed]
34. Wickner, R.B. A New Prion Controls Fungal Cell Fusion Incompatibility. *Proc. Natl. Acad. Sci. USA* **1997**, *94*, 10012–10014. [CrossRef] [PubMed]
35. Dos Reis, S.; Coulary-Salin, B.; Forge, V.; Lascu, I.; Bégueret, J.; Saupe, S.J. The HET-S Prion Protein of the Filamentous Fungus Podospora Anserina Aggregates in Vitro into Amyloid-like Fibrils. *J. Biol. Chem.* **2002**, *277*, 5703–5706. [CrossRef] [PubMed]
36. Melckebeke, V.; Wasmer, C.; Lange, A.; Ab, E.; Loquet, A.; Böckmann, A.; Meier, B.H. Atomic-Resolution Three-Dimensional Structure of HET-S(218–289) Amyloid Fibrils by Solid-State NMR. *J. Am. Chem. Soc.* **2010**, *132*, 13765–13775. [CrossRef] [PubMed]
37. Ritter, C.; Maddelein, M.-L.; Siemer, A.B.; Lührs, T.; Ernst, M.; Meier, B.H.; Saupe, S.J.; Riek, R. Correlation of Structural Elements and Infectivity of the HET-S Prion. *Nature* **2005**, *435*, 844–848. [CrossRef] [PubMed]
38. Balguerie, A.; Dos Reis, S.; Ritter, C.; Chaignepain, S.; Coulary-Salin, B.; Forge, V.; Bathany, K.; Lascu, I.; Schmitter, J.M.; Riek, R.; et al. Domain Organization and Structure-Function Relationship of the HET-S Prion Protein of Podospora Anserina. *EMBO J.* **2003**, *22*, 2071–2081. [CrossRef] [PubMed]
39. Greenwald, J.; Buhtz, C.; Ritter, C.; Kwiatkowski, W.; Choe, S.; Maddelein, M.-L.; Ness, F.; Cescau, S.; Soragni, A.; Leitz, D.; et al. The Mechanism of Prion Inhibition by HET-S. *Mol. Cell* **2010**, *38*, 889–899. [CrossRef] [PubMed]
40. Liebman, S.W.; Chernoff, Y.O. Prions in Yeast. *Genetics* **2012**, *191*, 1041–1072. [CrossRef] [PubMed]
41. Mizuno, N.; Baxa, U.; Steven, A.C. Structural Dependence of HET-S Amyloid Fibril Infectivity Assessed by Cryoelectron Microscopy. *Proc. Natl. Acad. Sci. USA* **2011**, *108*, 3252–3257. [CrossRef] [PubMed]
42. Wan, W.; Wille, H.; Stöhr, J.; Baxa, U.; Prusiner, S.B.; Stubbs, G. Degradation of fungal prion HET-s(218–289) induces formation of a generic amyloid fold. *Biophys. J.* **2012**, *102*, 2339–2344. [CrossRef] [PubMed]
43. Wan, W.; Stubbs, G. Fungal Prion HET-S as a Model for Structural Complexity and Self-Propagation in Prions. *Proc. Natl. Acad. Sci. USA* **2014**, *111*, 5201–5206. [CrossRef] [PubMed]
44. Kajava, A.V.; Steven, A.C. β-Rolls, β-Helices, and Other β-Solenoid Proteins. *Adv. Protein Chem.* **2006**, *73*, 55–96. [PubMed]
45. Kajava, A.V.; Steven, A.C. The Turn of the Screw: Variations of the Abundant β-Solenoid Motif in Passenger Domains of Type V Secretory Proteins. *J. Struct. Biol.* **2006**, *155*, 306–315. [CrossRef] [PubMed]
46. Kajava, A.V.; Baxa, U.; Steven, A.C. Beta Arcades: Recurring Motifs in Naturally Occurring and Disease-Related Amyloid Fibrils. *FASEB J.* **2010**, *24*, 1311–1319. [CrossRef] [PubMed]
47. Hennetin, J.; Jullian, B.; Steven, A.C.; Kajava, A.V. Standard Conformations of β-Arches in β-Solenoid Proteins. *J. Mol. Biol.* **2006**, *358*, 1094–1105. [CrossRef] [PubMed]
48. Jenkins, J.; Pickersgill, R. The Architecture of Parallel β-Helices and Related Folds. *Prog. Biophys. Mol. Biol.* **2001**, *77*, 111–175. [CrossRef]
49. Yoder, M.D.; Lietzke, S.E.; Jurnak, F. Unusual Structural Features in the Parallel β-Helix in Pectate Lyases. *Structure* **1993**, *1*, 241–251. [CrossRef]
50. Henrissat, B.; Heffron, S.E.; Yoder, M.D.; Lietzke, S.E.; Jurnak, F. Functional Implications of Structure-Based Sequence Alignment of Proteins in the Extracellular Pectate Lyase Superfamily. *Plant Physiol.* **1995**, *107*, 963–976. [CrossRef] [PubMed]
51. Bryan, A.W.; Starner-Kreinbrink, J.L.; Hosur, R.; Clark, P.L.; Berger, B. Structure-Based Prediction Reveals Capping Motifs That Inhibit β-Helix Aggregation. *Proc. Natl. Acad. Sci. USA* **2011**, *108*, 11099–11104. [CrossRef] [PubMed]
52. Kondo, H.; Hanada, Y.; Sugimoto, H.; Hoshino, T.; Garnham, C.P.; Davies, P.L.; Tsuda, S. Ice-Binding Site of Snow Mold Fungus Antifreeze Protein Deviates from Structural Regularity and High Conservation. *Proc. Natl. Acad. Sci. USA* **2012**, *109*, 9360–9365. [CrossRef] [PubMed]
53. Nelson, R.; Sawaya, M.R.; Balbirnie, M.; Madsen, A.Ø.; Riekel, C.; Grothe, R.; Eisenberg, D. Structure of the Cross-β Spine of Amyloid-like Fibrils. *Nature* **2005**, *435*, 773–778. [CrossRef] [PubMed]
54. Sawaya, M.R.; Sambashivan, S.; Nelson, R.; Ivanova, M.I.; Sievers, S.A.; Apostol, M.I.; Thompson, M.J.; Balbirnie, M.; Wiltzius, J.J.W.; McFarlane, H.T.; et al. Atomic Structures of Amyloid Cross-β Spines Reveal Varied Steric Zippers. *Nature* **2007**, *447*, 453–457. [CrossRef] [PubMed]

55. Riek, R.; Eisenberg, D.S. The Activities of Amyloids from a Structural Perspective. *Nature* **2016**, *539*, 227–235. [CrossRef] [PubMed]

56. Wiltzius, J.J.W.; Landau, M.; Nelson, R.; Sawaya, M.R.; Apostol, M.I.; Goldschmidt, L.; Soriaga, A.B.; Cascio, D.; Rajashankar, K.; Eisenberg, D. Molecular Mechanisms for Protein-Encoded Inheritance. *Nat. Struct. Mol. Biol.* **2009**, *16*, 973–978. [CrossRef] [PubMed]

57. Goedert, M.; Spillantini, M.G.; Jakes, R.; Rutherford, D.; Crowther, R.A. Multiple Isoforms of Human Microtubule-Associated Protein Tau: Sequences and Localization in Neurofibrillary Tangles of Alzheimer's disease. *Neuron* **1989**, *3*, 519–526. [CrossRef]

58. Mandelkow, E.M.; Mandelkow, E. Tau in Alzheimer's Disease. *Trends Cell Biol.* **1998**, *8*, 425–427. [CrossRef]

59. Crowther, R.A. Straight and Paired Helical Filaments in Alzheimer Disease Have a Common Structural Unit. *Proc. Natl. Acad. Sci. USA* **1991**, *88*, 2288–2292. [CrossRef] [PubMed]

60. Šimić, G.; Babić Leko, M.; Wray, S.; Harrington, C.; Delalle, I.; Jovanov-Milošević, N.; Bažadona, D.; Buée, L.; de Silva, R.; Di Giovanni, G.; et al. Tau Protein Hyperphosphorylation and Aggregation in Alzheimer's Disease and Other Tauopathies, and Possible Neuroprotective Strategies. *Biomolecules* **2016**, *6*, 1–28. [CrossRef] [PubMed]

61. Kirschner, D.A.; Abraham, C.; Selkoe, D.J. X-ray diffraction from intraneuronal paired helical filaments and extraneuronal amyloid fibers in Alzheimer disease indicates cross-beta conformation. *Proc. Natl. Acad. Sci. USA* **1986**, *83*, 503–507. [CrossRef] [PubMed]

62. Schweers, O.; Schönbrunn-Hanebeck, E.; Marx, A.; Mandelkow, E. Structural Studies of Tau-Protein and Alzheimer Paired Helical Filaments Show No Evidence for Beta-Structure. *J. Biol. Chem.* **1994**, *269*, 24290–24297. [PubMed]

63. von Bergen, M.; Barghorn, S.; Biernat, J.; Mandelkow, E.M.; Mandelkow, E. Tau aggregation is driven by a transition from random coil to beta sheet structure. *Biochim. Biophys. Acta* **2005**, *1739*, 158–166. [CrossRef] [PubMed]

64. Margittai, M.; Langen, R. Template-Assisted Filament Growth by Parallel Stacking of Tau. *Proc. Natl. Acad. Sci. USA* **2004**, *101*, 10278–10283. [CrossRef] [PubMed]

65. Baxa, U. Structural Basis of Infectious and Non-Infectious Amyloids. *Curr. Alzheimer Res.* **2008**, *5*, 308–318. [CrossRef] [PubMed]

66. Lührs, T.; Ritter, C.; Adrian, M.; Riek-Loher, D.; Bohrmann, B.; Döbeli, H.; Schubert, D.; Riek, R. 3D structure of Alzheimer's amyloid-beta(1–42) fibrils. *Proc. Natl. Acad. Sci. USA* **2005**, *102*, 17342–17347. [CrossRef] [PubMed]

67. Schmidt, M.; Rohou, A.; Lasker, K.; Yadav, J.K.; Schiene-Fischer, C.; Fändrich, M.; Grigorieff, N. Peptide Dimer Structure in an Aβ(1–42) Fibril Visualized with Cryo-EM. *Proc. Natl. Acad. Sci. USA* **2015**, *112*, 11858–11863. [CrossRef] [PubMed]

68. Riek, R. The Three-Dimensional Structures of Amyloids. *Cold Spring Harb. Perspect. Biol.* **2017**, *9*, 1–12. [CrossRef] [PubMed]

69. Goldsbury, C.S.; Wirtz, S.; Müller, S.A.; Sunderji, S.; Wicki, P.; Aebi, U.; Frey, P. Studies on the in Vitro Assembly of Aβ1-40: Implications for the Search for a Beta Fibril Formation Inhibitors. *J. Struct. Biol.* **2000**, *130*, 217–231. [CrossRef] [PubMed]

70. Sachse, C.; Xu, C.; Wieligmann, K.; Diekmann, S.; Grigorieff, N.; Fändrich, M. Quaternary Structure of a Mature Amyloid Fibril from Alzheimer's Aβ(1–40) Peptide. *J. Mol. Biol.* **2006**, *362*, 347–354. [CrossRef] [PubMed]

71. Rodriguez, J.A.; Ivanova, M.I.; Sawaya, M.R.; Cascio, D.; Reyes, F.E.; Shi, D.; Sangwan, S.; Guenther, E.L.; Johnson, L.M.; Zhang, M.; et al. Structure of the Toxic Core of α-Synuclein from Invisible Crystals. *Nature* **2015**, *525*, 486–490. [CrossRef] [PubMed]

72. Emamzadeh, F.N. Alpha-Synuclein Structure, Functions, and Interactions. *J. Res. Med. Sci.* **2016**, *21*, 29. [CrossRef] [PubMed]

73. Spillantini, M.G.; Schmidt, M.L.; Lee, V.M.-Y.; Trojanowski, J.Q.; Jakes, R.; Goedert, M. Alpha-Synuclein in Lewy Bodies. *Nature* **1997**, *388*, 839–840. [CrossRef] [PubMed]

74. Roeters, S.J.; Iyer, A.; Pletikapić, G.; Kogan, V.; Subramaniam, V.; Woutersen, S. Evidence for Intramolecular Antiparallel Beta-Sheet Structure in Alpha-Synuclein Fibrils from a Combination of Two-Dimensional Infrared Spectroscopy and Atomic Force Microscopy. *Sci. Rep.* **2017**, *7*, 41051. [CrossRef] [PubMed]

75. Hutchinson, E.G.; Thornton, J.M. The Greek Key Motif: Extraction, Classification and Analysis. *Protein Eng.* **1993**, *6*, 233–245. [CrossRef] [PubMed]

76. Giasson, B.I.; Murray, I.V.J.; Trojanowski, J.Q.; Lee, V.M.Y. A Hydrophobic Stretch of 12 Amino Acid Residues in the Middle of α-Synuclein Is Essential for Filament Assembly. *J. Biol. Chem.* **2001**, *276*, 2380–2386. [CrossRef] [PubMed]

77. Tzotzos, S.; Doig, A.J. Amyloidogenic Sequences in Native Protein Structures. *Protein Sci.* **2010**, *19*, 327–348. [CrossRef] [PubMed]

78. Zweckstetter, M.; Requena, J.R.; Wille, H. Elucidating the structure of an infectious protein. *PLoS Pathog.* **2017**, *13*, e1006229. [CrossRef] [PubMed]

79. Peralta, M.D.R.; Karsai, A.; Ngo, A.; Sierra, C.; Fong, K.T.; Hayre, N.R.; Mirzaee, N.; Ravikumar, K.M.; Kluber, A.J.; Chen, X.; et al. Engineering Amyloid Fibrils from β-Solenoid Proteins for Biomaterials Applications. *ACS Nano* **2015**, *9*, 449–463. [CrossRef] [PubMed]

80. Helmus, J.J.; Surewicz, K.; Surewicz, W.K.; Jaroniec, C.P. Conformational flexibility of Y145Stop human prion protein amyloid fibrils probed by solid-state nuclear magnetic resonance spectroscopy. *J. Am. Chem. Soc.* **2010**, *132*, 2393–2403. [CrossRef] [PubMed]

81. Helmus, J.J.; Surewicz, K.; Apostol, M.I.; Surewicz, W.K.; Jaroniec, C.P. Intermolecular alignment in Y145Stop human prion protein amyloid fibrils probed by solid-state NMR spectroscopy. *J. Am. Chem. Soc.* **2011**, *133*, 13934–13937. [CrossRef] [PubMed]

pathogens

MDPI

Review

Evolution of Diagnostic Tests for Chronic Wasting Disease, a Naturally Occurring Prion Disease of Cervids

Nicholas J. Haley [1],* and Jürgen A. Richt [2]

[1]   Department of Microbiology and Immunology, Arizona College of Osteopathic Medicine,
    Midwestern University, Glendale, AZ 85308, USA
[2]   College of Veterinary Medicine, Kansas State University (KSU), Manhattan, KS 66506, USA;
    jricht@vet.k-state.edu
*   Correspondence: nicholas.j.haley@gmail.com

Received: 30 June 2017; Accepted: 1 August 2017; Published: 5 August 2017

Abstract: Since chronic wasting disease (CWD) was first identified nearly 50 years ago in a captive mule deer herd in the Rocky Mountains of the United States, it has slowly spread across North America through the natural and anthropogenic movement of cervids and their carcasses. As the endemic areas have expanded, so has the need for rapid, sensitive, and cost effective diagnostic tests—especially those which take advantage of samples collected antemortem. Over the past two decades, strategies have evolved from the recognition of microscopic spongiform pathology and associated immunohistochemical staining of the misfolded prion protein to enzyme-linked immunoassays capable of detecting the abnormal prion conformer in postmortem samples. In a history that parallels the diagnosis of more conventional infectious agents, both qualitative and real-time amplification assays have recently been developed to detect minute quantities of misfolded prions in a range of biological and environmental samples. With these more sensitive and semi-quantitative approaches has come a greater understanding of the pathogenesis and epidemiology of this disease in the native host. Because the molecular pathogenesis of prion protein misfolding is broadly analogous to the misfolding of other pathogenic proteins, including Aβ and α-synuclein, efforts are currently underway to apply these in vitro amplification techniques towards the diagnosis of Alzheimer's disease, Parkinson's disease, and other proteinopathies. Chronic wasting disease—once a rare disease of Colorado mule deer—now represents one of the most prevalent prion diseases, and should serve as a model for the continued development and implementation of novel diagnostic strategies for protein misfolding disorders in the natural host.

Keywords: prion; cervids; PMCA; RT-QuIC; diagnosis

---

1. Background and Introduction

Chronic wasting disease (CWD) is a naturally occurring transmissible spongiform encephalopathy (TSE) known to affect a range of cervid species, including white-tailed and mule deer (*Odocoileus virginianus* and *Odocoileus hemionus*), North American elk (wapiti, *Cervus elaphus elaphus*), moose (*Alces alces*), and reindeer (*Rangifer tarandus*) [1–3]. Since its initial discovery nearly 50 years ago in Northern Colorado and Southern Wyoming, the disease has been reported in 22 additional states, 2 Canadian provinces, South Korea, and very recently in Norway (see https://www.nwhc.usgs.gov/disease_information/chronic_wasting_disease/ for a current map of the geographic extent of CWD in North America). As with other TSEs, including scrapie of sheep, bovine spongiform encephalopathy (BSE), and human variant and sporadic Creutzfeldt-Jakob disease (CJD), CWD is characterized by central nervous system pathology mediated by an abnormally folded isoform of the normal cellular prion protein (PrP^res when referring to the misfolded variant or PrP^{Sc} when referring to the infectious isoform specifically, and PrP^C, respectively). The primary structure of PrP^C, dictated by the host's prion protein gene (*PRNP*), plays a vital role in intra- and inter-species susceptibility, reducing susceptibility in animals with specific alleles and serving as the basis for the "species barrier", limiting the disease almost exclusively to cervids [4–9]. The molecular pathogenesis of prion diseases like CWD shares many common traits with other protein misfolding disorders, including Alzheimer's disease and Parkinson's disease, and while most prion diseases are decreasing or stable in prevalence, the ever-expanding range of CWD makes it a tempting model system for the broad development of novel diagnostic approaches for these proteinopathies.

In its present range, CWD has been found among both farmed and free-ranging cervids [2]. Although most evidence is anecdotal, both farmed and free-ranging animals have played a role in the progressive spread of the disease across North America and to South Korea [10–12]. The recent discovery of CWD in Norway is perplexing, with wildlife managers scrambling to determine not only the extent of infection, but also its source—whether arising in situ or imported in some form from North America [3,13]. While the natural or anthropogenic movement of animals may play the most prominent role in the spread of CWD, the movement of animal carcasses has likely also been involved in dissemination [14,15]; the role of animal byproducts and bodily fluids is less clear, although tissues and bodily fluids including deboned muscle [16] and fat [17], antler velvet [18], saliva [19,20], feces [21], and urine [19] have proven infectious under experimental conditions.

Making matters more difficult is the protracted nature of the disease—whereby several years may pass between preliminary exposure and the onset of clinical symptoms. Not unlike the pathogenesis of rabies infection in animals, the infectious PrP^{Sc} protein must make its way from the periphery, most likely following oral exposure, to the central nervous system. The spread to peripheral excretory tissues, either concurrent with or pursuant to CNS infection, permits the shedding of de novo misfolded PrP^{Sc} into the environment, perpetuating the transmission cycle. Most research has shown that the appearance of the misfolded PrP^{Sc} isoform in peripheral tissues and the onset of shedding may take place months or perhaps years before the appearance of clinical signs. Preclinical peripheral accumulation of prions and shedding in bodily fluids greatly contribute to the imperceptible movement of disease to CWD-free areas via infected animals and those byproducts described above.

As a result of disease expansion and the risks that the movement of clinical or pre-clinical animals, their carcasses, and byproducts may play in transmission, some urgency has been placed on the development of diagnostic approaches which are rapid, sensitive, cost effective, and can make use of samples collected either postmortem or antemortem. Paralleling the history of more conventional infectious agents, the evolution of prion diagnostic strategies has progressed first from the identification of characteristic microscopic pathologic changes [22], to antibody-antigen dependent detection systems [23–26], and eventually to the advent of techniques for the isolation [27] and amplification [28–32] of the building blocks of stored biological information—in the case of TSEs, the very structure of the prion protein itself.

Building on these approaches, new strategies are being developed to allow for the quantification of prion burden in a tissue, body fluid, or environmental sample. Perhaps a loftier goal, the development of in vitro techniques which may allow for strain discrimination would be tremendously helpful in identifying the source of recent or historic introductions of the disease across North American and now Scandinavia. As these approaches are implemented and refined for the detection of CWD, they will likewise lead to suitable diagnostic tests to meet objectives for the diagnosis of prions and other protein misfolding disorders.

2. The History of CWD Diagnostics

Roughly 12 years passed between the early clinical recognition of chronic wasting disease in the 1960s and its definitive grouping within the rapidly growing category of transmissible spongiform encephalopathies soon to be recognized globally as "prion" diseases [1]. The original clinical descriptions of CWD in mule deer are still appropriate today—a syndrome of slowly progressive neurologic dysfunction, behavioral changes, polyuria, polydipsia and hypersalivation, dysphagia and occasional aspiration pneumonia, and ultimately, death [2,33,34]. Like many other TSEs, postmortem diagnoses were based primarily on characteristic neurohistopathologic changes in the gray matter at all levels of the CNS—the spinal cord, mesencephalon, diencephalon, and both cerebellar and cerebral cortices. At the heart of the clinical signs, pathognomonic central nervous system lesions consisted of microcavitation of the neuropil, intracytoplasmic vacuolization, astrocytic hypertrophy and hyperplasia, and neuronal degeneration. Cached amongst these CNS lesions: amyloid plaques, best observed with Congo red or Bodian silver staining. Although the nature and origin of these plaques were unknown at the time, they were a consistent finding across TSEs of both animals and man.

With the definitive identification of the agent responsible for prion diseases, an abnormally folded and hardy conformer of the cellular prion protein [35,36], very specific immunoassays would be developed that could be used on a range of platforms, including fresh and fixed tissues. In cases where these immunoassays were not sensitive enough, bioassay in susceptible hosts—occasionally requiring serial passage—became the *de facto* testing method for infectivity. Each of these has served its respective fields—diagnostic medicine and research, for more than 20 years (Figure 1).

2.1. Immunohistochemistry, Western Blotting, and Enzyme Immunoassay

The initial discovery of the agents responsible for TSEs enabled the further development of diagnostic approaches beyond basic clinical and microscopic histopathological descriptions. The isolation of a misfolded cellular protein, found exclusively in the brains of TSE-infected animals and solely capable of inducing disease [36], permitted the development of an array of diagnostic assays dependent on the sensitivity and specificity of antibody-antigen interactions. These assays, including western blotting [23], immunohistochemistry [25], and enzyme immunoassay (EIA) [26], capable of distinguishing the normally folded cellular prion protein (PrP^C) and the misfolded, infectious isoform (PrP^{res}/PrP^{Sc}), are still considered the "gold standard" diagnostic approaches for CWD and other prion diseases (Figure 2).

The primary characteristic of the misfolded prion protein, PrP^{res}, which these assays take advantage of—its resistance to harsh conditions including acid treatment and enzymatic protease digestion, allowed for the reliable detection of infected individuals with high specificity. The amyloid plaques initially identified with routine histochemical staining in the CNS were found to be intensively immunoreactive to serum prepared from rabbits inoculated with hamster scrapie amyloid [25]. Brain homogenates from infected deer were also found to have protease-resistant remnants of immunoreactive prion amyloid when analyzed by SDS-PAGE and immune-dot blotting [23]. Although the presence of the protease-resistant core of the infectious prion protein is common to all prion diseases, its localization in the CNS and its immunoreactive banding pattern on western blot were found to help distinguish one prion agent from another [2]. The immunoreactive plaques observed in the CNS of deer with CWD are considered florid in nature, for example, whereas those found in cattle with bovine spongiform encephalopathy appear more granular. The western blotting pattern of CWD is a triplicate of di-, mono-, and unphosphorylated PrP^{res} protein bands 21–27 kD in size, with the diphosporylated band being the most intense [29]. The banding appearance of bovine spongiform encephalopathy PrP^{res}, in contrast, reveals a triplicate ranging from 17–28 kD, with di- and monophosphorylated bands frequently of equal intensities [37].

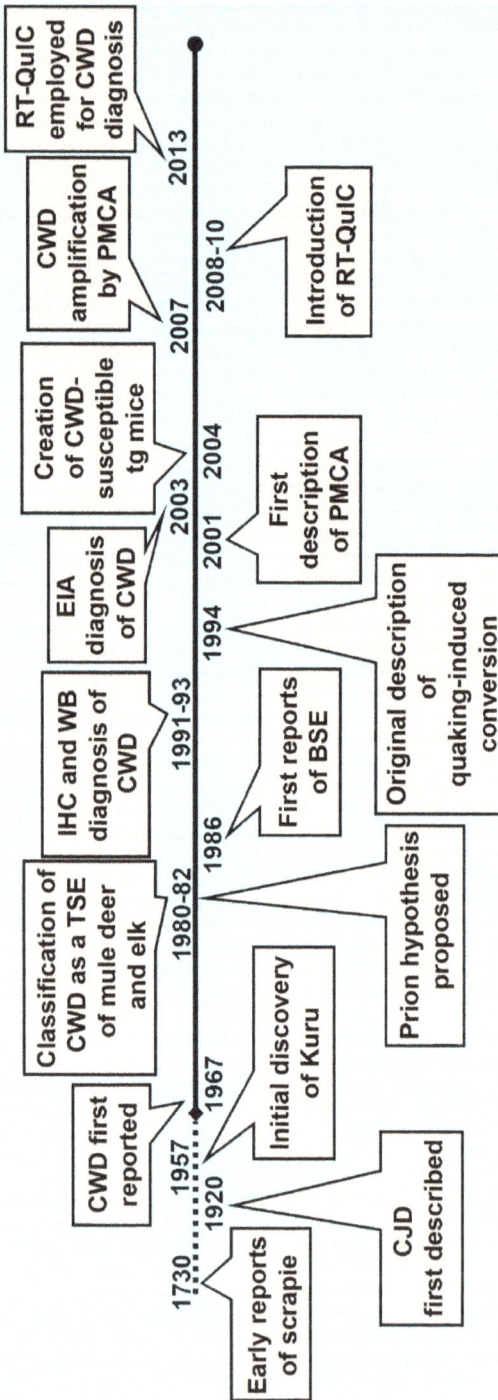

Figure 1. History of diagnostic developments for chronic wasting disease (CWD) and other transmissible spongiform encephalopathies (TSEs). CJD: Creutzfeldt-Jakob Disease; BSE: bovine spongiform encephalopathy; IHC: immunohistochemistry; WB: western blotting; EIA: enzyme immunoassay; PMCA: protein misfolding cyclic amplification; RT-QuIC: real time quaking-induced conversion.

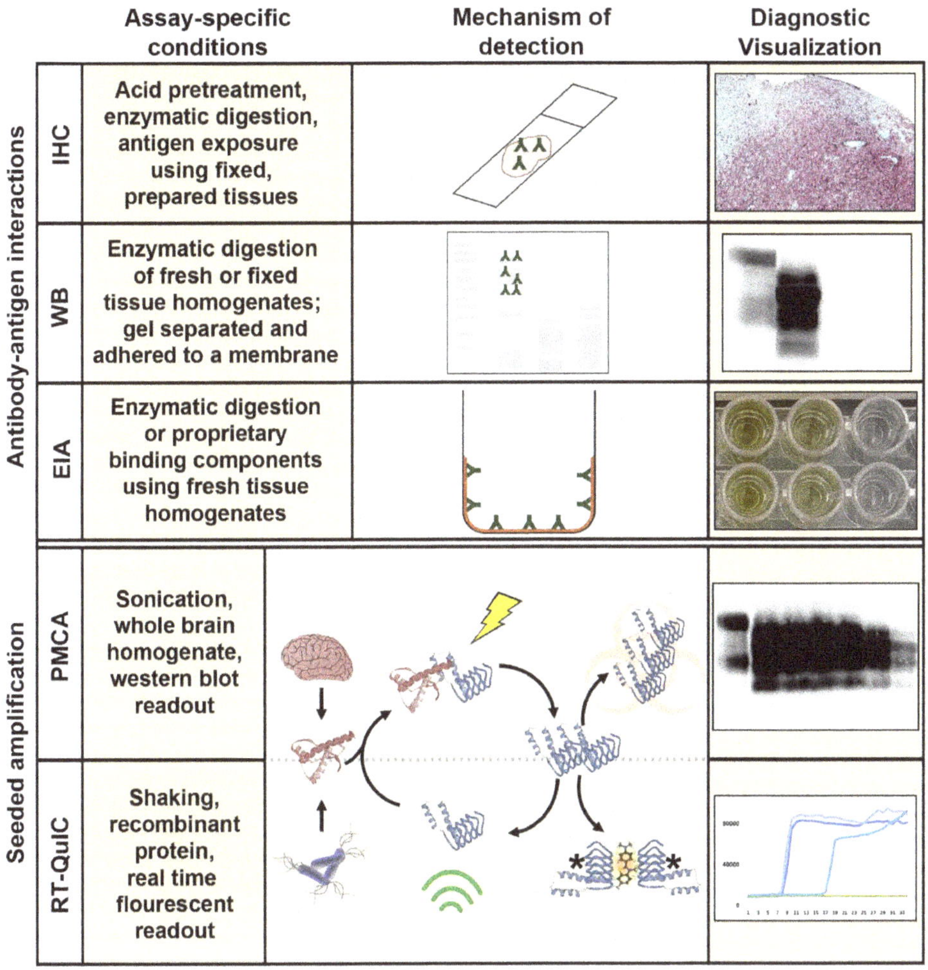

Figure 2. Summary of conventional CWD diagnostic strategies and seeded amplification methods for amplifying CWD prions in vitro. Distinguishing conditions for each assay, as well illustrative mechanisms of detection and representative diagnostic results are presented. IHC: immunohistochemistry; WB: western blotting; EIA: enzyme immunoassay; PMCA: protein misfolding cyclic amplification; RT-QuIC: real time quaking-induced conversion; * denotes that the structure of amplified products arising from recombinant PrP in RT-QuIC may be different than that produced by PMCA, potentially explaining the loss of infectivity seen with RT-QuIC.

From the advancements made with various immunoassays, more sensitive approaches to CWD diagnosis quickly evolved. Postmortem immunohistochemical studies of samples collected in the field and from experimental challenge studies have highlighted several target tissues as early harbingers of CWD infection, most importantly the dorsal motor nucleus of the vagus (DMNV) in the obex region of the brainstem and the medial retropharyngeal lymph nodes (RLN)—which are still considered the "gold standard" postmortem diagnostic tissues for regulatory diagnosis [38–40]. In deer, the RLN becomes positive before the DMNV, with rare exception [41], making it the most sensitive target tissue in this species. Elk, in contrast, may be DMNV positive without evidence of infection found in the RLN; as a result, both tissues should be examined in these animals [42]. In both species, progressive deposition of PrP^res in the DMNV and other regions of the brain has allowed diagnosticians to estimate the stage of infection through subjective scoring approaches [43–46]. Tonsilar tissue, interestingly, was one of the first tissues showing evidence of immunodeposition following exposure, and has been used experimentally to identify infected animals antemortem [39,47–49]. Later studies found that lymphoid tissue in the caudal rectum may also serve as a prognosticator for CNS infection, providing further opportunities for antemortem diagnosis [46,50].

Over the course of these diagnostic field and experimental studies, the growing geographical extent of the disease was examined [51–55], and evidence was uncovered in both deer and elk which showed that the host's prion gene (*PRNP*) sequence may modulate susceptibility [56–62]. Animals with several alleles harboring coding mutations, including 225S → F in mule deer [60], 132M → L in elk [61], and 96G → S in white-tailed deer [62], were underrepresented among animals found to be infected, and were therefore thought to have a lower relative risk of infection compared to their wild-type counterparts. Later studies more clearly demonstrated that cervids with these alleles were susceptible, though may have a more protracted incubation period than those with wild-type alleles [57]. Collectively, these introductory studies allowed researchers to estimate the geographical boundaries of CWD-endemic areas and assemble a preliminary picture of disease pathogenesis in cervids with a range of genetic backgrounds.

As efforts to better characterize CWD pathogenesis, especially routes of transmission, continued, it became necessary to pursue alternate strategies for prion detection in those biological samples thought to be involved. Immunohistochemistry was not a practical approach for bodily fluids and excreta like blood, saliva, urine, or feces. The presumably low levels of PrP^res in these samples also made identification difficult using conventional western blotting and EIA. The experimental exposure of susceptible species, then, became the most practical (albeit time consuming) mechanism for assessing infectivity in body fluids.

2.2. Bioassay

The early experiments characterizing the transmissibility of CWD, and later uncovering potential transmission routes, required an extensive reliance on both natural and experimental hosts. Initial studies in natural hosts—mule deer and elk [39,63,64]—were used to demonstrate that animal to animal contact and environmental contamination played very important roles in disease transmission. More granular studies in white-tailed deer, addressing the roles of specific bodily fluids and cellular components, soon showed that saliva and blood carried high levels of infectivity [20]; the roles of urine and feces at that time were less clear. Not long after, the development of transgenic murine models, susceptible to CWD, allowed for a more thorough examination of body fluids, greater consistency within and across experiments, and even permitted the titration of infectivity [21,65]. Transgenic mice helped further illuminate the role of specific blood fractions [66], and offered greater sensitivity in identifying infectivity in both feces and urine [19], as well as in the tissues of animals inoculated with these and other biological samples through secondary passage experiments [67]. While still widely used today, biological models for diagnostic purposes are extremely impractical for obvious reasons, including ethical considerations, costs, and prolonged incubation periods.

The development of antibody-antigen dependent assays (western blotting, IHC, and EIA) allowed for a better understanding of the pathogenesis of CWD and other prion diseases, and helped to identify the most appropriate tissues to collect and evaluate postmortem. As a result, the above described conventional testing strategies have helped elucidate the ever-growing range of CWD in North America and beyond, and have been used to identify cervid hosts with varying levels of susceptibility linked to the *PRNP* gene. Bioassays, meanwhile, have helped uncover the likely routes of transmission in bodily fluids, especially saliva—which could prove useful in developing antemortem tests. However, additional in vitro approaches, which could mimic the misfolding process that occurs in vivo, would need to be developed to allow for a more sensitive detection of infectivity in body fluids and other diagnostically appropriate samples.

3. The Present State of CWD Diagnostics

With the pioneering work of immunological and bioassay studies, much has been learned about the pathogenesis, transmission, and, equally important, the geographic distribution of CWD and other prion diseases. Although immunological tests were very *specific* for prion infection, concerns arose early on that these assays were not *sensitive* enough—suspicions often supported by bioassay findings [26,67,68]. Indeed, it is common practice to report CWD test results as "Not Detected", instead of "Negative", to acknowledge the so far unmeasured insensitivity of IHC, western blotting, and EIA. Because of ethical, practical, and monetary considerations, attention was turned from bioassay to other methods which might allow more rapid, sensitive, and cost-effective detection of CWD and other prion infections in vitro, using techniques and approaches common to the diagnosis of other infectious agents—including cell culture and various amplification techniques.

Concurrent with the development of more sensitive techniques for identifying CWD infected cervids, efforts have been made to shift the diagnostic focus in deer and elk from postmortem to antemortem detection. With the frequent movement of farmed and wild cervids and their byproducts across North America and beyond, it is becoming increasingly important to develop screening programs to prevent the introduction of CWD into new areas. Currently, farmed cervid herds in both the United States and Canada may enroll in voluntary herd health programs which facilitate the interstate or interprovincial sale of animals [69,70]. These programs typically require meticulous inventories and a consistent postmortem testing history and are commonly more stringent than the limitations placed on wildlife relocations—however they are not fail safe. In both farmed and wild cervids, antemortem testing prior to animal movement may add another layer of security to prevent the spread of CWD.

While progress has been made on assay development and antemortem testing strategies, some limitations remain. Bodily fluids have been shown to be infectious, and could therefore be used as a diagnostic sample—but little is known about the kinetics of shedding in bodily fluids over the course of disease. Easily accessible peripheral tissues (e.g., tonsil) have high diagnostic sensitivity late in the course of disease, but fall short when animals are in earlier stages. Lastly, a specific host genetic background, which has been linked to reduced susceptibility and/or delayed disease progression, may complicate detection in either bodily fluids or peripheral tissues (Figure 3). With a better understanding of CWD pathogenesis in all susceptible species and genetic backgrounds, the gains that have been made in sampling and testing approaches can more effectively be applied to improve both test sensitivity and specificity.

3.1. Amplification Assays for the Detection of Ultra-Low Levels of CWD Prions

Of the in vitro assays currently in development for detection of CWD prions, amplification assays are by far the ones getting the most attention [28,31]. At their very basic level, these assays take advantage of the proclivity of PrP^{Sc} to induce a conformational change in a normal cellular prion protein substrate (PrP^C). They may make use of the high levels of PrP^C found in the brains of transgenic mice, for example, or they can rely on bacterial expression systems to produce large amounts of recombinant PrP^C for use as a conversion substrate. Amyloid fibril disruption and generation of new prion "seeds" for amplification may be accomplished by simple shaking or through sonication. The readouts of the assays may require blotting techniques to visually detect amplified aggregates of PrP^{res}, paralleling conventional gel-based PCR, or they may take advantage of fluorescent molecules which bind to growing amyloid fibers, allowing a readout similar to real-time, quantitative PCR. In each case, the objective of these techniques is to amplify low levels of misfolded proteins in vitro which may be present in a sample, to levels which can be readily observed by more traditional methods (Figure 2).

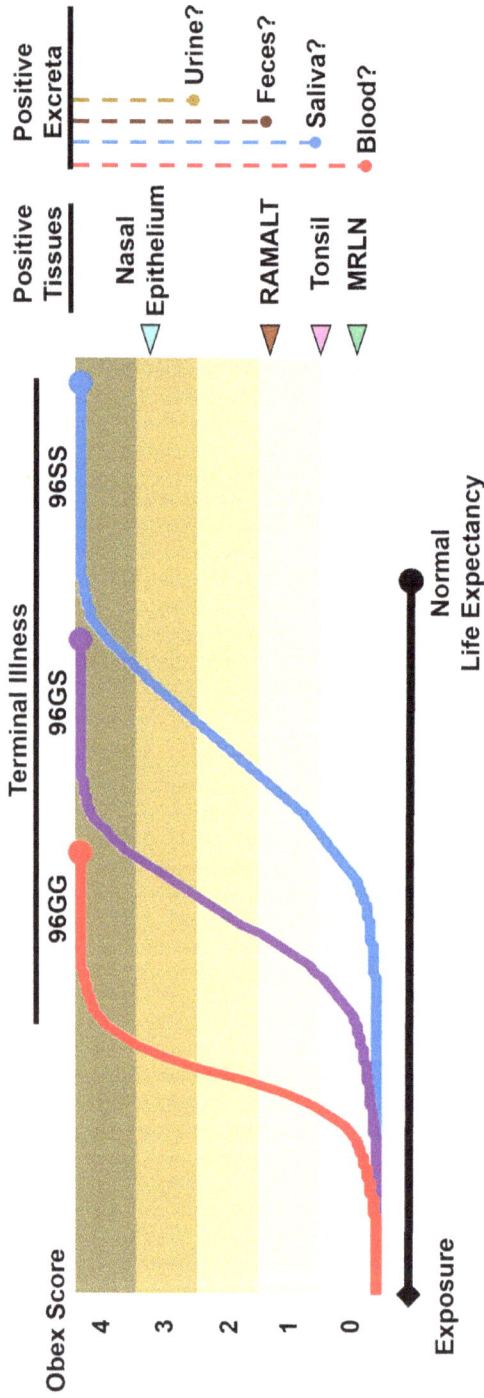

Figure 3. Model of the pathogenesis of chronic wasting disease in white-tailed deer with different *PRNP* backgrounds, with special attention on the diagnostic sensitivity of peripheral tissues and excreta. The disease seems to progress at different rates in animals with differing *PRNP* sequences, which affects the time points at which peripheral tissues may become positive by conventional or experimental diagnostic assays. Shedding in excreta is less well characterized, and the kinetics of prionemia, prionsialia, prionochezia, and prionuria (infectious prions in the blood, saliva, feces, and urine, respectively) may fluctuate during the course of infection. White-tailed deer with 96GG, GS, and SS *PRNP* sequences are considered, though the model would similarly apply to deer and elk with other variants of the *PRNP* gene. Obex scoring is a subjective, semi-quantitative method for visually estimating the amount of PrP^{res} deposition in the obex using immunohistochemistry, and has been used in studies in both deer and elk [44–46]. Data presented here have been compiled from several studies [19,39,42–19,71–74]. RAMALT: recto-anal mucosa associated lymphoid tissues; MRLN: medial retropharyngeal lymph node.

3.2. Protein Misfolding Cyclic Amplification

The first of these amplification techniques to be adapted for use with CWD, which helped to lay the groundwork for future developments in CWD amplification-based diagnostics, was the protein misfolding cyclic amplification assay (PMCA) [29,67]. This assay requires, most importantly, a cellular prion protein substrate derived from brain homogenates of susceptible, or potentially susceptible, hosts. For detection of CWD, very often these homogenates are derived from transgenic cervidized mice, which may express high levels of white-tailed deer or elk PrP^C, providing an abundance of substrate for in vitro conversion. The brains are commonly homogenized in phosphate buffered saline with a range of protease inhibitors and surfactants, to which the CWD-harboring sample, or "seed", is added and allowed to incubate at 37 °C for 24–48 h. The samples are sonicated intermittently to fragment the growing amyloid chain. These new amyloid fragments may then serve as seeds for further conversion reactions. After each experiment, the seed-substrate preparations may be treated with protease and evaluated by western blot for the resistant conformer, or they may be passaged into a new preparation of brain homogenate, in the case of "serial" PMCA (sPMCA) [73,75]. Serial PMCA, not unlike nested PCR, may involve up to ten passages or more of amplification over the course of several weeks in an attempt to achieve even greater sensitivity than conventional PMCA.

Several modifications have been described which improve the sensitivity of PMCA or sPMCA, including the addition of plastic beads or putative cofactors [76,77]. Some researchers have essentially hybridized PMCA with the quaking induced conversion assay described below, and applied an electrical current in an effort to improve sensitivity [78]. To detect the misfolded protein, many permutations still rely on protease treatment which destroys the normal cellular protein, and potentially some protease-sensitive isoforms of the infectious proteins, ultimately reducing sensitivity. To circumvent protease treatment, one group reported using a surround optical fiber immunoassay (SOFIA) to specifically identify the disease-associated form of the prion protein using immunocapture in combination with laser-induced fluorescence [79,80]. Each of these modified approaches have shown potential for the detection of exquisitely low levels of CWD prions, perhaps down to the attagram level—potentially at the cost of reduced specificity as is commonly seen in other diagnostics using extended PCR or nested PCR protocols [77].

Variations of the PMCA assay have been used to explore various areas of CWD pathogenesis, e.g., to assess the potential for infectivity in body fluids and other excreta [19,77,81,82], and to detect low levels of misfolded protein in soil [83], water [84], and plant samples [85]. Notably, the CWD seeds generated in vitro by sPMCA have proven infectious to some degree in susceptible hosts [86], indicating that the technique may accurately model what occurs in vivo; therein lays its advantage among amplification strategies. Neither PMCA nor any of its derivatives have, however, been used extensively in field studies which would allow researchers to test the true sensitivity and specificity against conventional IHC or EIA. Ultimately, four important considerations continued to drive the development of new diagnostic amplification techniques beyond PMCA: (1) the ethical concerns raised by continued use of animal hosts for a PrP^C conversion substrate; (2) the need for an assay which could detect all potential infectious conformers of the prion protein, including protease-sensitive forms; (3) time constraints commonly required in field surveillance; and (4) the need for a technically simple assay with a practical read-out, one which could more easily allow for quantification.

3.3. Quaking Induced Conversion

Many of the considerations described above would be met by a conceptually similar technique developed nearly in parallel—the quaking induced conversion assay, or QuIC [31,87]. Importantly, this approach makes use of recombinant PrP^C, an approach which has two distinct advantages over traditional PMCA: first, the protein substrate can be quickly and consistently produced in any cellular expression system, commonly *E. coli*, and second, it allows for the rapid design of substrates tailored to the researcher's needs, without the complicated intermediate steps needed to generate transgenic mice. Commonly, a truncated form of the Syrian hamster PrP protein is used as a substrate, however a number of cervid and non-cervid recombinant substrates have been developed for the detection of CWD and other prions of both animals and humans [88,89].

The QuIC technique seemingly went unnoticed by those researching CWD, until modifications, including the incorporation of a fluorescent dye and a plate reader capable of stringent shaking protocols, allowed it to evolve into a format that satisfied each of the considerations which had hindered the widespread implementation of PMCA [30]. As with PMCA, the shaking is believed to disrupt growing amyloid fibrils and multiply the number of seeds available for further amyloid formation. The fluorescent dye, commonly thioflavin T, is thought to intercalate within the growing amyloid fibril. When bound to amyloid, thioflavin T exhibits a different emission spectrum than when free in solution, permitting the user to monitor amyloid amplification in real time. Like qPCR, this consolidates the assay read-out into a technically simple, unambiguous amplification curve which may additionally allow some level of quantification.

Current permutations of the real time QuIC (RT-QuIC) assay monitor changes in fluorescence every 15–60 min, over periods of time ranging from 24–96 h or more. As with sPMCA, longer RT-QuIC protocols allow for amplification of lower levels of misfolded prions, while concurrently risking spontaneous misfolding and decreased specificity. Under these different protocols, RT-QuIC has been used to examine the initial steps of CWD tissue invasion [74], quantify the levels of misfolded protein in bodily fluids [71], and evaluate inter- and intra-species susceptibility to CWD in vitro [89]. It has also been blindly evaluated in parallel with PMCA, IHC, and EIA [32,44,45,77], allowing a direct comparison between RT-QuIC and conventional diagnostic approaches. Generally, these studies have shown RT-QuIC is at least as sensitive as IHC or EIA.

The strengths of RT-QuIC lie in its consistency, malleability, rapidity and ease of interpretation. Because it relies solely on recombinant PrP^C as a conversion substrate, it is less capable of modeling the in vivo conversion process than PMCA. Importantly, the amplified products generated by RT-QuIC have not yet been shown to be infectious in vivo, as they have with PMCA. In fact, very few diagnostic approaches, short of viral or bacterial culture and isolation methods, are dependent on infectivity. Thus, neither of these caveats should prevent the implementation of RT-QuIC as a diagnostic approach for CWD or other prion diseases.

3.4. Tyramide Signal Amplification

While the goal of both PMCA and RT-QuIC is to amplify low levels of misfolded prions by seeded conversion, tyramide signal amplification instead magnifies the signaling mechanisms present in conventional assays, and has been used experimentally for CWD specifically to improve IHC detection in fixed tissues [74,90]. In the case of IHC, horseradish peroxidase-labeled antibodies bound to CWD prion antigen in situ activates the tyramide substrate which then accumulates in the immediate vicinity of the antibody, amplifying signal intensity up to 15-fold [91]. This technique has been used to more effectively track the early pathogenesis of experimental CWD in both transgenic mice and deer, though has not yet found its way into clinical applications.

3.5. Cervid Prion Cell Assay

Just as cell culture systems have been developed for the detection and diagnosis of a range of viruses and intracellular bacteria, cell lines have likewise been developed for the cultivation and quantification of infectious prions [27]. Researchers have inserted a variety of alternate PrP gene sequences into the mutable rabbit kidney epithelial RK13 cell line, which have rendered them susceptible to species-specific prion replication. In the cervid prion cell assay, or CPCA, expression of the elk PrP gene resulted in an RK13 line susceptible to CWD which permits the titration of an infectious dose comparable to bioassay in transgenic mice. Although the CPCA effectively decreased the time and cost required for bioassay, and models in vivo infection more closely than amplification assays, the culture of prions in susceptible cell lines still remains limited in practicality compared to PMCA and RT-QuIC. As viral isolation and bacterial cell culture remain staples of microbiological testing today, so may cell culture have a future in the diagnosis of CWD in cervids.

3.6. Sample Selection for Antemortem Testing

Past and present detection strategies have supported the work on CWD pathogenesis and demonstrated the kinetics of shedding in bodily fluids and excreta. Using amplification approaches, evidence of CWD prion presence has been reported in a range of bodily fluids [19,77,81,92–94], making them tempting targets for the development of novel diagnostic strategies. Studies in other model systems, including sheep and humans, have identified peripheral tissues which may also serve as a useful diagnostic sample and indicator of central nervous system infection [95–97]. Through these discoveries, antemortem testing for CWD is becoming increasingly more sensitive and reliable, and may someday prove useful for screening prior to animal movement (Table 1).

3.7. Bodily Fluids and Excreta

With many infectious diseases of veterinary and human importance, assays which utilize bodily fluids—especially blood—are considered ideal choices for a diagnostic test. CWD is not unique in this regard, and the primary focus has been on the development of a hematologic test to identify infected animals [93,94]. Very little is known about the kinetics of prionemia (prion infectivity in the blood), or the kinetics of prion shedding in other forms of excreta, and yet a number of primarily amplification-based studies have attempted to identify the misfolded protein in these samples. In many cases, these techniques have been developed using a very limited number of infected animals, and more importantly a limited number of negative controls [98,99]. Very rarely have the techniques been successfully applied to large field studies, although several laboratories continue to pursue testing of saliva and urine [71], blood [94], and fecal samples [72] collected from experimentally exposed animals or during depopulations of CWD-infected farmed deer and elk. These studies will eventually allow for more direct comparisons to be made with conventional postmortem testing and allow researchers to evaluate their sensitivity and specificity.

3.8. Accessible Peripheral Tissues

Several accessible tissues, including peripheral lymphoid and neuroepithelial tissues, have been identified which may help identify CWD-infected deer and elk antemortem or postmortem [44–46,100–102]. Each of these tissues offer both strengths and weaknesses in their diagnostic feasibilities, and need to be considered on a case by case basis in their application. For example, lymphoid tissues like tonsil—where CWD prions may accumulate early in the course of infection—have been found to be quite sensitive when compared to central lymphoid and nervous tissues collected postmortem. To that end, the direct sampling of medial retropharyngeal lymph nodes might be expected to offer near perfect sensitivity. The aforementioned tissues are, however, rather difficult to sample practically and repeatedly when compared to other, less sensitive peripheral tissues like recto-anal mucosal associated lymphoid tissue (RAMALT) [41]. Real-time QuIC analysis of olfactory neuroepithelial swabs, a relatively simple technique shown to be quite sensitive in the diagnosis of clinical Creutzfeldt-Jakob disease in humans, may only be effective in identifying deer and elk in the most terminal stages of CWD [44,45]. Accordingly, it should be remembered that irrespective of the sampling tissue and technique, or assay used, cases in the very early stages of infection may still test negative—making serial sampling indispensable for antemortem diagnosis. As more is learned about CWD pathogenesis and transmission, however, improvements in both tissue and body fluid sampling strategies will most certainly be made.

Table 1. A summary of published diagnostic approaches for chronic wasting disease. Data sets from larger, comprehensive studies with postmortem data are included in the table. Other smaller or incomplete studies are referenced elsewhere in this review. Shaded rows indicate currently approved post-mortem diagnostic approaches for CWD. MRLN: medial retropharyngeal lymph node; RAMALT: recto-anal mucosa associated lymphatic tissue; CSF: cerebrospinal fluid; IHC: immunohistochemistry; EIA: enzyme immunoassay; RT-QuIC: real time quaking-induced conversion; sPMCA: serial protein misfolding cyclic amplification; NA: not applicable; ND: not determined.

	Sample	Method	Number Positive Postmortem (Total Examined)	Sensitivity *	Specificity *	Reference	Sample Notes
Tissues	Brainstem (obex)	IHC	NA	NA	NA	NA	IHC and ELISA of brainstem and RLN are considered the "gold standard" postmortem diagnostic approach for CWD. In deer, RLN are generally considered more sensitive, while in elk it is recommended both tissues be evaluated to confirm a diagnosis.
		EIA	53 (1986)	92%	100%	[26]	
		IHC	NA	NA	NA	NA	
	MRLN	EIA	84 (2042)	99%	>99%	[26]	
		RT-QuIC	23 (1243)	100%	100%	[32]	Field samples, postmortem
3–8	Tonsil	IHC	100 (1150)	99%	100%	[49]	Field samples, postmortem
		sPMCA	30 (48)	ND †	ND †	[68]	Experimental animals, antemortem
	RAMALT	IHC	150 (561)	68%	>99%	[46]	Field samples, postmortem
		RT-QuIC	289 (409)	70%	94%	[45]	Field samples, antemortem
	Nasal brushings	RT-QuIC	289 (409)	16%	91%	[45]	Field samples, antemortem
Body Fluids/Excreta	Blood	RT-QuIC	16 (21)	93%	100%	[96]	Experimental animals, serial collection
	CSF	sPMCA	16 (37)	19%	100%	[79]	Field samples, postmortem
		RT-QuIC	26 (48)	50%	96%	[74]	Experimental animals, postmortem
	Saliva	RT-QuIC	18 (22)	78%	98%	[88]	Experimental animals, antemortem
	Urine	RT-QuIC	18 (22)	39%	100%	[88]	Experimental animals, antemortem
	Feces	sPMCA	5 (36)	ND †	ND †	[78]	Field samples, ante- and postmortem
		RT-QuIC	15 (25)	53%	100%	[97]	Field samples, antemortem

* The sensitivity and specificity of various approaches are compared to postmortem immunohistochemistry of the obex +/- retropharyngeal lymph nodes. † Sensitivity and specificity could not be calculated, since it was proposed that a number of samples in these studies were from CWD-positive animals which were IHC negative postmortem.

3.9. Sample Collection in Farmed and Free-Ranging Cervids

While post-mortem samples are relatively easy to collect on the necropsy floor or in the field, weather and equipment permitting, antemortem sampling presents its own unique challenges in both farmed and wild deer and elk. Farmed cervids are commonly collected in small groups, processed in modern handling systems and restrained in standard large animal squeeze chutes, which greatly facilitates the collection of accessible bodily fluid samples and rectal biopsies, for example [44,45,72]. More invasive biopsy collections from farmed cervids, including tonsil and retropharyngeal lymph node, requires deep sedation and anesthesia—a practice that is all but necessary for the collection of any samples from wild cervids [48,102]. Apart from the need for sedation or anesthesia, a more important factor limiting the efficiency at which wild cervids may be sampled is first finding and then capturing animals. Sampling in either group is not without risk, however, with farmed cervids occasionally suffering from severe injuries like broken limbs, and wild cervids at risk for the development of capture myopathy, an infrequent and often fatal syndrome resulting from the handling of wild cervids [103].

3.10. Genetic Background and Antemortem CWD Diagnostic Sensitivity

With the development of antemortem testing approaches came the discovery that an animal's genetic background could have a profound effect on antemortem diagnostic sensitivity. In both deer and elk, RAMALT and nasal brush testing in animals with the prototypical *PRNP* genotype (96GG in deer and 132MM in elk) have been found to have the highest diagnostic sensitivity [44–46]. Antemortem testing in animals with a *PRNP* genetic background considered less susceptible, for example, 96GS or SS in deer and 132ML or LL in elk, is significantly less sensitive. Taken together with the apparently reduced susceptibility in animals with specific *PRNP* sequences, the reduced sensitivity of peripheral tissues can be best explained by a slower disease progression in these genotypes. In support of this, less susceptible animals which were CWD-negative in peripheral tissues were more commonly found to be in earlier stages of the disease, implying that the appearance of detectable prions in these peripheral tissues may be dependent primarily on disease stage, and not genetic background [44–46]. Several unanswered questions remain, however: do sensitivity limitations apply to all peripheral tissues? Do they apply broadly to all diagnostic assays? How might bodily fluids be affected? Should we use this information to encourage cervid farmers to breed highly susceptible animals to afford regulators a greater test sensitivity, or should we encourage a shift towards more resistant animals to help slow or prevent the spread of CWD? Ongoing research and policy discussions will hopefully provide the answers needed to move forward.

As with well-described bacterial and viral diagnostic strategies, diagnostic approaches for CWD and other TSEs began with clinical and postmortem pathologic detection methods. These strategies quickly progressed to more sensitive molecular approaches, which sought to identify the agent using amplification techniques, and shifted the focus from postmortem to antemortem diagnosis. While not perfectly sensitive compared to postmortem testing, currently deployed amplification techniques for CWD have a comparable sensitivity to assays for other important diseases—most notably bovine tuberculosis in cervids, although the ramifications of CWD misdiagnosis may be far more consequential. The available prion amplification approaches importantly take advantage of the infectious prion's mechanisms for storing and reproducing information, just as PCR targets RNA and DNA molecules. In the case of TSEs, the ability to store and reproduce this information is imprinted in the structure of the abnormal prion protein itself. What other information may lay in this structure? Perhaps information which may encode strain, virulence, or zoonotic capacity? Can we identify a roadmap for pathogenesis or transmission in hosts with diverse genetic backgrounds? These are questions with absent or incomplete information, though with luck the tools currently in development will someday provide sufficient answers.

4. The Future of CWD Diagnostics

Detection capabilities for CWD and other infectious prions have progressed significantly over the past two decades, although there are still a number of areas requiring further research. First, demonstrating the improved sensitivity of prion amplification tests compared to conventional diagnostics is challenging, and will require well-structured experiments and well-defined samples in order for them to supplant immunohistochemistry or EIA. Second, there are critical gaps in epidemiologic studies which make it difficult to identify the source(s) of CWD introduced into previously naïve populations and to estimate environmental contamination in non-endemic areas. Finally, it should be remembered that while it is important to continue improving CWD diagnostics, it is equally important to translate these findings for the benefit human medicine, in the form of improved diagnostics for human prion diseases, Alzheimer's disease and other protein misfolding disorders.

4.1. Improving Current CWD Diagnostic Approaches

As the pathogenesis of CWD is further defined in susceptible species and genetic backgrounds, improvements in sampling strategies should be expected. For example, lymph node aspirates and oral swabs, which are commonly used to diagnose a range of diseases in veterinary and human medicine [104–106], could be suitable for early antemortem diagnosis when combined with the appropriate testing platform. Fecal samples collected in the field, in contrast, may allow for a more passive sampling strategy to identify populations with otherwise undetectably low levels of CWD prevalence. Effective tissue and body fluid sampling has developed slowly over the past few decades, and there is every expectation that it will continue to evolve into the future.

Testing strategies are likewise evolving, with ever increasing sensitivity being reported by PMCA and RT-QuIC and other novel diagnostic approaches currently in development. A perpetual hurdle to demonstrating advanced sensitivity is the difficulty in overcoming a "gold standard"—how should we interpret samples which are positive by an amplification assay like sPMCA or RT-QuIC, yet IHC or EIA negative? To illustrate this point, experimental longitudinal studies have shown that sPMCA can identify misfolded prions in the blood of transgenic mice with a known exposure history [99]; however, the amplification experiments were not performed blindly, and thus should be interpreted with caution. Other, appropriately blinded studies have shown that both sPMCA and RT-QuIC more readily identify CWD prions in terminal cervid samples compared when to IHC—however, because of their terminal nature, it is very difficult to prove these animals were truly infected without a confirmatory test like bioassay to support the diagnosis [32,67]. Ideally, studies seeking to demonstrate the enhanced sensitivity of prion amplification approaches should *prospectively* incorporate both a longitudinal and blinded strategy, with repeated sampling of animals with a known and unknown exposure history to demonstrate presence or absence of infection, verified by IHC or EIA, in animals initially diagnosed by experimental techniques. Appropriate negative controls, including tissue or bodily fluid controls from negative hosts, are critical, while inter-lab validation is also an important strategy to consider, especially when a limited number of samples are under evaluation. It remains to be seen how well the quantitative or semi-quantitative nature of assays like RT-QuIC may correlate to in vivo infectivity: at what point does amplification-based detection become biologically relevant? Experiments such as these are ongoing, and may soon provide insight into the true sensitivity and specificity of prion amplification assays, and, perhaps as importantly, the true sensitivity and specificity of conventional and "gold standard" diagnostic approaches.

4.2. Exploring New Frontiers in CWD Diagnostics

Along with ongoing improvements in current sampling and testing strategies, future efforts should continue to pursue new and uncharted areas in CWD diagnostic capabilities. Several studies have demonstrated the occurrence of a number of putative CWD strains circulating in the wild [107,108], and while strain-typing is commonplace for viral or bacterial agents, no currently available approach has been shown to allow for rapid discrimination of diverse CWD prion strains. Western blotting very crudely identifies differences between CWD and BSE, while clinical presentation helps to differentiate sporadic CJD from variant CJD, for example, but current diagnostic technologies do not confer the ability to differentiate CWD strains or specifically identify the sources of new CWD incursions. The amplification-based assays could most likely address this diagnostic gap, with preliminary research seemingly demonstrating that RT-QuIC could provide reliable information regarding human Creutzfeldt-Jakob disease isolates [109]. This technology may effectively translate to CWD strains, where the comparison of various amplification parameters of cervid isolates in different amplification substrates could be employed. The ability to differentiate CWD isolates would be extremely helpful in epidemiologic studies, by allowing apparently new epidemic foci to be traced to specific geographic locations or source herds. Perhaps new strains would be discovered, including isolates previously undetectable by currently available technologies. Strain discrimination and characterization would additionally provide evidence and insight into prion evolution and adaptation—critical information which could be incorporated into field studies and efforts to investigate host resistance, and possibly help predict vaccine utility.

With the quantitative abilities of RT-QuIC, approximate titration of prion burden in biologic or environmental samples may also be possible using CWD amplification assays [71]. Early studies of saliva and other body fluids have shown variable levels of prion seeding potential in samples collected at different time points during infection, and it may soon be possible to correlate levels of shedding to incubation periods and genetic background as well as secondary underlying disease—renal dysfunction or perhaps even viral or bacterial co-infections, for example. An understanding of prion burden in tissues may provide a more thorough understanding of CWD pathogenesis and disease staging, and permit diagnosticians to select more appropriate ante- or post-mortem tissues for sensitive diagnoses. The ability to assess environmental contamination will allow wildlife biologists to monitor disease movement more easily, while simultaneously affording estimates of reduced infectivity following environmental decontamination efforts.

Advancements in CWD testing will certainly benefit from the introduction of prion amplification assays into the diagnostic repertoire. Multi-dimensional assays like RT-QuIC, which provides a range of information including amplification rate and efficiency in mutable substrates, seem poised to shed light on CWD strains and biological or environmental burdens which will allow for more detailed studies into disease epidemiology and pathogenesis. The benefits that this work provides will not be limited to cervid health, however.

4.3. Realms beyond CWD Diagnosis

The TSEs are increasingly regarded as models for other protein misfolding disorders of the CNS and other organ systems, including Alzheimer's disease, Parkinson's disease, and chronic traumatic encephalopathy (CTE) [110–113]. The application of the lessons learned through the course of investigations into CWD and other TSEs to the diagnostic challenges presented by these increasingly common human neurologic disorders should also be considered. The fundamental mechanisms directing the propagation of prions are not unlike those responsible for the accumulation of Alzheimer's Aβ protein, or α-synuclein in Parkinson's disease, and the techniques introduced by sPMCA or RT-QuIC should be transferrable with modifications to substrate and reaction conditions [114]. Efforts are currently underway to assess this potential, with promising results in both tau- and synucleinopathies.

Among TSEs, CWD perhaps represents the ideal model system for developing and deploying these prion amplification tests, in that it uniquely represents a proteinopathy affecting natural populations and is the only TSE currently expanding in distribution. Sample selection will undoubtedly vary between distinct proteinopathies and host populations; however a structured implementation of amplification assays for CWD would certainly help lay the groundwork for advancements in naturally occurring protein misfolding disorders in humans.

The future of CWD diagnostics depends on continued progress in the understanding of disease pathogenesis and the identification of suitable antemortem samples, and most importantly refinement and implementation of amplification assays like RT-QuIC. The potential for these assays to discriminate CWD strains and quantify tissue and body fluid burden will provide invaluable information for both epidemiologic studies and risk assessments. Challenges in the diagnosis of naturally occurring human proteinopathies will be offset by opportunities to implement CWD diagnostic strategies, making the continued development of these assays essential for advancements in human health.

5. Conclusions

Chronic wasting disease, a prion disease of deer and elk first reported five decades ago, now represents the last of the TSEs for which transmission and dissemination remain unchecked. The tools available to diagnosticians for identifying infected animals have steadily progressed over that timeframe from clinical and pathological descriptions to antibody-antigen dependent immunoassays, and more recently have begun incorporating qualitative and quantitative prion amplification techniques. These tools have provided a deep understanding of disease pathogenesis and transmission, and allowed animal health professionals to monitor the expanding geographical presence of CWD. Sampling techniques have likewise evolved, with shifts from postmortem to antemortem approaches targeting peripheral tissues and body fluids, and may someday offer the ability to screen animals prior to movement or selectively identify animals for removal. In the future, CWD diagnostics may also offer hope for the rapid discrimination of strains and assessment of tissue burden and environmental contamination. Although CWD's role as the last remaining unmanaged TSE is a distinction neither agricultural nor wildlife professionals hold in high esteem, the discoveries over the past several decades have greatly assisted the continued development of assays directed toward protein misfolding disorders occurring in natural populations, and will ultimately benefit not just animal health but human health as well.

Acknowledgments: Author N.J.H. is supported, in part, by grants from the North American Deer Farmers Association and the North American Elk Breeders Association. Author J.A.R. has no funding acknowledgements to declare. The authors would like to thank the many research scientists, wildlife managers, agricultural agencies and producers who have contributed untold time and resources to help develop solutions for the CWD epidemic, a disease which is now multi-national in scale. It is only with their concerted efforts that progress will continue to be made in the diagnosis, prevention, and management of this destructive disease, and it is with their help that larger issues in human health—Alzheimer's disease, Parkinson's disease, and other protein misfolding disorders—may someday come with a more favorable prognosis. The authors would also like to thank Kristen Davenport and Emily Mertz for their candid reviews of the manuscript, substantial improvements were made through their assistance.

Conflicts of Interest: The authors declare no conflicts of interest.

References

1. Williams, E.S.; Young, S. Chronic wasting disease of captive mule deer: A spongiform encephalopathy. *J. Wildl. Dis.* **1980**, *16*, 89–98. [CrossRef] [PubMed]
2. Haley, N.J.; Hoover, E.A. Chronic Wasting Disease of Cervids: Current Knowledge and Future Perspectives. *Annu. Rev. Anim. Biosci.* **2015**, *3*, 305–325. [CrossRef] [PubMed]
3. Benestad, S.L.; Mitchell, G.; Simmons, M.; Ytrehus, B.; Vikoren, T. First case of chronic wasting disease in Europe in a Norwegian free-ranging reindeer. *Vet. Res.* **2016**, *47*, 88. [CrossRef] [PubMed]
4. Belay, E.D.; Maddox, R.A.; Williams, E.S.; Miller, M.W.; Gambetti, P.; Schonberger, L.B. Chronic wasting disease and potential transmission to humans. *Emerg. Infect. Dis.* **2004**, *10*, 977–984. [CrossRef] [PubMed]
5. Collins, S.J.; Lawson, V.A.; Masters, C.L. Transmissible spongiform encephalopathies. *Lancet* **2004**, *363*, 51–61. [CrossRef]
6. Kong, Q.; Huang, S.; Zou, W.; Vanegas, D.; Wang, M.; Wu, D.; Yuan, J.; Zheng, M.; Bai, H.; Deng, H.; et al. Chronic wasting disease of elk: Transmissibility to humans examined by transgenic mouse models. *J. Neurosci.* **2005**, *25*, 7944–7949. [CrossRef] [PubMed]

7. Wilson, R.; Plinston, C.; Hunter, N.; Casalone, C.; Corona, C.; Tagliavini, F.; Suardi, S.; Ruggerone, M.; Moda, F.; Graziano, S.; et al. Chronic wasting disease and atypical forms of bovine spongiform encephalopathy and scrapie are not transmissible to mice expressing wild-type levels of human prion protein. *J. Gen. Virol.* **2012**, *93 Pt 7*, 1624–1629. [CrossRef] [PubMed]

8. Perrott, M.R.; Sigurdson, C.J.; Mason, G.L.; Hoover, E.A. Mucosal transmission and pathogenesis of chronic wasting disease in ferrets. *J. Gen. Virol.* **2012**, *94 Pt 2*, 432–442. [CrossRef] [PubMed]

9. Heisey, D.M.; Mickelsen, N.A.; Schneider, J.R.; Johnson, C.J.; Langenberg, J.A.; Bochsler, P.N.; Keane, D.P.; Barr, D.J. Chronic wasting disease (CWD) susceptibility of several North American rodents that are sympatric with cervid CWD epidemics. *J. Virol.* **2009**, *4*, 210–215. [CrossRef] [PubMed]

10. Sohn, H.J.; Kim, J.H.; Choi, K.S.; Nah, J.J.; Joo, Y.S.; Jean, Y.H.; Ahn, S.W.; Kim, O.K.; Kin, D.Y.; Balachandran, A. A case of chronic wasting disease in an elk imported to Korea from Canada. *J. Vet. Med. Sci.* **2002**, *64*, 855–858. [CrossRef] [PubMed]

11. Williams, E.S.; Miller, M.W.; Kreeger, T.J.; Kahn, R.H.; Thorne, E.T. Chronic wasting disease of deer and elk: A review with recommendations for management. *J. Wildl. Manag.* **2002**, *66*, 551–563. [CrossRef]

12. Williams, E.S. Chronic wasting disease. *Vet. Pathol.* **2005**, *42*, 530–549. [CrossRef] [PubMed]

13. Stokstad, E. Norway plans to exterminate a large reindeer herd to stop a fatal infectious brain disease. *Science Magazine*, 3 August 2017.

14. Picard, J. Experts explain deer disease. *Oneida Daily Dispatch News*, 13 May 2005.

15. Watts, T. CWD leads to new regulations for taxidermists. *Oakland Press News*, 28 May 2009.

16. Angers, R.C.; Browning, S.R.; Seward, T.S.; Sigurdson, C.J.; Miller, M.W.; Hoover, E.A.; Telling, G.C. Prions in skeletal muscles of deer with chronic wasting disease. *Science* **2006**, *311*, 1117. [CrossRef] [PubMed]

17. Race, B.; Meade-White, K.; Race, R.; Chesebro, B. Prion infectivity in fat of deer with chronic wasting disease. *J. Virol.* **2009**, *83*, 9608–9610. [CrossRef] [PubMed]

18. Angers, R.C.; Seward, T.S.; Napier, D.; Green, M.; Hoover, E.; Spraker, T.; O'Rourke, K.; Balachandran, A.; Telling, G.C. Chronic wasting disease prions in elk antler velvet. *Emerg. Infect. Dis.* **2009**, *15*, 696–703. [CrossRef] [PubMed]

19. Haley, N.J.; Seelig, D.M.; Zabel, M.D.; Telling, G.C.; Hoover, E.A. Detection of CWD prions in urine and saliva of deer by transgenic mouse bioassay. *PLoS ONE* **2009**, *4*, e4848. [CrossRef] [PubMed]

20. Mathiason, C.K.; Powers, J.G.; Dahmes, S.J.; Osborn, D.A.; Miller, K.V.; Warren, R.J.; Mason, G.L.; Hays, S.A.; Hayes-Klug, J.; Seelig, D.M.; et al. Infectious prions in the saliva and blood of deer with chronic wasting disease. *Science* **2006**, *314*, 133–136. [CrossRef] [PubMed]

21. Tamguney, G.; Miller, M.W.; Wolfe, L.L.; Sirochman, T.M.; Glidden, D.V.; Palmer, C.; Lemus, A.; DeArmond, S.J.; Prusiner, S.B. Asymptomatic deer excrete infectious prions in faeces. *Nature* **2009**, *461*, 529–532. [CrossRef] [PubMed]

22. Bahmanyar, S.; Williams, E.S.; Johnson, F.B.; Young, S.; Gajdusek, D.C. Amyloid plaques in spongiform encephalopathy of mule deer. *J. Comp. Pathol.* **1985**, *95*, 1–5. [CrossRef]

23. Guiroy, D.C.; Williams, E.S.; Song, K.J.; Yanagihara, R.; Gajdusek, D.C. Fibrils in brain of Rocky Mountain elk with chronic wasting disease contain scrapie amyloid. *Acta Neuropathol.* **1993**, *86*, 77–80. [CrossRef] [PubMed]

24. Guiroy, D.C.; Williams, E.S.; Yanagihara, R.; Gajdusek, D.C. Immunolocalization of scrapie amyloid (PrP27-30) in chronic wasting disease of Rocky Mountain elk and hybrids of captive mule deer and white-tailed deer. *Neurosci. Lett.* **1991**, *126*, 195–198. [CrossRef]

25. Guiroy, D.C.; Williams, E.S.; Yanagihara, R.; Gajdusek, D.C. Topographic distribution of scrapie amyloid-immunoreactive plaques in chronic wasting disease in captive mule deer (*Odocoileus hemionus hemionus*). *Acta Neuropathol.* **1991**, *81*, 475–478. [CrossRef] [PubMed]

26. Hibler, C.P.; Wilson, K.L.; Spraker, T.R.; Miller, M.W.; Zink, R.R.; DeBuse, L.L.; Anderson, E.; Schweitzer, D.; Kennedy, J.A.; Baeten, L.A.; et al. Field validation and assessment of an enzyme-linked immunosorbent assay for detecting chronic wasting disease in mule deer (*Odocoileus hemionus*), white-tailed deer (*Odocoileus virginianus*), and Rocky Mountain elk (Cervus elaphus nelsoni). *J. Vet. Diagn. Investig.* **2003**, *15*, 311–319. [CrossRef] [PubMed]

27. Bian, J.; Napier, D.; Khaychuck, V.; Angers, R.; Graham, C.; Telling, G. Cell-based quantification of chronic wasting disease prions. *J. Virol.* **2010**, *84*, 8322–8326. [CrossRef] [PubMed]

28. Soto, C.; Saborio, G.P.; Anderes, L. Cyclic amplification of protein misfolding: Application to prion-related disorders and beyond. *Trends Neurosci.* **2002**, *25*, 390–394. [CrossRef]

29. Kurt, T.D.; Perrott, M.R.; Wilusz, C.J.; Wilusz, J.; Supattapone, S.; Telling, G.C.; Hoover, E.A. Efficient in vitro amplification of chronic wasting disease PrPRES. *J. Virol.* **2007**, *81*, 9605–9608. [CrossRef] [PubMed]

30. Wilham, J.M.; Orru, C.D.; Bessen, R.A.; Atarashi, R.; Sano, K.; Race, B.; Meade-White, K.D.; Taubner, L.M.; Timmes, A.; Caughey, B. Rapid end-point quantitation of prion seeding activity with sensitivity comparable to bioassays. *PLoS Pathog.* **2010**, *6*, e1001217. [CrossRef] [PubMed]

31. Atarashi, R.; Moore, R.A.; Sim, V.L.; Hughson, A.G.; Dorward, D.W.; Onwubiko, H.A.; Priola, S.A.; Caughey, B. Ultrasensitive detection of scrapie prion protein using seeded conversion of recombinant prion protein. *Nat. Methods* **2007**, *4*, 645–650. [CrossRef] [PubMed]

32. Haley, N.J.; Carver, S.; Hoon-Hanks, L.L.; Henderson, D.M.; Davenport, K.A.; Bunting, E.; Gray, S.; Trindle, B.; Galeota, J.; LeVan, I.; et al. Detection of chronic wasting disease in the lymph nodes of free-ranging cervids by real-time quaking-induced conversion. *J. Clin. Microbiol.* **2014**, *52*, 3237–3243. [CrossRef] [PubMed]

33. Miller, M.W.; Williams, E.S. Chronic wasting disease of cervids. *Curr. Top. Microbiol. Immunol.* **2004**, *284*, 193–214. [PubMed]

34. Sigurdson, C.J.; Aguzzi, A. Chronic wasting disease. *Biochim. Biophys. Acta* **2007**, *1772*, 610–618. [CrossRef] [PubMed]

35. Bolton, D.C.; McKinley, M.P.; Prusiner, S.B. Identification of a protein that purifies with the scrapie prion. *Science* **1982**, *218*, 1309–1311. [CrossRef] [PubMed]

36. Prusiner, S.B. Novel proteinaceous infectious particles cause scrapie. *Science* **1982**, *216*, 136–144. [CrossRef] [PubMed]

37. Masujin, K.; Okada, H.; Miyazawa, K.; Matsuura, Y.; Imamura, M.; Iwamaru, Y.; Murayama, Y.; Yokoyama, T. Emergence of a novel bovine spongiform encephalopathy (BSE) prion from an atypical H-type BSE. *Sci. Rep.* **2016**, *6*, 22753. [CrossRef] [PubMed]

38. Peters, J.; Miller, J.M.; Jenny, A.L.; Peterson, T.L.; Carmichael, K.P. Immunohistochemical diagnosis of chronic wasting disease in preclinically affected elk from a captive herd. *J. Vet. Diagn. Investig.* **2000**, *12*, 579–582. [CrossRef] [PubMed]

39. Sigurdson, C.J.; Williams, E.S.; Miller, M.W.; Spraker, T.R.; O'Rourke, K.I.; Hoover, E.A. Oral transmission and early lymphoid tropism of chronic wasting disease PrPres in mule deer fawns (*Odocoileus hemionus*). *J. Gen. Virol.* **1999**, *80 Pt 10*, 2757–2764. [CrossRef] [PubMed]

40. Keane, D.P.; Barr, D.J.; Keller, J.E.; Hall, S.M.; Langenberg, J.A.; Bochsler, P.N. Comparison of retropharyngeal lymph node and obex region of the brainstem in detection of chronic wasting disease in white-tailed deer (*Odocoileus virginianus*). *J. Vet. Diagn. Investig.* **2008**, *20*, 58–60. [CrossRef] [PubMed]

41. Keane, D.P.; Barr, D.J.; Bochsler, P.N.; Hall, S.M.; Gidlewski, T.; O'Rourke, K.I.; Spraker, T.R.; Samuel, M.D. Chronic wasting disease in a Wisconsin white-tailed deer farm. *J. Vet. Diagn. Investig.* **2008**, *20*, 698–703. [CrossRef] [PubMed]

42. Spraker, T.R.; Balachandran, A.; Zhuang, D.; O'Rourke, K.I. Variable patterns of distribution of PrP(CWD) in the obex and cranial lymphoid tissues of Rocky Mountain elk (Cervus elaphus nelsoni) with subclinical chronic wasting disease. *Vet. Rec.* **2004**, *155*, 295–302. [CrossRef] [PubMed]

43. Fox, K.A.; Jewell, J.E.; Williams, E.S.; Miller, M.W. Patterns of PrPCWD accumulation during the course of chronic wasting disease infection in orally inoculated mule deer (*Odocoileus hemionus*). *J. Gen. Virol.* **2006**, *87 Pt 11*, 3451–3461. [CrossRef] [PubMed]

44. Haley, N.J.; Siepker, C.; Hoon-Hanks, L.L.; Mitchell, G.; Walter, W.D.; Manca, M.; et al. Seeded Amplification of Chronic Wasting Disease Prions in Nasal Brushings and Recto-anal Mucosa-Associated Lymphoid Tissues from Elk by Real-Time Quaking-Induced Conversion. *J. Clin. Microbiol.* **2016**, *54*, 1117–1126. [CrossRef] [PubMed]

45. Haley, N.J.; Siepker, C.; Walter, W.D.; Thomsen, B.V.; Greenlee, J.J.; Lehmkuhl, A.D.; Richt, J.A. Antemortem Detection of Chronic Wasting Disease Prions in Nasal Brush Collections and Rectal Biopsy Specimens from White-Tailed Deer by Real-Time Quaking-Induced Conversion. *J. Clin. Microbiol.* **2016**, *54*, 1108–1116. [CrossRef] [PubMed]

46. Thomsen, B.V.; Schneider, D.A.; O'Rourke, K.I.; Gidlewski, T.; McLane, J.; Allen, R.W.; McIsaac, A.A.; Mitchell, G.B.; Keane, D.P.; Spraker, T.R.; et al. Diagnostic accuracy of rectal mucosa biopsy testing for chronic wasting disease within white-tailed deer (*Odocoileus virginianus*) herds in North America: Effects of age, sex, polymorphism at PRNP codon 96, and disease progression. *J. Vet. Diagn. Investig.* **2012**, *24*, 878–887. [CrossRef] [PubMed]

47. Sigurdson, C.J.; Barillas-Mury, C.; Miller, M.W.; Oesch, B.; van Keulen, L.J.; Langeveld, J.P.; Hoover, E.A. PrP(CWD) lymphoid cell targets in early and advanced chronic wasting disease of mule deer. *J. Gen. Virol.* **2002**, *83 Pt 10*, 2617–2628. [CrossRef] [PubMed]

48. Wild, M.A.; Spraker, T.R.; Sigurdson, C.J.; O'Rourke, K.I.; Miller, M.W. Preclinical diagnosis of chronic wasting disease in captive mule deer (*Odocoileus hemionus*) and white-tailed deer (*Odocoileus virginianus*) using tonsillar biopsy. *J. Gen. Virol.* **2002**, *83 Pt 10*, 2629–2634. [CrossRef] [PubMed]

49. Spraker, T.R.; O'Rourke, K.I.; Balachandran, A.; Zink, R.R.; Cummings, B.A.; Miller, M.W.; Powers, B.E. Validation of monoclonal antibody F99/97.6.1 for immunohistochemical staining of brain and tonsil in mule deer (*Odocoileus hemionus*) with chronic wasting disease. *J. Vet. Diagn. Investig.* **2002**, *14*, 3–7. [CrossRef] [PubMed]

50. Spraker, T.R.; VerCauteren, K.C.; Gidlewski, T.; Schneider, D.A.; Munger, R.; Balachandran, A.; O'Rourke, K.I. Antemortem detection of PrPCWD in preclinical, ranch-raised Rocky Mountain elk (Cervus elaphus nelsoni) by biopsy of the rectal mucosa. *J. Vet. Diagn. Investig.* **2009**, *21*, 15–24. [CrossRef] [PubMed]

51. Miller, M.W.; Williams, E.S.; McCarty, C.W.; Spraker, T.R.; Kreeger, T.J.; Larsen, C.T.; Thorne, E.T. Epizootiology of chronic wasting disease in free-ranging cervids in Colorado and Wyoming. *J. Wildl. Dis.* **2000**, *36*, 676–690. [CrossRef] [PubMed]

52. Williams, E.S.; Miller, M.W. Chronic wasting disease in deer and elk in North America. *Rev. Sci. Tech.* **2002**, *21*, 305–316. [CrossRef] [PubMed]

53. Joly, D.O.; Samuel, M.D.; Langenberg, J.A.; Blanchong, J.A.; Batha, C.A.; Rolley, R.E.; Keane, D.P.; Ribic, C.A. Spatial epidemiology of chronic wasting disease in Wisconsin white-tailed deer. *J. Wildl. Dis.* **2006**, *42*, 578–588. [CrossRef] [PubMed]

54. Jennelle, C.S.; Samuel, M.D.; Nolden, C.A.; Keane, D.P.; Barr, D.J.; Johnson, C.; Vanderloo, J.P.; Aiken, J.M.; Hamir, A.N.; Hoover, E.A. Surveillance for transmissible spongiform encephalopathy in scavengers of white-tailed deer carcasses in the chronic wasting disease area of Wisconsin. *J. Toxicol. Environ. Health A* **2009**, *72*, 1018–1024. [CrossRef] [PubMed]

55. Saunders, S.E.; Bartelt-Hunt, S.L.; Bartz, J.C. Occurrence, transmission, and zoonotic potential of chronic wasting disease. *Emerg. Infect. Dis.* **2012**, *18*, 369–376. [CrossRef] [PubMed]

56. Johnson, C.; Johnson, J.; Vanderloo, J.P.; Keane, D.; Aiken, J.M.; McKenzie, D. Prion protein polymorphisms in white-tailed deer influence susceptibility to chronic wasting disease. *J. Gen. Virol.* **2006**, *87 Pt 7*, 2109–2114. [CrossRef] [PubMed]

57. O'Rourke, K.I.; Spraker, T.R.; Zhuang, D.; Greenlee, J.J.; Gidlewski, T.E.; Hamir, A.N. Elk with a long incubation prion disease phenotype have a unique PrPd profile. *Neuroreport* **2007**, *18*, 1935–1938. [CrossRef] [PubMed]

58. Kelly, A.C.; Mateus-Pinilla, N.E.; Diffendorfer, J.; Jewell, E.; Ruiz, M.O.; Killefer, J.; Shelton, P.; Beissel, T.; Novakofski, J. Prion sequence polymorphisms and chronic wasting disease resistance in Illinois white-tailed deer (*Odocoileus virginianus*). *Prion* **2008**, *2*, 28–36. [CrossRef] [PubMed]

59. Robinson, S.J.; Samuel, M.D.; O'Rourke, K.I.; Johnson, C.J. The role of genetics in chronic wasting disease of North American cervids. *Prion* **2012**, *6*, 153–162. [CrossRef] [PubMed]

60. Jewell, J.E.; Conner, M.M.; Wolfe, L.L.; Miller, M.W.; Williams, E.S. Low frequency of PrP genotype 225SF among free-ranging mule deer (*Odocoileus hemionus*) with chronic wasting disease. *J. Gen. Virol.* **2005**, *86 Pt 8*, 2127–2134. [CrossRef] [PubMed]

61. O'Rourke, K.I.; Besser, T.E.; Miller, M.W.; Cline, T.F.; Spraker, T.R.; Jenny, A.L.; Wild, M.A.; Sebarth, G.L.; Williams, E.S. PrP genotypes of captive and free-ranging Rocky Mountain elk (Cervus elaphus nelsoni) with chronic wasting disease. *J. Gen. Virol.* **1999**, *80 Pt 10*, 2765–2769.

62. Johnson, C.; Johnson, J.; Clayton, M.; McKenzie, D.; Aiken, J. Prion protein gene heterogeneity in free-ranging white-tailed deer within the chronic wasting disease affected region of Wisconsin. *J. Wildl. Dis.* **2003**, *39*, 576–581. [CrossRef] [PubMed]

63. Miller, M.W.; Williams, E.S.; Hobbs, N.T.; Wolfe, L.L. Environmental sources of prion transmission in mule deer. *Emerg. Infect. Dis.* **2004**, *10*, 1003–1006. [CrossRef] [PubMed]
64. Miller, M.W.; Wild, M.A.; Williams, E.S. Epidemiology of chronic wasting disease in captive Rocky Mountain elk. *J. Wildl. Dis.* **1998**, *34*, 532–538. [CrossRef] [PubMed]
65. Browning, S.R.; Mason, G.L.; Seward, T.; Green, M.; Eliason, G.A.; Mathiason, C.; Miller, M.W.; Williams, E.S.; Hoover, E.; Telling, G.C. Transmission of prions from mule deer and elk with chronic wasting disease to transgenic mice expressing cervid PrP. *J. Virol.* **2004**, *78*, 13345–13350. [CrossRef] [PubMed]
66. Mathiason, C.K.; Hayes-Klug, J.; Hays, S.A.; Powers, J.; Osborn, D.A.; Dahmes, S.J.; Miller, K.V.; Warren, R.J.; Mason, G.L.; Telling, G.C.; et al. B cells and platelets harbor prion infectivity in the blood of deer infected with chronic wasting disease. *J. Virol.* **2010**, *84*, 5097–5107. [CrossRef] [PubMed]
67. Haley, N.; Mathiason, C.; Zabel, M.D.; Telling, G.C.; Hoover, E. Detection of sub-clinical CWD infection in conventional test-negative deer long after oral exposure to urine and feces from CWD+ deer. *PLoS ONE* **2009**, *4*, e7990. [CrossRef] [PubMed]
68. Haley, N.J.; Mathiason, C.; Carver, S.; Telling, G.C.; Zabel, M.C.; Hoover, E.A. Sensitivity of protein misfolding cyclic amplification vs. immunohistochemistry in antemortem detection of CWD infection. *J. Gen. Virol.* **2012**, *93*, 1141–1150. [CrossRef] [PubMed]
69. *Chronic Wasting Disease Program Standards*; United States Department of Agriculture: Ames, IA, USA, 2014.
70. Canadian Food Inspection Agency. *Accredited Veterinarian's Manual*; Canadian Food Inspection Agency: Ottawa, ON, Canada, 2017.
71. Henderson, D.M.; Davenport, K.A.; Haley, N.J.; Denkers, N.D.; Mathiason, C.K.; Hoover, E.A. Quantitative assessment of prion infectivity in tissues and body fluids by real-time quaking-induced conversion. *J. Gen. Virol.* **2015**, *96 Pt 1*, 210–219. [CrossRef] [PubMed]
72. Henderson, D.; Tennant, J.; Haley, N.; Denkers, N.D.; Mathiason, C.; Hoover, E. Detection of CWD Prion-seeding Activity in Deer and Elk Feces by Real-time Quaking Induced Conversion. *J. Gen. Virol.* **2017**, in press.
73. Haley, N.J.; Mathiason, C.K.; Carver, S.; Zabel, M.; Telling, G.C.; Hoover, E.A. Detection of CWD prions in salivary, urinary, and intestinal tissues of deer: Potential mechanisms of prion shedding and transmission. *J. Virol.* **2011**, *85*, 6309–6318. [CrossRef] [PubMed]
74. Hoover, C.E.; Davenport, K.A.; Henderson, D.M.; Denkers, N.D.; Mathiason, C.K.; Soto, C.; Zabel, M.D.; Hoover, E.A. Pathways of Prion Spread during Early Chronic Wasting Disease in Deer. *J. Virol.* **2017**, *91*, e00077-17. [CrossRef] [PubMed]
75. Castilla, J.; Saa, P.; Morales, R.; Abid, K.; Maundrell, K.; Soto, C. Protein misfolding cyclic amplification for diagnosis and prion propagation studies. *Methods Enzymol.* **2006**, *412*, 3–21. [PubMed]
76. Gonzalez-Montalban, N.; Makarava, N.; Ostapchenko, V.G.; Savtchenk, R.; Alexeeva, I.; Rohwer, R.G.; Baskakov, I.V. Highly efficient protein misfolding cyclic amplification. *PLoS Pathog.* **2011**, *7*, e1001277. [CrossRef] [PubMed]
77. Haley, N.J.; Van de Motter, A.; Carver, S.; Henderson, D.; Davenport, K.; Seelig, D.M.; Mathiason, C.; Hoover, E. Prion-seeding activity in cerebrospinal fluid of deer with chronic wasting disease. *PLoS ONE* **2013**, *8*, e81488. [CrossRef] [PubMed]
78. Park, J.H.; Choi, Y.G.; Park, S.J.; Choi, H.S.; Choi, E.K.; Kim, Y.S. Ultra-efficient Amplification of Abnormal Prion Protein by Modified Protein Misfolding Cyclic Amplification with Electric Current. *Mol. Neurobiol.* **2017**. [CrossRef] [PubMed]
79. Chang, B.; Gray, P.; Piltch, M.; Bulgin, M.S.; Sorensen-Melson, S.; Miller, M.W.; Davies, P.; Brown, D.R.; Coughlin, D.R.; Rubenstein, R.; et al. Surround optical fiber immunoassay (SOFIA): An ultra-sensitive assay for prion protein detection. *J. Virol. Methods* **2009**, *159*, 15–22. [CrossRef] [PubMed]
80. Rubenstein, R.; Chang, B.; Gray, P.; Piltch, M.; Bulgin, M.S.; Sorensen-Melson, S.; Miller, M.W. A novel method for preclinical detection of PrPSc in blood. *J. Gen. Virol.* **2010**, *91 Pt 7*, 1883–1892. [CrossRef] [PubMed]
81. Pulford, B.; Spraker, T.R.; Wyckoff, A.C.; Meyerett, C.; Bender, H.; Ferguson, A.; Wyatt, B.; Lockwood, K.; Powers, J.; Telling, C.G.; et al. Detection of PrPCWD in feces from naturally exposed Rocky Mountain elk (Cervus elaphus nelsoni) using protein misfolding cyclic amplification. *J. Wildl. Dis.* **2012**, *48*, 425–434. [CrossRef] [PubMed]

82. Nichols, T.A.; Spraker, T.R.; Gidlewski, T.; Powers, J.G.; Telling, G.C.; VerCauteren, K.C.; Zabel, M.D. Detection of prion protein in the cerebrospinal fluid of elk (Cervus canadensis nelsoni) with chronic wasting disease using protein misfolding cyclic amplification. *J. Vet. Diagn. Investig.* **2012**, *24*, 746–749. [CrossRef] [PubMed]

83. Saunders, S.E.; Shikiya, R.A.; Langenfeld, K.; Bartelt-Hunt, S.L.; Bartz, J.C. Replication efficiency of soil-bound prions varies with soil type. *J. Virol.* **2011**, *85*, 5476–5482. [CrossRef] [PubMed]

84. Nichols, T.A.; Pulford, B.; Wyckoff, A.C.; Meyerett, C.; Michel, B.; Gertig, K.; Hoover, E.A.; Jewell, J.E.; Telling, G.C.; Zabel, M.D. Detection of protease-resistant cervid prion protein in water from a CWD-endemic area. *Prion* **2009**, *3*, 171–183. [CrossRef] [PubMed]

85. Pritzkow, S.; Morales, R.; Moda, F.; Khan, U.; Telling, G.C.; Hoover, E.; Soto, C. Grass plants bind, retain, uptake, and transport infectious prions. *Cell Rep.* **2015**, *11*, 1168–1175. [CrossRef] [PubMed]

86. Meyerett, C.; Michel, B.; Pulford, B.; Spraker, T.R.; Nichols, T.A.; Johnson, T. In vitro strain adaptation of CWD prions by serial protein misfolding cyclic amplification. *Virology* **2008**, *382*, 267–276. [CrossRef] [PubMed]

87. Kocisko, D.A.; Come, J.H.; Priola, S.A.; Chesebro, B.; Raymond, G.J.; Lansbury, P.T.; Caughey, B. Cell-free formation of protease-resistant prion protein. *Nature* **1994**, *370*, 471–474. [CrossRef] [PubMed]

88. Atarashi, R.; Wilham, J.M.; Christensen, L.; Hughson, A.G.; Moore, R.A.; Johnson, L.M.; Onwubiko, H.A.; Priola, S.A.; Caughey, B. Simplified ultrasensitive prion detection by recombinant PrP conversion with shaking. *Nat. Methods* **2008**, *5*, 211–212. [CrossRef] [PubMed]

89. Davenport, K.A.; Henderson, D.M.; Bian, J.; Telling, G.C.; Mathiason, C.K.; Hoover, E.A. Insights into Chronic Wasting Disease and Bovine Spongiform Encephalopathy Species Barriers by Use of Real-Time Conversion. *J. Virol.* **2015**, *89*, 9524–9531. [CrossRef] [PubMed]

90. Seelig, D.M.; Mason, G.L.; Telling, G.C.; Hoover, E.A. Chronic Wasting Disease Prion Trafficking via the Autonomic Nervous System. *Am. J. Pathol.* **2011**, *179*, 1319–1328. [CrossRef] [PubMed]

91. Chao, J.; DeBiasio, R.; Zhu, Z.; Giuliano, K.A.; Schmidt, B.F. Immunofluorescence signal amplification by the enzyme-catalyzed deposition of a fluorescent reporter substrate (CARD). *Cytometry* **1996**, *23*, 48–53. [CrossRef]

92. Henderson, D.M.; Manca, M.; Haley, N.J.; Denkers, N.D.; Nalls, A.V.; Mathiason, C.K.; Caughey, B.; Hoover, E.A. Rapid Antemortem Detection of CWD Prions in Deer Saliva. *PLoS ONE* **2013**, *8*, e74377. [CrossRef] [PubMed]

93. Elder, A.M.; Henderson, D.M.; Nalls, A.V.; Wilham, J.M.; Caughey, B.W.; Hoover, E.A.; Kincaid, A.E.; Bartz, J.C.; Mathiason, C.K. In Vitro Detection of prionemia in TSE-Infected Cervids and Hamsters. *PLoS ONE* **2013**, *8*, e80203. [CrossRef] [PubMed]

94. Elder, A.M.; Henderson, D.M.; Nalls, A.V.; Hoover, E.A.; Kincaid, A.E.; Bartz, J.C.; Mathiason, C.K. Immediate and Ongoing Detection of Prions in the Blood of Hamsters and Deer following Oral, Nasal, or Blood Inoculations. *J. Virol.* **2015**, *89*, 7421–7424. [CrossRef] [PubMed]

95. Van Keulen, L.J.; Schreuder, B.E.; Meloen, R.H.; Mooij-Harkes, G.; Vromans, M.E.; Langeveld, J.P. Immunohistochemical detection of prion protein in lymphoid tissues of sheep with natural scrapie. *J. Clin. Microbiol.* **1996**, *34*, 1228–1231. [PubMed]

96. Gonzalez, L.; Dagleish, M.P.; Bellworthy, S.J.; Siso, S.; Stack, M.J.; Chaplin, M.J.; Balachandran, A. Postmortem diagnosis of preclinical and clinical scrapie in sheep by the detection of disease-associated PrP in their rectal mucosa. *Vet. Rec.* **2006**, *158*, 325–331. [CrossRef] [PubMed]

97. O'Rourke, K.I.; Baszler, T.V.; Besser, T.E.; Miller, J.M.; Cutlip, R.C.; Wells, G.A.; Ryder, S.J.; Parish, S.M.; Hamir, A.N.; Cockett, N.E.; et al. Preclinical diagnosis of scrapie by immunohistochemistry of third eyelid lymphoid tissue. *J. Clin. Microbiol.* **2000**, *38*, 3254–3259.

98. Castilla, J.; Saa, P.; Soto, C. Detection of prions in blood. *Nat. Med.* **2005**, *11*, 982–985. [CrossRef] [PubMed]

99. Saa, P.; Castilla, J.; Soto, C. Presymptomatic detection of prions in blood. *Science* **2006**, *313*, 92–94. [CrossRef] [PubMed]

100. Spraker, T.R.; Gidlewski, T.L.; Balachandran, A.; VerCauteren, K.C.; Creekmore, L.; Munger, R.D. Detection of PrP(CWD) in postmortem rectal lymphoid tissues in Rocky Mountain elk (Cervus elaphus nelsoni) infected with chronic wasting disease. *J. Vet. Diagn. Investig.* **2006**, *18*, 553–557. [CrossRef] [PubMed]

101. Wolfe, L.L.; Spraker, T.R.; Gonzalez, L.; Dagleish, M.P.; Sirochman, T.M.; Brown, J.C.; Jeffrey, M.; Miller, M.W. PrPCWD in rectal lymphoid tissue of deer (Odocoileus spp.). *J. Gen. Virol.* **2007**, *88 Pt 7*, 2078–2082. [CrossRef] [PubMed]
102. Monello, R.J.; Powers, J.G.; Hobbs, N.T.; Spraker, T.R.; O'Rourke, K.I.; Wild, M.A. Efficacy of antemortem rectal biopsies to diagnose and estimate prevalence of chronic wasting disease in free-ranging cow elk (Cervus elaphus nelsoni). *J. Wildl. Dis.* **2013**, *49*, 270–278. [CrossRef] [PubMed]
103. Dechen Quinn, A.C.; Williams, D.M.; Porter, W.F.; Fitzgerald, S.D.; Hynes, K. Effects of capture-related injury on postcapture movement of white-tailed deer. *J. Wildl. Dis.* **2014**, *50*, 250–258. [CrossRef] [PubMed]
104. Sharma, K.; Mewara, A.; Gupta, N.; Sharma, A.; Varma, S. Multiplex PCR in diagnosis of *M. tuberculosis* and *M. avium* co-infection from lymph node in an AIDS patient. *Indian J. Med. Microbiol.* **2015**, *33*, 151–153. [CrossRef] [PubMed]
105. Smith, A.J.; Robertson, D.; Tang, M.K.; Jackson, M.S.; MacKenzie, D.; Bagg, J. Staphylococcus aureus in the oral cavity: A three-year retrospective analysis of clinical laboratory data. *Br. Dent. J.* **2003**, *195*, 701–703. [CrossRef] [PubMed]
106. Warren, W.P.; Balcarek, K.; Smith, R.; Pass, R.F. Comparison of rapid methods of detection of cytomegalovirus in saliva with virus isolation in tissue culture. *J. Clin. Microbiol.* **1992**, *30*, 786–789. [PubMed]
107. Perrott, M.R.; Sigurdson, C.J.; Mason, G.L.; Hoover, E.A. Evidence for distinct chronic wasting disease (CWD) strains in experimental CWD in ferrets. *J. Gen. Virol.* **2011**, *93 Pt 1*, 212–221. [CrossRef] [PubMed]
108. Angers, R.C.; Kang, H.E.; Napier, D.; Browning, S.; Seward, T.; Mathiason, C.; Balachandran, A.; McKenzie, D.; Castilla, J.; Soto, C.; et al. Prion strain mutation determined by prion protein conformational compatibility and primary structure. *Science* **2010**, *328*, 1154–1158. [CrossRef] [PubMed]
109. Orru, C.D.; Groveman, B.R.; Raymond, L.D.; Hughson, A.G.; Nonno, R.; Zou, W.; Ghetti, B.; Gambetti, P.; Caughey, B. Bank Vole Prion Protein As an Apparently Universal Substrate for RT-QuIC-Based Detection and Discrimination of Prion Strains. *PLoS Pathog.* **2015**, *11*, e1004983.
110. Ayers, J.I.; Giasson, B.I.; Borchelt, D.R. Prion-like Spreading in Tauopathies. *Biol. Psychiatry* **2017**. [CrossRef] [PubMed]
111. Olanow, C.W. Do prions cause Parkinson disease?: The evidence accumulates. *Ann. Neurol.* **2014**, *75*, 331–333. [CrossRef] [PubMed]
112. Stopschinski, B.E.; Diamond, M.I. The prion model for progression and diversity of neurodegenerative diseases. *Lancet Neurol.* **2017**, *16*, 323–332. [CrossRef]
113. Edwards, G., 3rd; Moreno-Gonzalez, I.; Soto, C. Amyloid-beta and tau pathology following repetitive mild traumatic brain injury. *Biochem. Biophys. Res. Commun.* **2017**, *483*, 1137–1142. [CrossRef] [PubMed]
114. Schmitz, M.; Cramm, M.; Llorens, F.; Muller-Cramm, D.; Collins, S.; Atarashi, R.; Satoh, K.; Orru, C.D.; Groveman, B.R.; Zafar, S.; et al. The real-time quaking-induced conversion assay for detection of human prion disease and study of other protein misfolding diseases. *Nat. Protoc.* **2016**, *11*, 2233–2242. [CrossRef] [PubMed]

pathogens

MDPI

Review

The Structure of PrP^{Sc} Prions

Holger Wille [1] and Jesús R. Requena [2,*]

[1] Centre for Prions and Protein Folding Diseases & Department of Biochemistry, University of Alberta, Edmonton, AB T6G 2M8, Canada; wille@ualberta.ca

[2] CIMUS Biomedical Research Institute & Department of Medical Sciences, University of Santiago de Compostela-IDIS, 15782 Santiago de Compostela, Spain

* Correspondence: jesus.requena@usc.es; Tel.: +34-8818-15464

Received: 22 January 2018; Accepted: 3 February 2018; Published: 7 February 2018

Abstract: PrP^{Sc} (scrapie isoform of the prion protein) prions are the infectious agent behind diseases such as Creutzfeldt–Jakob disease in humans, bovine spongiform encephalopathy in cattle, chronic wasting disease in cervids (deer, elk, moose, and reindeer), as well as goat and sheep scrapie. PrP^{Sc} is an alternatively folded variant of the cellular prion protein, PrP^C, which is a regular, GPI-anchored protein that is present on the cell surface of neurons and other cell types. While the structure of PrP^C is well studied, the structure of PrP^{Sc} resisted high-resolution determination due to its general insolubility and propensity to aggregate. Cryo-electron microscopy, X-ray fiber diffraction, and a variety of other approaches defined the structure of PrP^{Sc} as a four-rung β-solenoid. A high-resolution structure of PrP^{Sc} still remains to be solved, but the four-rung β-solenoid architecture provides a molecular framework for the autocatalytic propagation mechanism that gives rise to the alternative conformation of PrP^{Sc}. Here, we summarize the current knowledge regarding the structure of PrP^{Sc} and speculate about the molecular conversion mechanisms that leads from PrP^C to PrP^{Sc}.

Keywords: PrP^{Sc}; prion structure; β-solenoid; cryo-electron microscopy; prion propagation; amyloid

1. Introduction

PrP^{Sc} was the first prion—i.e., infectious protein—to be discovered, and continues to be the quintessential prion, not only because of its historical preeminence, but also because of its association with a unique class of fatal diseases. PrP^{Sc} is the only prion known to date to have caused local epidemic and epizootic outbursts. Some of these have captured the attention of the public and even caused shockwaves of panic [1].

Ovine and caprine scrapie—fatal neurodegenerative ailments endemic in Europe—have been known for centuries, but it was not until the 1930s that its infectious nature was discovered [2]. Later, in the 1980s, PrP^{Sc} was identified as the infectious agent causing transmission of scrapie, and as the first prion ever, and was used to define the term "prion" [3,4]. Despite efforts to eradicate it, scrapie continues to be enzootic in Europe. However, it is not transmissible to humans, a phenomenon known as transmission barrier. The practice of industrial cannibalism resulted in PrP^{Sc} prions being recycled into cattle feed and causing bovine spongiform encephalopathy ("mad cow disease") [1,4,5]. The ensuing bovine spongiform encephalopathy (BSE) epizootic affected hundreds of thousands of animals throughout Europe in the 1980s to 2000s. In turn, BSE PrP^{Sc} prions transmitted to humans, causing transmissible variant Creutzfeldt–Jakob disease (vCJD). Fortunately, the barrier governing transmission of BSE PrP^{Sc} to humans is quite high, which limited the number of vCJD cases to about 200, whereas millions of individuals are likely to have been exposed to BSE PrP^{Sc} by the oral route [1]. While cases of vCJD have subsided, retrospective histopathological analyses of tonsil and appendix samples suggest that thousands of individuals harbor PrP^{Sc} in their bodies, although a very long incubation time has prevented the appearance of clinical disease so far [6]. Very long incubation times have also been observed for Kuru. Kuru was an epidemic caused by human PrP^{Sc} transmitted orally through ritual cannibalism among the Fore people from Papua New Guinea beginning in the 1950s [7]. In this instance, an initial case of sporadic Creutzfeldt–Jakob disease (CJD) is suspected to have triggered the localized epidemic. CJD PrP^{Sc} is known to have been transmitted from humans to humans iatrogenically, through the reutilization of improperly decontaminated neurosurgical instruments, dura mater grafts, or treatment with cadaveric growth hormone containing traces of PrP^{Sc} [4,5]. Also, at least three cases of transmission of vCJD PrP^{Sc} through blood transfusion have been documented [8].

Another example of widespread infectious transmission of PrP^{Sc} prions is chronic wasting disease (CWD), which affects various cervid species and which is very contagious resulting in efficient horizontal transmission. CWD was first detected in the state of Colorado (USA) and has since spread through very extensive areas of North America [9], where it appears to be becoming enzootic. More recently, six cases have surfaced in Norway [10]. It is not currently known whether these are related to North American CWD or arose spontaneously in Norwegian moose, reindeer, and red deer populations.

All in all, while substantial experimental evidence is accruing to suggest that other misfolded proteins such as Aβ, tau or α-synuclein might be prions or at least feature prion-like behavior of the affected proteins [11–13], PrP^{Sc} prions stand out as truly infectious, at times highly contagious, and disease-causing pathogens that command close attention.

Yet, how can a protein become infectious? Classic infectious agents reproduce because they contain nucleic acids, biomolecules that can be copied and therefore amplified. More specifically, what is copied is the primary structure of these nucleic acids, whether DNA or RNA. In contrast, propagation of prions, and more specifically, of PrP^{Sc} prions, involves reproduction not of their primary, but of their secondary, tertiary, and quaternary structures, i.e., their conformation [4]. PrP^{Sc} coerces PrP^{C}, a glycosylphosphatidylinositol-anchored (GPI-anchored) membrane protein with the same primary but different secondary, tertiary, and quaternary structures, to adopt the PrP^{Sc} conformation. This likely involves complete unfolding of PrP^{C}, first, followed by refolding through a series of molecular events in which PrP^{Sc} acts as a physical template (vide infra). To fully understand this process at the molecular level, it is essential to first know the structure of PrP^{Sc}. This review presents a comprehensive summary of what we currently know about the structure of PrP^{Sc} and how its structure might encode a possible mechanism for its conformational replication. This mechanism also provides hints to explain the strain and transmission barrier phenomena, crucial in the epidemiology and epizootiology of PrP^{Sc}.

2. The Architecture of PrP^{Sc} Prions

The structure of PrP^{Sc} is based on a four-rung β-solenoid architecture (Figure 1), as was revealed recently by cryo-electron microscopy and three-dimensional (3D) reconstructions [14]. By analyzing 3D reconstructions from individual PrP^{Sc} amyloid fibrils, and by taking the molecular density of PrP^{Sc} into consideration [15], it was possible to determine the average molecular height of each PrP^{Sc} molecule along the fibril axis as ~17.7 Å [14]. Individual measurements ranged from 16.1 to 19.25 Å, while a four-rung β-solenoid architecture would be expected to have a height of 19.2 Å (=4 × 4.8 Å). A single particle approach, which was used to average data from a much larger number of PrP^{Sc} amyloid fibril segments, produced molecular height peaks around 20 and 40 Å [14]. The former was in good agreement with the results obtained from individual amyloid fibrils, while the latter measurement suggested a larger assembly unit along the fibril axis, encompassing two monomers in a potential head-to-head/tail-to-tail configuration (Figure 2). While the cryo-electron microscopy data helped to decipher the overall architecture of PrP^{Sc} as a four-rung β-solenoid, the resolution was not sufficient to resolve the structure in atomic details.

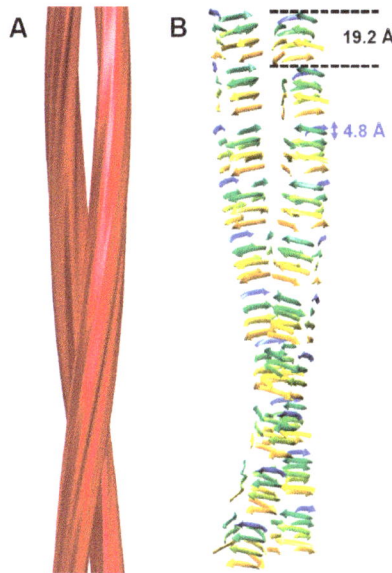

Figure 1. Four-rung β-solenoid architecture of PrP^{Sc}. (**A**) Three-dimensional reconstruction of a PrP^{Sc} amyloid fibril with two protofilaments. (**B**) Cartoon representation of a four-rung β-solenoid architecture drawn to approximate the 3D reconstruction in (**A**). The 4.8 Å spacing of individual β-strands running perpendicular to the fibril axis is indicated, as is the 19.2 Å height of an individual PrP^{Sc} molecule. Figure adapted from Vázquez-Fernández et al. PLoS Pathog. 2016, 12, e1005835 [14].

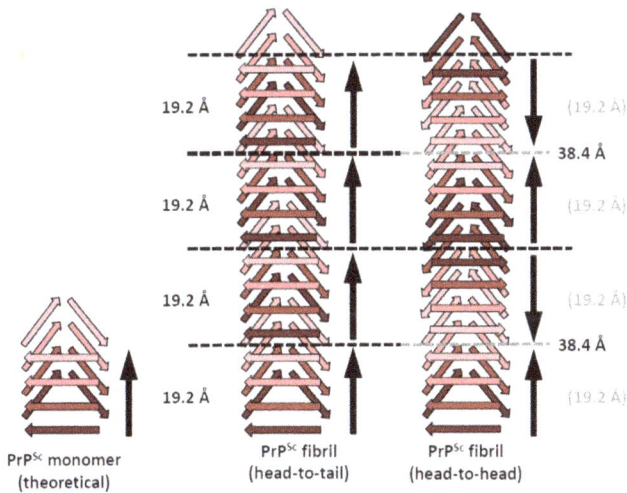

Figure 2. Schematic representations of possible head-to-tail and head-to-head (tail-to-tail) architectures for PrP^{Sc} amyloid fibrils. The different architectures would lead to a polar fibril in the case of head-to-tail stacking, while a head-to-head (tail-to-tail) architecture would give rise to a non-polar fibril. The ~40 Å signal that was obtained with the single particle image processing approach [14] would favor a head-to-head arrangement, since there is no straightforward mechanism by which a head-to-tail arrangement would produce such a spacing.

A β-solenoidal core was originally proposed by one of us as the key architectural element of PrP[Sc], based on electron crystallography studies of 2D crystals from the N-terminally truncated PrP[Sc] (PrP27-30) [16,17]. By comparing 2D projection maps from PrP27-30 2D crystals with those of an even smaller "mini-prion", PrP[Sc]106 [18], the structure of PrP[Sc] was constrained to contain a β-helix or β-solenoid structure at its core. At the time, it was assumed that PrP[Sc] would contain residual α-helix structure at the C-terminus, an interpretation which is not supported by more recent experimental evidence [14,19,20].

X-ray fiber diffraction from brain-derived PrP27-30 and PrP[Sc] amyloid fibrils gave a series of meridional diffraction signals at 9.6, 6.4, and 4.8 Å, which correspond to the second, third, and fourth order diffraction, respectively, of a 19.2 Å β-sheet structure [21]. The equatorial diffraction signatures were equally informative, in that a prominent ~10 Å signal, which is characteristic for generic stacked β-sheet amyloid structures, including those present in recombinant PrP amyloid fibrils [21,22], was absent. This absence is a strong indicator that the underlying architecture is that of a β-solenoid, as demonstrated by comparison with diffraction results obtained from the HET-s amyloid [23], which has been shown by solid state NMR to contain a two-rung β-solenoid structure [24]. All together, the X-ray fiber diffraction results provided clear evidence that the structure of PrP[Sc] contains a four-rung β-solenoid fold at its core [21]. Subsequent analyses indicated that shorter fragments of PrP could either adopt a generic stacked β-sheet structure or shorter β-solenoids [25]. The shortest form of PrP that could support the formation of transmissible prions, PrP[Sc]106, was also found to contain a four-rung β-solenoid fold [18,25]. A number of other studies have obtained high-resolution structures from short, PrP-based peptides, often adopting "steric zippers" or related structures, but those short peptides have no biological relevance and lack the structural complexities that characterize PrP[Sc].

Moreover, the β-solenoidal structure agrees with a number of structural restraints gathered through the years with a variety of biophysical and biochemical methods. Fourier-transform infrared (FTIR) spectroscopy, and circular dichroism spectroscopy (CD), had demonstrated a high β-sheet content of PrP[Sc] and its N-terminally truncated variant, PrP27-30 [19,26–29]. More specifically, FTIR-based estimates of β-sheet in PrP27-30 range from 43–61% [26–28]. For some time, the FTIR data were interpreted to imply that PrP[Sc] and PrP27-30 contained a substantial fraction of α-helical structure. However, Smirnovas et al. have shown that the ~1660 cm^{-1} band in the FTIR spectra of PrP[Sc] and PrP27-30, which had been attributed to α-helices based on calibration using globular proteins, is also present in the spectrum of amyloid fibrils formed by recombinant prion protein (recPrP) known to exhibit a parallel in-register β-structure and to be completely devoid of α-helices [19]. Furthermore, the ~1660 cm^{-1} FTIR band overlaps with bands in the same region arising from turns and coils. Therefore, it can be safely concluded that the FTIR-based data do not support the presence of α-helices in PrP[Sc] [20]. To sum up, all these studies suggest that PrP27-30 consists of about 50% β-strands and 50% random coil loops; this fits very well a four-rung β-solenoid with short β-strands connected by loops. It should be noted that the HET-s(218–289) prion, whose structure conforms to a two-rung β-solenoid and is probably quite similar to PrP[Sc], contains ~53% of β-strands connected by ~47% loops and turns [24].

The extremely compact nature of the β-solenoid core, spanning virtually from the N- to the C-terminus of PrP27-30 (~90 to 230) is also in good agreement with the known resistance of PrP27-30 to protease digestion. Classically, when PrP[Sc] of most strains is subjected to proteinase K (PK) treatment, its N-terminal residues, up to position 86/98 depending on the strain, is readily digested. This portion is therefore believed to retain the completely unfolded secondary structure that it exhibits in PrP[C] [4]. It is noteworthy that this region is totally dispensable for infectivity, and therefore it can be considered as not being part of the "prionic domain" of PrP[Sc]. Nevertheless, insertions and deletions in the N-terminal domain are known to cause familial prion diseases [30]. The C-terminal domain, which corresponds to the β-solenoid part of PrP[Sc], forms the core of the PrP[Sc] amyloid fibril from which the N-terminal 23-86/98 "tail" projects into the medium.

On the other hand, the existence of connecting loops, would explain the presence of small quantities of smaller PK-resistant fragments besides PrP27-30 [31–34]. Any β-solenoidal protein subjected to treatment with a relatively non-specific protease such as PK would be expected to undergo partial cleavage at the more flexible, less compact coils connecting the β-strands, while these would be relatively spared [35,36]. The extraordinary resilience of PrP^{Sc} to PK observed in early studies [4,37] suggests that the PrP27-30 β-solenoid is extremely compact, i.e., its connecting loops and turns are likely very tightly packed against the β-strands that make up the solenoid core. This would result in the scarcity of PK cleavages at these sites as compared to the fast, extensive, and complete digestion of the unfolded N-terminal stretch. However, secondary cleavages within the β-solenoid core have indeed been progressively unveiled in many PrP^{Sc} samples, particularly as a wider variety of antibodies has been used to probe the PK digests [31–34]. Furthermore, a number of PrP^{Sc} strains with increased susceptibility to PK have been identified, including Variably Protease Sensitive Prionopaty (VPSP) PrP^{Sc} [38,39], or PrP^{Sc} from spontaneously ill transgenic bank voles overexpressing PrP with the 109I polymorphism [40,41]. Also, shorter PK-resistant PrP^{Sc} fragments have been identified as being characteristic of many "atypical" PrP^{Sc} strains [39,42]. It is noteworthy that many of these strains of PrP^{Sc} are characteristically cleaved by PK at a position around ~150–153 [39]. The same cleavage has been identified as one of the most prominent secondary internal cleavages within PrP^{Sc} of "classic" strains [31–34,36]. This strongly suggests that the ~150–153 region corresponds to an important loop, perhaps connecting different rungs of the solenoid, a hinge of sorts.

Minor differences in threading (vide infra) and/or lateral packing of loops would obviously result in very significant changes in susceptibility of different strains of PrP^{Sc} to proteolysis.

3. Other Models of PrP^{Sc}

Before confluent X-ray fiber diffraction and cryo-EM studies defined the β-solenoid as the basic architectural element of PrP^{Sc}, a number of structural models were put forward. We have extensively reviewed them and their shortcomings [20]. Here, we will just refer to a recent one, which has received more attention due to its similarity to recently solved amyloid structures: the parallel in-register intermolecular beta-sheet model (PIRIBS) [43]. In this model, each molecule of PrP stacks on top of the preceding molecule perfectly in register. Hence, a single molecule of PrP contributes just 4.8 Å in height to the rise of a PrP amyloid fibril. Given the size of PrP, a single molecule would have to cover both "protofilaments" in the observed amyloid fibrils. As a consequence, the "protofilaments" that can be observed in electron micrographs of PrP^{Sc} amyloid [14,21,44] would not exist as separate entities and represent merely imaging artifacts. Interestingly, X-ray fiber diffraction suggested a PIRIBS-like conformation for recombinant PrP amyloid that was found to be non-infectious [21,22].

The PIRIBS model is incompatible with the height measurements from both the X-ray fiber diffraction experiments [21], and the cryo EM observations [14]. Both had independently indicated the height of a molecule of PrP^{Sc} to be 19.2 Å (see above). Furthermore, the dense stacking of the PIRIBS model cannot accommodate the bulk of the glycosylation side chains [45], which, due to their bulk, take up more space than is available in a tight packing as the PIRIBS model requires (Figure 3).

4. Implications of the Structure of PrP^{Sc} for Its Propagation

While only the overall architecture of PrP^{Sc} has been deciphered and important structural elements are still undefined (Figure 1), it is now possible, for the first time, to formulate a sound hypothesis about how PrP^{Sc} prions propagate [46]. A β-solenoid has inherent templating capabilities: its upper- and lowermost rungs contain "unpaired" β-strands that can propagate their hydrogen-bonding pattern into any amyloidogenic peptide they encounter [47]. In fact, the edge strands of native, soluble proteins that contain a β-solenoid have evolved to be capped by loops or α-helices that block unregulated β-sheet propagation. Furthermore, when these capping structures are eliminated by means of protein-engineering techniques, the resulting "decapped" β-solenoids become unstable and undergo edge-to-edge-driven oligomerization [48]. Therefore, the upper- and lowermost β-solenoid rungs in PrP^{Sc} can template an incoming, unfolded PrP molecule, and mold it into an additional β-solenoid rung (Figure 3). Once this supplementary β-solenoid rung is formed, it offers a fresh, "sticky" surface that can continue templating the remaining, unfolded portion of the incoming PrP molecule, until a second rung is generated. This process can be repeated two more times until the entire length of the incoming PrP polypeptide chain has been molded into four newly formed rungs, thus completing a new four-rung β-solenoid structure (Figure 4). The newly formed upper- or lower-most rungs can now serve as a fresh templating surface for a new incoming unfolded PrP molecule, in a process that can proceed ad infinitum. As already mentioned above, the presence of bulky carbohydrate chains in the incoming unfolded PrP molecule must certainly impose constraints to the templating process; in turn, the GPI moiety present both in the incoming PrP molecule and in the PrP^{Sc} template likely anchoring them to the cell membrane and/or endocytic vesicle milieu, probably impose constraints relative to the cellular location(s) in which conversion takes place. Templating must necessarily be based on either a head-to-tail or a head-to-head/tail-to-tail orientation. In the former case, templating of β-sheets would involve heterotypic contacts between different parts of the molecule, while the latter would involve homotypic contacts. The structural arguments and data favoring one or the other of these two possibilities will be discussed in the following section.

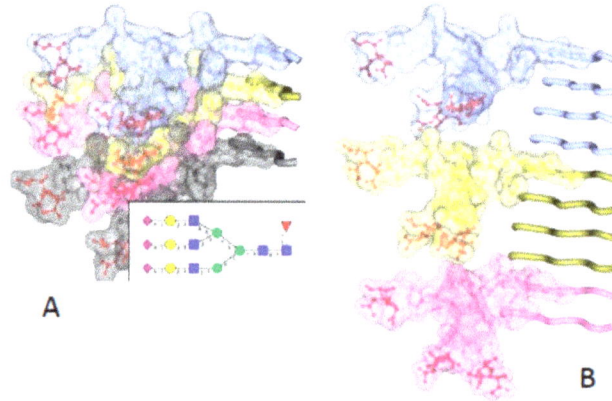

Figure 3. The bulky, N-linked glycans impose spatial constraints on folding patterns of PrP^{Sc} and rule out a flat, in-register stacking. Cross β-sheet structures carrying tri-antennary N-glycans (shown in inset) on each successive β-strand (**A**) or every fourth β-strand (**B**). Polypeptide chains are represented in tube form, whereas the N-glycans are shown as a combination of ball-and-stick and volume representations. Each PrP molecule with corresponding N-glycan is rendered a different color. Sialic acid residues are colored in red; N-glycan electrostatic surfaces are semi-transparent. To model the dimension of cross-β structures, the Authors of the original figure (see below) adapted the parallel beta-sheet model from PDB database entry 2RNM, an NMR structure for HET-s(218–289) prion in its amyloid form [24]. The structure of a tri-antennary N-linked glycan was taken from PDB entry 3QUM, a crystal structure of human prostate specific antigen (PSA) [50]. Adapted from Baskakov, I.V. and Katorcha, E. Frontiers in Neuroscience 2016, 10, 358 [45].

How are strain and transmission barrier properties encoded in the β-solenoid architecture of PrP^{Sc}? Slightly different threading, resulting in slight differences in the amino acid composition of the β-strands and loops are an obvious source of variability giving rise to different variations of the main β-solenoid theme, as already noted by Langedijk and colleagues [49]. These variations, affecting the topography of the upper- and lowermost rungs, would have an obvious impact on the templating properties of a given β-solenoid variant (strain), as depicted in Figure 3. Thus, the presence in the templating surfaces of charged or bulky amino acid residues might impose restrictions to their ability to receive and template a PrP chain of a given amino acidic sequence. This is particularly evident with respect to the sequences at the N- and C-termini of the unfolded PrP substrate, the first stretches that need to adapt to these templating surfaces; but also for the rest of the sequence, as every new rung template generates a fresh templating surface with its own steric and charge constraints. Ultimately, a higher resolution structure of PrP^{Sc} should allow to fully understand these properties of PrP^{Sc} at the molecular level.

It is noteworthy that the molecular forces responsible for templating, namely, hydrogen-bonding, charge interactions, aromatic stacking, and steric constraints, are essentially the same as those operating during DNA replication. However, they lack its exquisite precision and the complex proofreading mechanisms that provide the flexibility of nucleic acid replication. The higher complexity of the PrP^{Sc} structure, as compared to that of nucleic acids, will require particular efforts to achieve a complete understanding of all the molecular aspects associated with propagation of PrP^{Sc} prions.

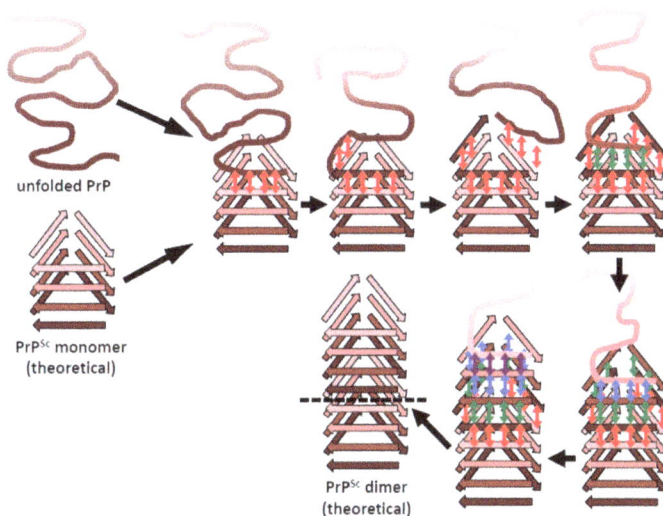

Figure 4. The β-solenoid architecture of PrP^{Sc} suggests a mechanism to template incoming molecules of unfolded PrP onto the existing β-solenoid fold to generate a copy of itself. For simplicity, this mechanistic model is based on a head-to-tail arrangement. An incoming molecule of unfolded PrP would interact with an uncapped β-solenoid surface and adopt a β-strand conformation by forming backbone hydrogen-bonds (red arrows). Once the first rung of a nascent β-solenoid configuration has been formed, it would self-template successive rungs of β-solenoid structure using the same mechanism (green, blue, and purple arrows, respectively). Once the fourth and final rung has been templated, a new molecule of PrP^{Sc} is formed and the original template surface has been re-created. Any mutations facilitating unfolding of PrP^{C} would lead to increased chances of propagation events.

5. Head-to-Head or Head-to-Tail Stacking?

The four-rung β-solenoid architecture of PrP^{Sc} implies two possible stacking modes: a head-to-tail stacking, resulting in polar fibril assemblies, or a head-to-head stacking, which would produce non-polar fibril assemblies (Figure 2). A head-to-head stacking mode would result from a templating process in which the first rung formed is in-register with respect to the templating rung, whereas subsequently formed rungs are not. This initial in-register stacking seems a very elegant and simple option. However, it adds the complication of successive PrP^{Sc} subunits with opposite handedness/twist, which would add an unusual level of complexity to the structure of PrP^{Sc}. At the moment, the experimental evidence of roughly four-nanometer axial repeats in single particle averages from PrP^Sc fibrils slightly favor a head-to-head stacking [14], but the evidence is not strong enough to unequivocally resolve the question. Furthermore, it should be noted that a similar "vertical pairing" signal characteristically appears in Fast Fourier Transforms from Het-s prion fibers [15], in which templating and stacking are unequivocally known to be head-to-tail [24].

In contrast, a head-to-tail stacking would entail templating between heterotypic sequences, lacking the elegance and simplicity that in-register stacking/templating would provide. On the other hand, this propagation mode would not result in the opposite handedness/twist problems, as discussed for head-to-head templating. Intrinsically, the head-to-tail stacking would result in a prominent 19.2 Å periodicity, based on the molecular height of PrP^Sc (see above). Any larger spacings along the fibril axis would require alternative explanations, which cannot be provided based on currently available data.

6. Concluding Remarks and Outlook

Elucidation of the architecture of PrP^Sc allows at last to understand the molecular underpinnings of the propagation of this lethal prion. It is noteworthy that the mechanisms involved are not too different from those at play during replication of DNA: hydrogen bonding and steric fitting. However, while for DNA all the templating information can be deduced from and therefore stored in its primary structure, and can therefore be seen as "digital", for PrP^Sc it involves secondary and tertiary and maybe even quaternary structure levels, and therefore it can be viewed as "analog". Hence, it is not surprising that nucleic acids and not prions have been selected as the main elements of heredity, although prebiotic templating of amyloid has been suggested as an early precursor of cellular life [51]. The basic understanding of the four-rung β-solenoid also hints at a possible explanation of the strain and transmission barrier phenomena: strains are likely to correspond to minor variations in the threading of the solenoid, while transmission barriers are likely the consequence of steric hindrances arising from differences in the sequence of the incoming and templating PrP molecules. However, to fully understand these phenomena it will be necessary to refine our current understanding of the structure of PrP^Sc through higher resolution data. Further studies to that end, using improved cryo-EM techniques and NMR applied to recombinant PrP^Sc are being currently carried out in our and other laboratories.

Acknowledgments: The authors would like to acknowledge support from the Alberta Prion Research Institute (grants APRI 201600012 and 201600029) and the Alberta Livestock and Meat Agency (grant ALMA 2016A001R) (all to H.W.), and from the European Commission Grant FP7222887 "Priority," Spanish Ministry of Education Grant BFU2006-04588/BMC, and Spanish Ministry of Economy Grant BFU2013-48436-C2-1-P (all to J.R.R.). We thank Ilia Baskakov and Elizaveta Katorcha, University of Maryland, for allowing us to adapt and use Figure 3.

Author Contributions: H.W. and J.R.R. conceived the ideas expressed in this review and jointly wrote the paper.

Conflicts of Interest: The authors declare no conflict of interest.

References

1. Requena, J.R.; Kristensson, K.; Korth, C.; Zurzolo, C.; Simmons, M.; Aguilar-Calvo, P.; Aguzzi, A.; Andreoletti, O.; Benestad, S.L.; Böhm, R.; et al. The Priority position paper: Protecting Europe's food chain from prions. *Prion* **2016**, *10*, 165–181. [CrossRef] [PubMed]

2. Cuille, J.; Chelle, P.L. Pathologie animale—La maladie dite tremblante du mouton est-elle inoculable? *C. R. Hebd. Seances Acad. Sci.* **1936**, *203*, 1552–1554.

3. Prusiner, S.B. Novel proteinaceous infectious particles cause scrapie. *Science* **1982**, *216*, 136–144. [CrossRef] [PubMed]

4. Prusiner, S.B. Prions. *Proc. Natl. Acad. Sci. USA* **1998**, *95*, 13363–13383. [CrossRef] [PubMed]

5. Aguzzi, A.; Calella, A.M. Prions: Protein aggregation and infectious diseases. *Physiol. Rev.* **2009**, *89*, 1105–1152. [CrossRef] [PubMed]

6. De Marco, M.F.; Linehan, J.; Gill, O.N.; Clewley, J.P.; Brandner, S. Large-scale immunohistochemical examination for lymphoreticular prion protein in tonsil specimens collected in Britain. *J. Pathol.* **2010**, *222*, 380–387. [CrossRef] [PubMed]

7. Alpers, M.P. The epidemiology of kuru: Monitoring the epidemic from its peak to its end. *Philos. Trans. R. Soc. Lond. B Biol. Sci.* **2008**, *363*, 3707–3713. [CrossRef] [PubMed]
8. Urwin, P.J.; Mackenzie, J.M.; Llewelyn, C.A.; Will, R.G.; Hewitt, P.E. Creutzfeldt-Jakob disease and blood transfusion: Updated results of the UK Transfusion Medicine Epidemiology Review Study. *Vox Sang.* **2016**, *110*, 310–316. [CrossRef] [PubMed]
9. Hannaoui, S.; Schatzl, H.M.; Gilch, S. Chronic wasting disease: Emerging prions and their potential risk. *PLoS Pathog.* **2017**, *13*, e1006619. [CrossRef] [PubMed]
10. Benestad, S.L.; Mitchell, G.; Simmons, M.; Ytrehus, B.; Vikøren, T. First case of chronic wasting disease in Europe in a Norwegian free-ranging reindeer. *Vet. Res.* **2016**, *47*, 88. [CrossRef] [PubMed]
11. Jucker, M.; Walker, L.C. Self-propagation of pathogenic protein aggregates in neurodegenerative diseases. *Nature* **2013**, *501*, 45–51. [CrossRef] [PubMed]
12. Prusiner, S.B. A unifying role for prions in neurodegenerative diseases. *Science* **2012**, *336*, 1511–1513. [CrossRef] [PubMed]
13. Castilla, J.; Requena, J.R. Prion-like diseases: Looking for their niche in the realm of infectious diseases. *Virus Res.* **2015**, *207*, 1–4. [CrossRef] [PubMed]
14. Vázquez-Fernández, E.; Vos, M.R.; Afanasyev, P.; Cebey, L.; Sevillano, A.M.; Vidal, E.; Rosa, I.; Renault, L.; Ramos, A.; Peters, P.J.; et al. The structural architecture of an infectious mammalian prion using electron cryomicroscopy. *PLoS Pathog.* **2016**, *8*, e1005835. [CrossRef] [PubMed]
15. Mizuno, N.; Baxa, U.; Steven, A.C. Structural dependence of HET-s amyloid fibril infectivity assessed by cryoelectron microscopy. *Proc. Natl. Acad. Sci. USA* **2011**, *108*, 3252–3257. [CrossRef] [PubMed]
16. Wille, H.; Michelitsch, M.D.; Guenebaut, V.; Supattapone, S.; Serban, A.; Cohen, F.E.; Agard, D.A.; Prusiner, S.B. Structural studies of the scrapie prion protein by electron crystallography. *Proc. Natl. Acad. Sci. USA* **2002**, *99*, 3563–3568. [CrossRef] [PubMed]
17. Govaerts, C.; Wille, H.; Prusiner, S.B.; Cohen, F.E. Evidence for assembly of prions with left-handed beta-helices into trimers. *Proc. Natl. Acad. Sci. USA* **2004**, *101*, 8342–8347. [CrossRef] [PubMed]
18. Supattapone, S.; Bosque, P.; Muramoto, T.; Wille, H.; Aagaard, C.; Peretz, D.; Nguyen, H.O.; Heinrich, C.; Torchia, M.; Safar, J.; et al. Prion protein of 106 residues creates an artificial transmission barrier for prion replication in transgenic mice. *Cell* **1999**, *96*, 869–878. [CrossRef]
19. Smirnovas, V.; Baron, G.S.; Offerdahl, D.K.; Raymond, G.J.; Caughey, B.; Surewicz, W.K. Structural organization of brain-derived mammalian prions examined by hydrogen-deuterium exchange. *Nat. Struct. Mol. Biol.* **2011**, *18*, 504–506. [CrossRef] [PubMed]
20. Requena, J.R.; Wille, H. The structure of the infectious prion protein: Experimental data and molecular models. *Prion* **2014**, *8*, 60–66. [CrossRef] [PubMed]
21. Wille, H.; Bian, W.; McDonald, M.; Kendall, A.; Colby, D.W.; Bloch, L.; Ollesch, J.; Borovinskiy, A.L.; Cohen, F.E.; Prusiner, S.B.; et al. Natural and synthetic prion structure from X-ray fiber diffraction. *Proc. Natl. Acad. Sci. USA* **2009**, *106*, 16990–16995. [CrossRef] [PubMed]
22. Ostapchenko, V.G.; Sawaya, M.R.; Makarava, N.; Savtchenko, R.; Nilsson, K.P.; Eisenberg, D.; Baskakov, I.V. Two amyloid States of the prion protein display significantly different folding patterns. *J. Mol. Biol.* **2010**, *400*, 908–921. [CrossRef] [PubMed]
23. Wan, W.; Wille, H.; Stöhr, J.; Baxa, U.; Prusiner, S.B.; Stubbs, G. Degradation of fungal prion HET-s (218–289) induces formation of a generic amyloid fold. *Biophys. J.* **2012**, *102*, 2339–2344. [CrossRef] [PubMed]
24. Wasmer, C.; Lange, A.; Van Melckebeke, H.; Siemer, A.B.; Riek, R.; Meier, B.H. Amyloid fibrils of the HET-s(218–289) prion form a β solenoid with a triangular hydrophobic core. *Science* **2008**, *319*, 1523–1526. [CrossRef] [PubMed]
25. Wan, W.; Wille, H.; Stöhr, J.; Kendall, A.; Bian, W.; McDonald, M.; Tiggelaar, S.; Watts, J.C.; Prusiner, S.B.; Stubbs, G. Structural studies of truncated forms of the prion protein PrP. *Biophys. J.* **2015**, *108*, 548–1554. [CrossRef] [PubMed]
26. Caughey, B.W.; Dong, A.; Bhat, K.S.; Ernst, D.; Hayes, S.F.; Caughey, W.S. Secondary structure analysis of the scrapie-associated protein PrP 27-30 in water by infrared spectroscopy. *Biochemistry* **1991**, *30*, 7672–7680. [CrossRef] [PubMed]
27. Caughey, B.; Raymond, G.J.; Bessen, R.A. Strain-dependent differences in beta-sheet conformations of abnormal prion protein. *J. Biol. Chem.* **1998**, *273*, 32230–32235. [CrossRef] [PubMed]

28. Pan, K.M.; Baldwin, M.; Nguyen, J.; Gasset, M.; Serban, A.; Groth, D.; Mehlhorn, I.; Huang, Z.; Fletterick, R.J.; Cohen, F.E.; et al. Conversion of alpha-helices into beta-sheets features in the formation of the scrapie prion proteins. *Proc. Natl. Acad. Sci. USA* **1993**, *90*, 10962–10966. [CrossRef] [PubMed]
29. Safar, J.; Roller, P.P.; Gajdusek, D.C.; Gibbs, C.J., Jr. Conformational transitions, dissociation, and unfolding of scrapie amyloid (prion) protein. *J. Biol. Chem.* **1993**, *268*, 20276–20284. [PubMed]
30. Schmitz, M.; Dittmar, K.; Llorens, F.; Gelpi, E.; Ferrer, I.; Schulz-Schaeffer, W.J.; Zerr, I. Hereditary human prion diseases: An update. *Mol. Neurobiol.* **2016**, *54*, 4138–4149. [CrossRef] [PubMed]
31. Zou, W.Q.; Capellari, S.; Parchi, P.; Sy, M.S.; Gambetti, P.; Chen, S.G. Identification of novel proteinase K-resistant C-terminal fragments of PrP in Creutzfeldt-Jakob disease. *J. Biol. Chem.* **2003**, *278*, 40429–40436. [CrossRef] [PubMed]
32. Zanusso, G.; Farinazzo, A.; Prelli, F.; Fiorini, M.; Gelati, M.; Ferrari, S.; Righetti, P.G.; Rizzuto, N.; Frangione, B.; Monaco, S. Identification of distinct N-terminal truncated forms of prion protein in different Creutzfeldt-Jakob disease subtypes. *J. Biol. Chem.* **2004**, *279*, 38936–38942. [CrossRef] [PubMed]
33. Sajnani, G.; Pastrana, M.A.; Dynin, I.; Onisko, B.; Requena, J.R. Scrapie prion protein structural constraints obtained by limited proteolysis and mass spectrometry. *J. Mol. Biol.* **2008**, *382*, 88–98. [CrossRef] [PubMed]
34. Vázquez-Fernández, E.; Alonso, J.; Pastrana, M.A.; Ramos, A.; Stitz, L.; Vidal, E.; Dynin, I.; Petsch, B.; Silva, C.J.; Requena, J.R. Structural organization of mammalian prions as probed by limited proteolysis. *PLoS ONE* **2012**, *7*, e50111. [CrossRef] [PubMed]
35. Hubbard, S.J. The structural aspects of limited proteolysis of native proteins. *Biochim. Biophys. Acta* **1998**, *1382*, 191–206. [CrossRef]
36. Silva, C.J.; Vázquez-Fernández, E.; Onisko, B.; Requena, J.R. Proteinase K and the structure of PrPSc: The good, the bad and the ugly. *Virus Res.* **2015**, *207*, 120–126. [CrossRef] [PubMed]
37. Bolton, D.C.; McKinley, M.P.; Prusiner, S.B. Identification of a protein that purifies with the scrapie prion. *Science* **1982**, *218*, 1309–1311. [CrossRef] [PubMed]
38. Gambetti, P.; Dong, Z.; Yuan, J.; Xiao, X.; Zheng, M.; Alshekhlee, A.; Castellani, R.; Cohen, M.; Barria, M.A.; Gonzalez-Romero, D.; et al. A novel human disease with abnormal prion protein sensitive to protease. *Ann. Neurol.* **2008**, *63*, 697–708. [CrossRef] [PubMed]
39. Pirisinu, L.; Nonno, R.; Esposito, E.; Benestad, S.L.; Gambetti, P.; Agrimi, U.; Zou, W.Q. Small ruminant Nor98 prions share biochemical features with human Gerstmann-Straussler-Scheinker disease and variably protease-sensitive prionopathy. *PLoS ONE* **2013**, *8*, e66405. [CrossRef] [PubMed]
40. Watts, J.C.; Giles, K.; Stöhr, J.; Oehler, A.; Bhardwaj, S.; Grillo, S.K.; Patel, S.; DeArmond, S.J.; Prusiner, S.B. Spontaneous generation of rapidly transmissible prions in transgenic mice expressing wild-type bank vole prion protein. *Proc. Natl. Acad. Sci. USA* **2012**, *109*, 3498–3503. [CrossRef] [PubMed]
41. Watts, J.C.; Giles, K.; Bourkas, M.E.; Patel, S.; Oehler, A.; Gavidia, M.; Bhardwaj, S.; Lee, J.; Prusiner, S.B. Towards authentic transgenic mouse models of heritable PrP prion diseases. *Acta Neuropathol.* **2016**, *132*, 593–610. [CrossRef] [PubMed]
42. Götte, D.R.; Benestad, S.L.; Laude, H.; Zurbriggen, A.; Oevermann, A.; Seuberlich, T. Atypical scrapie isolates involve a uniform prion species with a complex molecular signature. *PLoS ONE* **2011**, *6*, e27510. [CrossRef] [PubMed]
43. Groveman, B.R.; Dolan, M.A.; Taubner, L.M.; Kraus, A.; Wickner, R.B.; Caughey, B. Parallel in-register intermolecular β-sheet architectures for prion-seeded prion protein (PrP) amyloids. *J. Biol. Chem.* **2014**, *289*, 24129–24142. [CrossRef] [PubMed]
44. Sim, V.L.; Caughey, B. Ultrastructures and strain comparison of under-glycosylated scrapie prion fibrils. *Neurobiol. Aging* **2009**, *30*, 2031–2042. [CrossRef] [PubMed]
45. Baskakov, I.V.; Katorcha, E. Multifaceted role of sialylation in prion diseases. *Front. Neurosci.* **2016**, *10*, 358. [CrossRef] [PubMed]
46. Vázquez-Fernández, E.; Young, H.S.; Requena, J.R.; Wille, H. The structure of mammalian prions and their aggregates. *Int. Rev. Cell Mol. Biol.* **2017**, *329*, 277–301. [PubMed]
47. Richardson, J.S.; Richardson, D.C. Natural beta-sheet proteins use negative design to avoid edge-to-edge aggregation. *Proc. Natl. Acad. Sci. USA* **2002**, *99*, 2754–2759. [CrossRef] [PubMed]
48. Bryan, A.W.; Starner-Kreinbrink, J.L.; Hosur, R.; Clark, P.L.; Berger, B. Structure-based prediction reveals capping motifs that inhibit β-helix aggregation. *Proc. Natl. Acad. Sci. USA* **2011**, *108*, 11099–11104. [CrossRef] [PubMed]

49. Langedijk, J.P.; Fuentes, G.; Boshuizen, R.; Bonvin, A.M. Two-rung model of a left-handed beta-helix for prions explains species barrier and strain variation in transmissible spongiform encephalopathies. *J. Mol. Biol.* **2006**, *360*, 907–920. [CrossRef] [PubMed]

50. Stura, E.A.; Muller, B.H.; Bossus, M.; Michel, S.; Jolivet-Reynaud, C.; Ducancel, F. Crystal structure of human prostate-specific antigen in a sandwich antibody complex. *J. Mol. Biol.* **2011**, *414*, 530–544. [CrossRef] [PubMed]

51. Greenwald, J.; Friedmann, M.P.; Riek, R. Amyloid aggregates arise from amino acid condensations under prebiotic conditions. *Angew. Chem. Int. Ed.* **2016**, *55*, 11609–11613. [CrossRef] [PubMed]

Perspective

The Evolutionary unZIPping of a Dimerization Motif—A Comparison of ZIP and PrP Architectures

Jian Hu [1,2], Holger Wille [3,4] and Gerold Schmitt-Ulms [5,6,*]

[1] Department of Chemistry, Michigan State University, East Lansing, MI 48824, USA; hujian1@msu.edu
[2] Department of Biochemistry and Molecular Biology, Michigan State University, East Lansing, MI 48824, USA
[3] Department of Biochemistry, University of Alberta, Edmonton, AB T6G 2M8, Canada; wille@ualberta.ca
[4] Centre for Prions and Protein Folding Diseases, University of Alberta, Edmonton, AB T6G 2M8, Canada
[5] Tanz Centre for Research in Neurodegenerative Diseases, University of Toronto, Toronto, ON M5T 2S8, Canada
[6] Department of Laboratory Medicine and Pathobiology, University of Toronto, Toronto, ON M5S 1A8, Canada
* Correspondence: g.schmittulms@utoronto.ca; Tel.: +1-416-507-6864

Received: 5 December 2017; Accepted: 22 December 2017; Published: 29 December 2017

Abstract: The cellular prion protein, notorious for its causative role in a range of fatal neurodegenerative diseases, evolved from a Zrt-/Irt-like Protein (ZIP) zinc transporter approximately 500 million years ago. Whilst atomic structures for recombinant prion protein (PrP) from various species have been available for some time, and are believed to stand for the structure of PrP^C, the first structure of a ZIP zinc transporter ectodomain was reported only recently. Here, we compare this ectodomain structure to structures of recombinant PrP. A shared feature of both is a membrane-adjacent helix-turn-helix fold that is coded by a separate exon in the respective ZIP transporters and is stabilized by a disulfide bridge. A 'CPALL' amino acid motif within this cysteine-flanked core domain appears to be critical for dimerization and has undergone stepwise regression in fish and mammalian prion proteins. These insights are intriguing in the context of repeated observations of PrP dimers. Other structural elements of ZIP transporters and PrP are discussed with a view to distilling shared versus divergent biological functions.

Keywords: ZIP metal ion transporter; prion protein; dimerization; evolution

1. Introduction

Prion proteins are notorious for their central role in fatal neurodegenerative diseases in a subset of mammalian species, including humans [1–3]. In prion diseases, the cellular prion protein (PrP^C) undergoes structural rearrangements to a β-sheet-rich conformer termed PrP^{Sc} (named after Scrapie in sheep, the first known prion disease). That essential role of PrP in these disorders was demonstrated by showing that the knockout of the prion protein gene (*Prnp*) renders mice refractory to acquiring the disease [4]. A side-product of mouse *Prnp* knockout studies undertaken concomitantly in several laboratories was the identification of a paralog of the prion gene, termed Doppel (*Dpl*). *Dpl* maps to a genomic region C-terminal to *Prnp* and, consequently, was determined to have arisen from a gene duplication event [5,6]. Further genomic sequence analyses revealed that the Shadoo (Sho) gene coded for an additional prion protein paralog [7,8]. The ancestry of this small gene family was enigmatic until 2009 when PrP was initially observed to interact with a subset of zinc transporters of the Zrt-/Irt-like Protein (ZIP) family [9], and subsequent bioinformatic analyses revealed PrP and ZIP transporters to meet several criteria that establish common ancestry [10]. The evolutionary relationship was particularly apparent in comparisons of PrP and ZIP ectodomain sequences in fish genomes, which exhibit a degree of sequence similarity and identity previously reported in pair-wise sequence comparisons of PrP and Dpl, or PrP and Sho. In contrast to ZIPs, which are multi-spanning transmembrane proteins, the prion protein is anchored in the membrane by a glycosylphosphatidylinositol (GPI) anchor, a shift in topology also observed in other protein families [11,12]. Consistent with the view that the prion protein founder gene represented a truncated ZIP gene, such a shift in topology can be experimentally induced when a gene coding for a transmembrane protein is truncated at the 3' end of its first transmembrane domain [13].

Whereas the biology of the prion protein in health and disease has been extensively studied and reviewed [3,14,15], considerably less is known about ZIP transporters, which are coded by members of the solute carrier 39a (Slc39a) gene family. In humans and mice, this family comprises 14 genes, whose gene products appear to be tasked with the import of zinc and other divalent cations into the cytosol, either from the extracellular space or from intracellular compartments. Autosomal recessive mutations in ZIP4 and ZIP13 genes have been linked to *Acrodermatitis enteropathica*, a rare zinc deficiency syndrome [16], and a form of Ehlers-Danlos syndrome characterized by a skeletal dysplasia that mainly affects the spine, and also causes developmental deformations of the hands [17]. The latter symptoms speak to an emerging pattern of ZIP-dependent phenotypes that point toward roles of these proteins in specific morphogenetic programs. In particular, members of the so-called LIV1 subfamily of ZIPs, featuring ectodomains with homology to PrP [10], stand out in this way (Figure 1A). For instance, ZIP6 and ZIP10, the ZIP transporters most closely related to PrP, were shown to contribute to the mammalian oocyte-to-egg transition [18]. Moreover, the morpholino-based knockdown of ZIP6 or ZIP10 caused an embryonic arrest in zebrafish that exhibited characteristics of an impaired epithelial-to-mesenchymal transition (EMT) [19,20]. Test embryos exhibited a phenotype reminiscent of a similar impairment to that observed following the knockdown of PrP in the same paradigm [21]. The striking overlap in ZIP6- and ZIP10-dependent phenotypes was recently resolved by data, which clarified that these proteins form a functional heteromer [19,22]. The ability of these ZIPs to interact directly may also account for their original appearance amongst a short list of PrP interacting proteins [9]. This theory assumes that PrP inherited structural features responsible for interactions amongst these ZIPs from a common ancestor. It is currently unclear whether direct or third-party interactions underlie shared links of PrP and ZIPs to EMT. Recent work in NMuMG cells, a mammalian EMT model, put a spotlight on the neural cell adhesion molecule (NCAM1), by showing that not just PrP [23–25] but also the ZIP6-ZIP10 heteromer [22] predominantly interacts with this cell adhesion molecule. However, although both PrP and ZIP6-ZIP10 affect post-translational modifications on NCAM1 during its involvement in EMT, they do so in different and perhaps complementary ways. More specifically, whereas signaling downstream of PrP was shown to control the transcription of the sialyltransferase (ST8SIA2) that mediates NCAM1 polysialylation [26], the ZIP6-ZIP10 heteromer appeared to control NCAM1 phosphorylation at a specific cluster of cytosolic phosphoacceptor sites through its recruitment of GSK3 [22].

What are the structural features that govern these similarities and differences between PrP and its closest ZIP family members? Until recently, high-resolution structural data were only available for recombinantly expressed PrP [27–29], but not for ZIPs. According to these data, prion proteins from various species are composed of a disordered *N*-terminal domain and a folded *C*-terminal domain characterized by three α-helices and a short two-stranded β-sheet. The two most *C*-terminal α-helices form a conserved helix-turn-helix fold that is stabilized by an internal disulfide bridge and can be post-translationally modified by up to two complex *N*-glycans. Attempts to solve the structure of a fish prion protein, which presumably would be more closely related to ZIP ectodomain structures, were not met with success [30]. Finally, the first high-resolution crystal structure of a ZIP ectodomain [31,32] became available in the past couple of years. This was followed by the discovery of a separate structure of a prokaryotic ZIP containing the transmembrane domain conserved in the ZIP family [32]. Although the ZIP ectodomain structure is from ZIP4, a relatively distant PrP relative [10], the sequence of its membrane-adjacent domain is sufficiently similar to ZIP6 and ZIP10 to be of interest in this context. Here we describe this structure, compare it to PrP, and discuss its significance for understanding the biology and evolution of mammalian prion proteins.

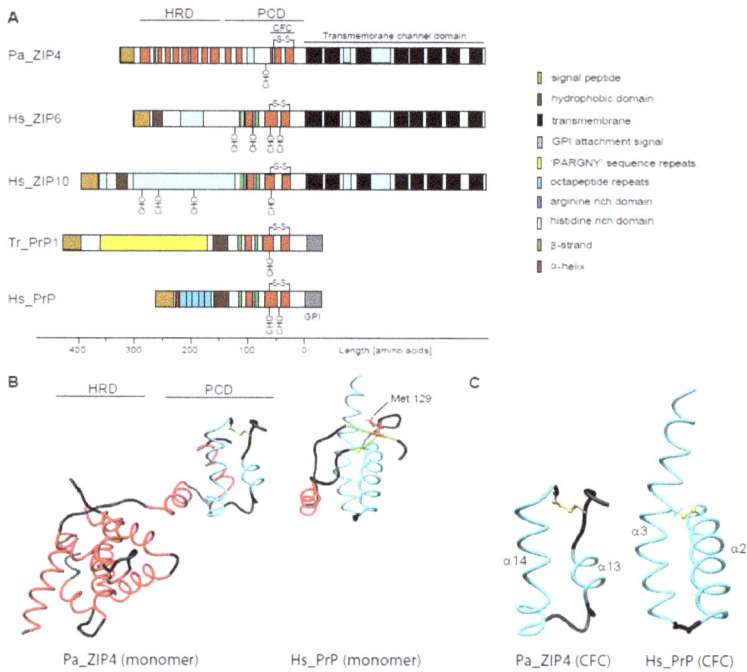

Figure 1. Comparison of molecular architectures of Zrt-/Irt-like Proteins (ZIP)s and prion protein (PrP) molecules, comprising either helix-rich domains (HRDs) or disordered domains at their *N*-terminus and a separate module consisting of helix-turn-helix motifs adjacent to the plasma membrane. (**A**) Domain organization of selected members of the ZIP/PrP protein family. The depiction of all proteins was centered on the predicted site of their insertion into the outer face of the plasma membrane. Pa, *Pteropus alecto*; Hs, *Homo sapiens*; Tr, *Takifugu rubripes*. (**B**) Side-by-side views of the bat (Pa) ZIP4 ectodomain monomer (PDB: 4x82; Zhang et al., 2016; Nat Commun) and the human PrP NMR structure (PDB: 1qm0; Zahn et al., 2000; PNAS). Homologous α-helices are colored in cyan. Just for reference, the rendering also depicts the position of the side-chain of methionine-129 within the first short β-strand in PrP^C. (**C**) Comparison of CFC domains of bat ZIP4 (PDB: 4x82) and human PrP (PDB: 1qm0). As the available bat ZIP4 ectodomain structure does not contain residues *C*-terminally to the CFC, its structure ends on the respective cysteine, but the PrP structure extends to the *C*-terminus.

2. Main

2.1. Comparison of Molecular Architectures of ZIP and Prion Proteins

The ZIP4 ectodomain (from *Pteropus alecto*, the fruit bat) was solved at a resolution of 2.85 Å and was shown to be composed of two structurally independent subdomains. Its *N*-terminal subdomain of 156 amino acids, bearing no sequence homology to PrP, folds into a globular cluster of nine α-helices, termed helix-rich domain (HRD). The HRD is connected through a flexible linker to a folded *C*-terminal domain comprising amino acid residues 192–322. The latter featured two helix-turn-helix folds composed of helices α10–α11 and α13–α14 that were connected to each other by a disordered histidine-rich segment (residues 232–255) and a short α12 helix (Figure 1B). A striking feature of the ZIP4 ectodomain is that it was crystallized as a dimer, held together by extensive interactions among the hydrophobic residues on a large interface including the 'PAL' sequence motifs present in the middle of helix α14 of interacting protomers. Hence, it was suggested to refer to the *C*-terminal subdomain as the PAL-motif containing domain (PCD). When comparing the ZIP4 PCD to PrP, the most apparent shared characteristic is the C-terminal helix-turn-helix fold, which in PrP comprises its helices α2 and α3 (often referred to as helices B and C). However, helix α2 in PrP is longer than α13 in ZIP4, thereby spanning homologous sequence segments that encompass ZIP4 PCD helices α12 plus α13 (Figure 1C). More N-terminal sequence elements present in mammalian prion proteins, including helix α1 (helix A) and the short two-stranded β-sheet observed in this sequence region, cannot be aligned to structural elements within ZIP4. While these elements are well-conserved within mammalian PrP, they fall within a sequence segment that is highly divergent not only in ZIPs but also in fish prion proteins. The diversity of this region can be attributed to the presence of repetitive elements that are prone to contract and expand throughout evolution but may also indicate that this region is not essential for a shared core function of members of this protein family. In the ZIP4 PCD, a relatively short and disordered histidine-rich fragment (31 amino acids) is present between the pair of helix-turn-helix folds. The corresponding region is expanded in the subbranch of ZIPs closer related to PrP, encompassing 89 and 232 residues in human ZIP6 and ZIP10, respectively. As the respective N-terminal domain of mammalian PrP^C is known to be highly disordered, one may speculate that N-terminal regions of related ZIPs might also not acquire a particular fold. This may indeed be the case for ZIPs 6 and 10, whose histidine-rich fragments, preceding the second helix-turn-helix fold (α13 and α14 in ZIP4 PCD), are predicted to be unstructured, and which have no sequences that correspond to the HRD in ZIP4.

2.2. Disappearance of CPAL Motif in the Prion Protein Subbranch of ZIP-PrP Protein Family

The ZIP4 ectodomain structure draws attention to the membrane-adjacent helix-turn-helix fold, which appears to be a core element shared between homologous members of the ZIP/PrP protein family (Figure 1C). Extensive sequence comparisons of ZIP genes across a multitude of organisms revealed this sequence segment to have evolved approximately 750 million years ago in a ZIP ancestral gene around the time when metazoans evolved [33]. The approximate timing of this event can be inferred from the presence of homologous exons in the genomes of basal metazoan organisms, including *Trichoplax adhaerens*, a marine organism with a body plan lacking organs. Whereas the exon-intron structure of ZIP genes is generally quite diverse, the genomic segment coding for this core region is invariably flanked by introns in metazoans (Figure 2A). Consequently, the absence of these flanking introns in all PrP sequences represented one of the strongest pieces of evidence in support of the conclusion that the prion protein founder gene must have evolved by retroinsertion of a spliced transcript of an ancient ZIP gene [33]. Additional elements by which this segment is recognizable include a pair of highly conserved cysteines (hence its designation as the cysteine-flanked core (CFC) domain) and the aforementioned PAL sequence. Close comparison of PrP and ZIP ortholog sequences bearing the PAL motif establish that only the proline in this motif is highly conserved, the second position is very often occupied by a small amino acid (A/G), and the third position by a hydrophobic one (mostly L/I/V). The motif is often preceded by a cysteine residue and is followed by another hydrophobic amino acid (mostly L/I/V). A regression of this extended CPALL motif appears to have occurred throughout PrP evolution, i.e., fish prion proteins lack the cysteine residue, which is present in its closest ZIP relatives, and a full replacement of the PAL motif with a sequence stretch characterized by polar amino acids occurred in mammalian PrP genes (Figure 2B). The CFC also frequently encompasses Nx(T/S) glycan acceptor sites, which, in a subset of prion proteins [34] and close ZIP relatives [35], were shown to be glycan-occupied, but nonetheless cannot be considered a core feature of the CFC domain. There is another variation to the CFC that can be observed in a subset of members of the ZIP/PrP family, namely a relatively large sequence insertion at a site predicted to form the 'turn' within the helix-turn-helix fold. Such insertions are found in ZIP orthologs of some insects (e.g., the mosquito *Anopheles aegypti*) and a subset of fish prion proteins (e.g., PrP2 from pufferfish, *Takifugu rubripes*). Consistent with the interpretation that the respective region in the native folds of these proteins must be able to accommodate relatively large additional structures, the second N-glycan acceptor site in mammalian PrP also maps to this 'turn'. One may speculate that the lack of the PAL motif in mammalian PrP is linked to the insertion of this second N-glycan acceptor site. For example, it may either functionally replace the PAL motif (see below) or offer some other unknown evolutionary fitness adaptation.

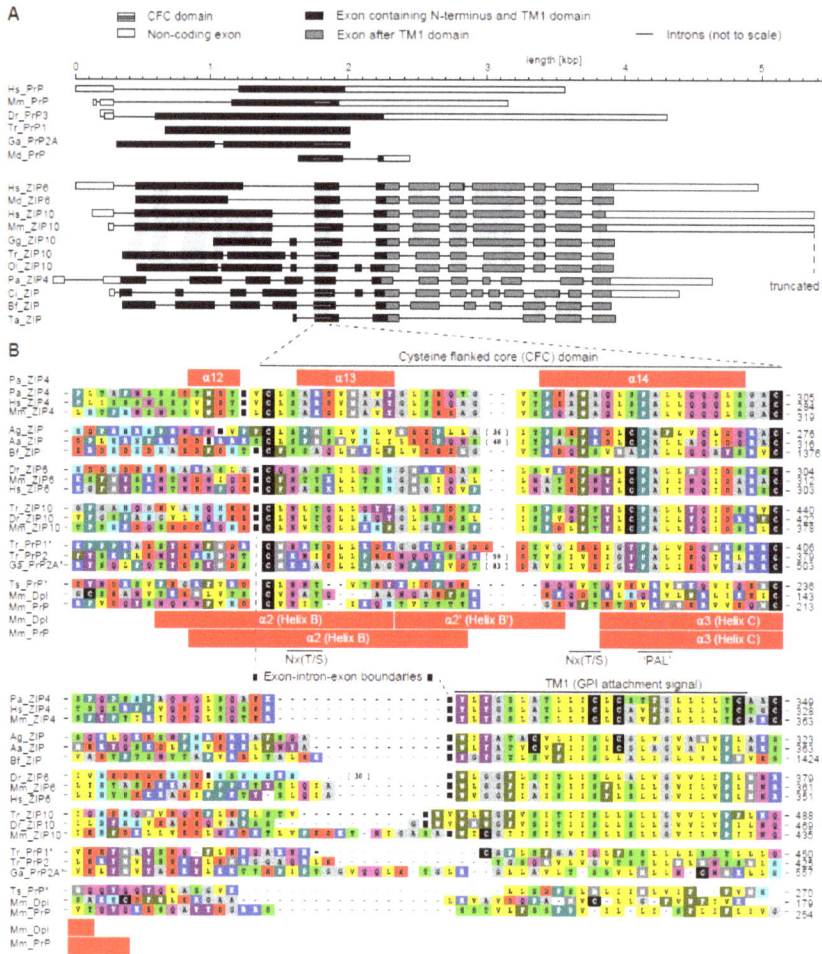

Figure 2. The cysteine-flanked core (CFC) domain represents an ancient intron-flanked module, from which the PAL motif gradually disappeared in the prion protein subbranch of this protein family. (**A**) Intron–exon gene architecture of selected ZIP and PrP sequences. The organization of most prion protein genes resembles the one shown for human PrP. Examples in this panel represent a wide breadth of available architectures. (**B**) Multiple sequence alignment of selected CFC sequences illustrates the gradual loss of the extended CPALL motif in fish and mammalian prion proteins. The selection of sequences was made with a view to best illustrate trends and features described in the main text. Note the absence of the cysteine preceding the PAL motif in fish PrP sequences, and the replacement of the PAL motif in mammalian PrP. Species abbreviations: Hs, *Homo sapiens*; Mm, *Mouse musculus*; Dr, *Danio rerio*; Tr, *Takifugu rubripes*; Ga, *Gasterosteus aculateus*; Md, *Mouse domestica*; Gg, *Gallus gallus*; Ol, *Oryzias latipes*; Pa, *Pteropus alecto*; Ci, *Ciona intestinalis*; Bf, *Branchiostoma florida*; Ta, *Trichoplax adhaerens*; Ag, *Anopheles gambiae*; Aa, *Anopheles aegypti*; Ts, *Trachemys scripta*. Both panels are derivatives of original Figures 2 and 3 in reference [33] used under CC BY.

2.3. The CPAL Motif Represents a Dimerization Interface

In the context reviewed here, perhaps the single most interesting insight revealed by the ZIP4 ectodomain structure is the discovery of the ZIP4 dimer interface as being centered on the hydrophobic PAL motif (Figure 3A). Several lines of evidence suggested for some time that ZIP proteins comprising a CFC domain can assemble into dimers in vitro, and may also exist as dimers in vivo. For example, recombinant ZIP5 and ZIP13 ectodomains were observed to purify as homodimers [36,37]. Additional combinatorial complexity of ZIP protein structures may exist due to the formation of heterodimers. The aforementioned ZIP6–ZIP10 heteromer is the first example of this phenomenon (note that current data are consistent with the latter complex representing a heterodimer. However, a rigorous validation of the stoichiometry of this interaction has not been made, hence its tentative designation as a heteromer [19,22]). As all of the above proteins share a homologous PAL motif, it is to be expected that this motif will also be central to their dimerization interface. When the recombinant ZIP4 construct was truncated C-terminal to its helix α11, the expressed protein no longer eluted as a dimer on size exclusion chromatography, indicating that the CFC is indeed essential for dimerization [31]. Moreover, a serine-to-cysteine replacement of the amino acid that precedes the ZIP4 PAL motif, thereby mimicking the predicted dimerization interface of the ZIPs that naturally harbor a cysteine in this position, caused the mutant to migrate in a Western blot as an SDS-resistant dimer (Figure 3B). This finding is intriguing, as it suggests the enhanced stability of the mutant dimer is due to the respective cysteine residues in the interacting monomers forming a disulfide bridge *in trans*. Consistent with this interpretation, the protein shifted to monomeric molecular weight, when the same samples were subjected to disulfide bridge reducing conditions prior to their SDS-PAGE separation. Additional evidence supporting this dimerization model came from the biochemical characterization of the ectodomain of ZIP14 (from *Pteropus alecto*), which naturally possesses a cysteine residue immediately preceding the conserved proline residue. Again, as for the aforementioned artificial ZIP4 mutant, a disulfide bond-mediated dimer was observed in the SDS-PAGE analysis, and a cysteine-to-serine substitution led to a monomeric species under non-reducing condition. These results corroborate the conclusion that the PAL motif-centered dimerization represents a universal model for ZIPs containing this motif in their ectodomains.

Whereas there seems to be good agreement on the predominant existence of ZIPs comprising a CFC domain as dimers, the same cannot be said for PrP, although PrP dimers were repeatedly reported. For instance, a 60 kDa PrP dimer was initially observed in murine neuroblastoma cells expressing hamster PrP [38]. Extensive subsequent recombinant work with various PrP constructs, which most often relied on in vitro refolding steps, revealed the protein to give rise to monomeric NMR structures [28]. The folding repertoire of PrP is more complex though, as the protein could be observed to crystallize as a monomer or dimer [39]. Curiously, subtle differences in the protein sequence (e.g., single amino acid mutations or the presence of the M129V polymorphism) were critical for whether the protein crystallized as a non-swapped or swapped dimer, and seemed to determine the specific architecture of the dimer interface [38,40]. Although the aforementioned studies corroborated the notion that PrP can adopt a surprising diversity of conformational states, the in vivo relevance of these states is difficult to gauge. Cell-based studies with N-glycosylated wild-type PrP^{C} observed a monomer–dimer equilibrium on the basis of crosslinking experiments, size exclusion chromatography, and enzyme-linked immunosorbent assay analysis [41]. Similarly, expression of cattle, hamster and human PrP in baculovirus-transduced insect cells led to dimeric PrP [42]. It is currently not understood if differences in the folding behavior of PrP in bacterial or eukaryotic cells merely reflects a need for a eukaryotic cellular protein folding environment, indicates that the attachment of N-glycans is critical for dimer formation, or has other causes. That dimerization might play a critical role in PrP's biogenesis and trafficking was corroborated by fusing it to the dimerization domain of the FK506 binding protein (Fv). When dimerization was induced by the addition of an Fv dimerization ligand, the authors observed profound increases in the levels of PrP that reached the cell surface [43,44].

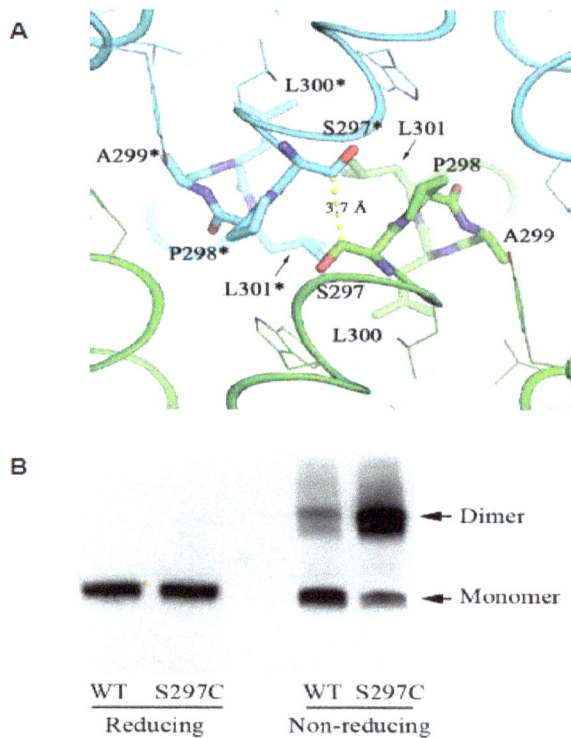

Figure 3. The CPAL motif present in LIV-1 and fish PrP sequences represents a dimerization interface. (**A**) Structural model of human ZIP4-ECD dimerization interface based on the crystal structure of *Pteropus alecto* ZIP4-ECD. The two protomers are colored in green and cyan, respectively. The residues in the 'SPAL' sequence are shown in stick mode. The '*' symbol indicates the residues from the other protomer. The yellow dashed line shows the distance between the C_β atoms of the two interacting S297 residues. (**B**) Validation of the dimerization interface in (**A**) by substitution of S297 with a cysteine residue in the full-length human ZIP4 expressed in HEK293T cells. Both panels are derivatives of the original Figure 3C, D in reference [31] used under CC BY.

Currently available data on dimerization motifs for mammalian PrP indicate that these are non-homologous to the ZIP4 dimer interface discussed above and instead rely on structural features not shared with this ZIP transporter. It remains to be seen to which extent the structures of more closely related ZIPs will be informative for understanding this aspect of the biology of mammalian PrP.

To date, the smallest extensively studied PrP deletion construct known to convert and transmit prion-like disease in mice was of 106 amino acids (Δ23–88, Δ141–176) [45]. This expression product comprised the entire helix CFC domain, which in human begins at amino acid 179. However, although the CFC is a critical component of disease-associated prion proteins, and appears to be retained in their proteinase-resistant core of 27 to 30 kDa (PrP27-30), the study of ZIPs may not help us elucidate their structures (the reader is reminded that even fish PrP or the mammalian paralogs Dpl and Sho have not been shown to undergo prion-like conversion). Rather, we anticipate that comparative analyses of PrP and ZIPs may continue to help elucidate the physiological function of PrP^C, and may hold a key to understanding the molecular mechanisms by which PrP affects its next neighbors.

Based on the striking conservation of the PAL motif in fish PrP sequences, our prediction is that the latter not only exist naturally as dimers, but also adapt the interface seen in ZIP4. Dramatic effects of point mutations on the ability of ZIP4 to fold and reach the cell surface [16,31] and a strong interaction of the ZIP6–ZIP10 complex with calreticulin (an ER resident chaperone) suggest that the assembly of ZIPs might be intricate. It then may not come as a surprise that fish PrP was observed to be refractory to recombinant protein refolding protocols that repeatedly produced high-quality NMR structures for mammalian prion proteins [30]. It is tempting to speculate that a move to a eukaryotic expression system, possibly augmented by a tyrosine-to-cysteine replacement of the tyrosine preceding the 'PAL' motif present in fish prion sequences, will lead to better-behaved fish PrP expression products [30].

3. Conclusions

Studying the physiological function of homologous segments within ZIPs may hold a key to understanding elusive aspects of the biology of PrP. Nonetheless, because ZIP4 is a relatively distant PrP relative, there are limits to the extent insights into its biology will be informative for understanding PrP. However, this overall approach should become more valuable once structures for the ZIP6–ZIP10 heteromer become available. Already, this general comparative approach has precipitated research that revealed a role of PrP in EMT [26]. It also was instrumental for understanding that the PrP–NCAM1 interaction is unlikely to have evolved following the split of PrP sequences from its ZIP relatives, but was most likely inherited by PrP and the ZIP6–ZIP10 complex from a common ancestor [22]. Currently missing are insights into the molecular workings of PrP and the molecular mechanisms by which its presence influences its interactors. We anticipate that major advances in this direction will be made once the ZIP6–ZIP10 complex can be functionally interrogated to dissect how structural elements within this complex affect its function.

Acknowledgments: Work on this project was supported through funding made available by the Canadian Institutes for Health Research (grant number 324987, G.S.-U.), a research team award by the Alberta Prion Research Institute (grant number 201600028, H.W. and G.S.-U.), and a NIH R01 operating grant (grant number R01GM115373, J.H.). We also gratefully acknowledge generous ongoing support by the Borden Rosiak family (G.S.-U.) and the Arnold Irwin family (G.S.-U.).

Author Contributions: All authors contributed to conceiving, writing and editing the paper.

Conflicts of Interest: The authors declare no conflict of interest.

Abbreviations

Aa	*Anopheles aegypti*
Ag	*Anopheles gambiae*
AE	*Acrodermatitis enteropathica*
Bf	*Branchiostoma florida*
CDF	cation diffusion facilitator
CFC	cysteine-flanked core
Ci	*Ciona intestinalis*
CTD	carboxy-terminal domain
Dm	*Drosophila melanogaster*
Dr	*Danio rerio*
ECD	extracellular domain
Ga	*Gasterosteus aculateus*
Gg	*Gallus gallus*
GPI	glycosylphosphatidylinositol
HRD	helix-rich domain
Hs	*Homo sapiens*
LZT	LIV-1 subfamily of ZIP zinc transporters
Md	*Monodelphis domestica*
Mm	*Mus musculus*
Ol	*Oryzias latipes*
ORF	open reading frame

Pa	*Pteropus alecto*
PCD	PAL-motif containing domain
PL	prion-like
PrP^C	cellular prion protein
SLC	solute carrier
Ta	*Trichoplax adhaerens*
TM	transmembrane
Tr	*Takifugu rubripes*
Ts	*Trachemys scripta*

References

1. Prusiner, S.B. Novel proteinaceous infectious particles cause scrapie. *Science* **1982**, *216*, 136–144. [CrossRef] [PubMed]
2. Prusiner, S.B. Prions. *Proc. Natl. Acad. Sci. USA* **1998**, *95*, 13363–13383. [CrossRef] [PubMed]
3. Prusiner, S.B. Cell biology. A unifying role for prions in neurodegenerative diseases. *Science* **2012**, *336*, 1511–1513. [CrossRef] [PubMed]
4. Sailer, A.; Bueler, H.; Fischer, M.; Aguzzi, A.; Weissmann, C. No propagation of prions in mice devoid of PrP. *Cell* **1994**, *77*, 967–968. [CrossRef]
5. Moore, R.C.; Lee, I.Y.; Silverman, G.L.; Harrison, P.M.; Strome, R.; Heinrich, C.; Karunaratne, A.; Pasternak, S.H.; Chishti, M.A.; Liang, Y.; et al. Ataxia in prion protein (PrP)-deficient mice is associated with upregulation of the novel PrP-like protein doppel. *J. Mol. Biol.* **1999**, *292*, 797–817. [CrossRef] [PubMed]
6. Silverman, G.L.; Qin, K.; Moore, R.C.; Yang, Y.; Mastrangelo, P.; Tremblay, P.; Prusiner, S.B.; Cohen, F.E.; Westaway, D. Doppel is an N-glycosylated, glycosylphosphatidylinositol-anchored protein. Expression in testis and ectopic production in the brains of Prnp(0/0) mice predisposed to Purkinje cell loss. *J. Biol. Chem.* **2000**, *275*, 26834–26841. [PubMed]
7. Premzl, M.; Sangiorgio, L.; Strumbo, B.; Marshall Graves, J.A.; Simonic, T.; Gready, J.E. Shadoo, a new protein highly conserved from fish to mammals and with similarity to prion protein. *Gene* **2003**, *314*, 89–102. [CrossRef]
8. Westaway, D.; Daude, N.; Wohlgemuth, S.; Harrison, P. The PrP-like proteins Shadoo and Doppel. *Top. Curr. Chem.* **2011**, *305*, 225–256. [PubMed]
9. Watts, J.C.; Huo, H.; Bai, Y.; Ehsani, S.; Jeon, A.H.; Shi, T.; Daude, N.; Lau, A.; Young, R.; Xu, L.; et al. Interactome analyses identify ties of PrP and its mammalian paralogs to oligomannosidic N-glycans and endoplasmic reticulum-derived chaperones. *PLoS Pathog.* **2009**, *5*, e1000608. [CrossRef]
10. Schmitt-Ulms, G.; Ehsani, S.; Watts, J.C.; Westaway, D.; Wille, H. Evolutionary descent of prion genes from the ZIP family of metal ion transporters. *PLoS ONE* **2009**, *4*, e7208. [CrossRef] [PubMed]
11. Hatinen, T.; Holm, L.; Airaksinen, M.S. Loss of neurturin in frog—Comparative genomics study of GDNF family ligand-receptor pairs. *Mol. Cell Neurosci.* **2007**, *34*, 155–167. [CrossRef] [PubMed]
12. Airaksinen, M.S.; Holm, L.; Hatinen, T. Evolution of the GDNF family ligands and receptors. *Brain Behav. Evol.* **2006**, *68*, 181–190. [CrossRef] [PubMed]
13. Bell, L.M.; Solomon, K.R.; Gold, J.P.; Tan, K.N. Cytoplasmic tail deletion of T cell receptor (TCR) beta-chain results in its surface expression as glycosylphosphatidylinositol-anchored polypeptide on mature T cells in the absence of TCR-alpha. *J. Biol. Chem.* **1994**, *269*, 22758–22763. [PubMed]
14. Aguzzi, A.; Baumann, F.; Bremer, J. The Prion's Elusive Reason for Being. *Annu. Rev. Neurosci.* **2008**, *31*, 439–477. [CrossRef] [PubMed]
15. Castle, A.R.; Gill, A.C. Physiological Functions of the Cellular Prion Protein. *Front. Mol. Biosci.* **2017**, *4*, 19. [CrossRef] [PubMed]
16. Wang, F.; Kim, B.E.; Dufner-Beattie, J.; Petris, M.J.; Andrews, G.; Eide, D.J. Acrodermatitis enteropathica mutations affect transport activity, localization and zinc-responsive trafficking of the mouse ZIP4 zinc transporter. *Hum. Mol. Genet.* **2004**, *13*, 563–571. [CrossRef] [PubMed]

17. Giunta, C.; Elcioglu, N.H.; Albrecht, B.; Eich, G.; Chambaz, C.; Janecke, A.R.; Yeowell, H.; Weis, M.; Eyre, D.R.; Kraenzlin, M.; et al. Spondylocheiro dysplastic form of the Ehlers-Danlos syndrome—An autosomal-recessive entity caused by mutations in the zinc transporter gene SLC39A13. *Am. J. Hum. Genet.* **2008**, *82*, 1290–1305. [CrossRef] [PubMed]

18. Kong, B.Y.; Duncan, F.E.; Que, E.L.; Kim, A.M.; O'Halloran, T.V.; Woodruff, T.K. Maternally-derived zinc transporters ZIP6 and ZIP10 drive the mammalian oocyte-to-egg transition. *Mol. Hum. Reprod.* **2014**, *20*, 1077–1089. [CrossRef] [PubMed]

19. Taylor, K.M.; Muraina, I.; Brethour, D.; Schmitt-Ulms, G.; Nimmanon, T.; Ziliotto, S.; Kille, P.; Hogstrand, C. Zinc transporter ZIP10 forms a heteromer with ZIP6 which regulates embryonic development and cell migration. *Biochem. J.* **2016**, *473*, 2531–2544. [CrossRef] [PubMed]

20. Yamashita, S.; Miyagi, C.; Fukada, T.; Kagara, N.; Che, Y.S.; Hirano, T. Zinc transporter LIVI controls epithelial-mesenchymal transition in zebrafish gastrula organizer. *Nature* **2004**, *429*, 298–302. [CrossRef] [PubMed]

21. Malaga-Trillo, E.; Solis, G.P.; Schrock, Y.; Geiss, C.; Luncz, L.; Thomanetz, V.; Stuermer, C.A. Regulation of embryonic cell adhesion by the prion protein. *PLoS Biol.* **2009**, *7*, e55. [CrossRef] [PubMed]

22. Brethour, D.; Mehrabian, M.; Williams, D.; Wang, X.; Ghodrati, F.; Ehsani, S.; Rubie, E.A.; Woodgett, J.R.; Sevalle, J.; Xi, Z.; et al. A ZIP6-ZIP10 heteromer controls NCAM1 phosphorylation and integration into focal adhesion complexes during epithelial-to-mesenchymal transition. *Sci. Rep.* **2017**, *7*, 40313. [CrossRef] [PubMed]

23. Schmitt-Ulms, G.; Legname, G.; Baldwin, M.A.; Ball, H.L.; Bradon, N.; Bosque, P.J.; Crossin, K.L.; Edelman, G.M.; DeArmond, S.J.; Cohen, F.E.; et al. Binding of neural cell adhesion molecules (N-CAMs) to the cellular prion protein. *J. Mol. Biol.* **2001**, *314*, 1209–1225. [CrossRef] [PubMed]

24. Slapsak, U.; Salzano, G.; Amin, L.; Abskharon, R.N.; Ilc, G.; Zupancic, B.; Biljan, I.; Plavec, J.; Giachin, G.; Legname, G. The N Terminus of the Prion Protein Mediates Functional Interactions with the Neuronal Cell Adhesion Molecule (NCAM) Fibronectin Domain. *J. Biol. Chem.* **2016**, *291*, 21857–21868. [CrossRef] [PubMed]

25. Santuccione, A.; Sytnyk, V.; Leshchyns'ka, I.; Schachner, M. Prion protein recruits its neuronal receptor NCAM to lipid rafts to activate p59fyn and to enhance neurite outgrowth. *J. Cell Biol.* **2005**, *169*, 341–354. [CrossRef] [PubMed]

26. Mehrabian, M.; Brethour, D.; Wang, H.; Xi, Z.; Rogaeva, E.; Schmitt-Ulms, G. The Prion Protein Controls Polysialylation of Neural Cell Adhesion Molecule 1 during Cellular Morphogenesis. *PLoS ONE* **2015**, *10*, e0133741. [CrossRef] [PubMed]

27. Riek, R.; Hornemann, S.; Wider, G.; Billeter, M.; Glockshuber, R.; Wuthrich, K. NMR structure of the mouse prion protein domain PrP(121-321). *Nature* **1996**, *382*, 180–182. [CrossRef] [PubMed]

28. Wuthrich, K.; Riek, R. Three-dimensional structures of prion proteins. *Adv. Protein Chem.* **2001**, *57*, 55–82. [PubMed]

29. Calzolai, L.; Lysek, D.A.; Perez, D.R.; Guntert, P.; Wuthrich, K. Prion protein NMR structures of chickens, turtles, and frogs. *Proc. Natl. Acad. Sci. USA* **2005**, *102*, 651–655. [CrossRef] [PubMed]

30. Christen, B.; Wuthrich, K.; Hornemann, S. Putative prion protein from Fugu (*Takifugu rubripes*). *FEBS J.* **2008**, *275*, 263–270. [CrossRef] [PubMed]

31. Zhang, T.; Sui, D.; Hu, J. Structural insights of ZIP4 extracellular domain critical for optimal zinc transport. *Nat. Commun.* **2016**, *7*, 11979. [CrossRef] [PubMed]

32. Zhang, T.; Liu, J.; Fellner, M.; Zhang, C.; Sui, D.; Hu, J. Crystal structures of a ZIP zinc transporter reveal a binuclear metal center in the transport pathway. *Sci. Adv.* **2017**, *3*, e1700344. [CrossRef] [PubMed]

33. Ehsani, S.; Tao, R.; Pocanschi, C.L.; Ren, H.; Harrison, P.M.; Schmitt-Ulms, G. Evidence for retrogene origins of the prion gene family. *PLoS ONE* **2011**, *6*, e26800. [CrossRef] [PubMed]

34. Rudd, P.M.; Endo, T.; Colominas, C.; Groth, D.; Wheeler, S.F.; Harvey, D.J.; Wormald, M.R.; Serban, H.; Prusiner, S.B.; Kobata, A.; et al. Glycosylation differences between the normal and pathogenic prion protein isoforms. *Proc. Natl. Acad. Sci. USA* **1999**, *96*, 13044–13049. [CrossRef] [PubMed]

35. Ehsani, S.; Salehzadeh, A.; Huo, H.; Reginold, W.; Pocanschi, C.L.; Ren, H.; Wang, H.; So, K.; Sato, C.; Mehrabian, M.; et al. LIV-1 ZIP ectodomain shedding in prion-infected mice resembles cellular response to transition metal starvation. *J. Mol. Biol.* **2012**, *422*, 556–574. [CrossRef] [PubMed]

36. Pocanschi, C.L.; Ehsani, S.; Mehrabian, M.; Wille, H.; Reginold, W.; Trimble, W.S.; Wang, H.; Yee, A.; Arrowsmith, C.H.; Bozoky, Z.; et al. The ZIP5 Ectodomain Co-Localizes with PrP and May Acquire a PrP-Like Fold That Assembles into a Dimer. *PLoS ONE* **2013**, *8*, e72446. [CrossRef] [PubMed]

37. Bin, B.H.; Fukada, T.; Hosaka, T.; Yamasaki, S.; Ohashi, W.; Hojyo, S.; Miyai, T.; Nishida, K.; Yokoyama, S.; Hirano, T. Biochemical characterization of human ZIP13 protein: A homo-dimerized zinc transporter involved in the Spondylocheiro dysplastic Ehlers-Danlos syndrome. *J. Biol. Chem.* **2011**, *286*, 40255–40265. [CrossRef] [PubMed]

38. Priola, S.A.; Caughey, B.; Wehrly, K.; Chesebro, B. A 60-kDa prion protein (PrP) with properties of both the normal and scrapie-associated forms of PrP. *J. Biol. Chem.* **1995**, *270*, 3299–3305. [CrossRef] [PubMed]

39. Knaus, K.J.; Morillas, M.; Swietnicki, W.; Malone, M.; Surewicz, W.K.; Yee, V.C. Crystal structure of the human prion protein reveals a mechanism for oligomerization. *Nat. Struct. Biol.* **2001**, *8*, 770–774. [CrossRef] [PubMed]

40. Lee, S.; Antony, L.; Hartmann, R.; Knaus, K.J.; Surewicz, K.; Surewicz, W.K.; Yee, V.C. Conformational diversity in prion protein variants influences intermolecular beta-sheet formation. *EMBO J.* **2010**, *29*, 251–262. [CrossRef] [PubMed]

41. Meyer, R.K.; Lustig, A.; Oesch, B.; Fatzer, R.; Zurbriggen, A.; Vandevelde, M. A monomer-dimer equilibrium of a cellular prion protein (PrPC) not observed with recombinant PrP. *J. Biol. Chem.* **2000**, *275*, 38081–38087. [CrossRef] [PubMed]

42. Hundt, C.; Gauczynski, S.; Leucht, C.; Riley, M.L.; Weiss, S. Intra- and interspecies interactions between prion proteins and effects of mutations and polymorphisms. *Biol. Chem.* **2003**, *384*, 791–803. [CrossRef] [PubMed]

43. Beland, M.; Roucou, X. Homodimerization as a molecular switch between low and high efficiency PrP C cell surface delivery and neuroprotective activity. *Prion* **2013**, *7*, 170–174. [CrossRef] [PubMed]

44. Beland, M.; Motard, J.; Barbarin, A.; Roucou, X. PrP(C) homodimerization stimulates the production of PrPC cleaved fragments PrPN1 and PrPC1. *J. Neurosci.* **2012**, *32*, 13255–13263. [CrossRef] [PubMed]

45. Supattapone, S.; Bosque, P.; Muramoto, T.; Wille, H.; Aagaard, C.; Peretz, D.; Nguyen, H.-O.B.; Heinrich, C.; Torchia, M.; Safar, J.; et al. Prion protein of 106 residues creates an artificial transmission barrier for prion replication in transgenic mice. *Cell* **1999**, *96*, 869–878. [CrossRef]

pathogens

MDPI

Review

Recombinant PrP and Its Contribution to Research on Transmissible Spongiform Encephalopathies

Jorge M. Charco [1], Hasier Eraña [1], Vanessa Venegas [1], Sandra García-Martínez [1], Rafael López-Moreno [1], Ezequiel González-Miranda [1], Miguel Ángel Pérez-Castro [1] and Joaquín Castilla [1,2,*]

[1] CIC bioGUNE, Parque Tecnológico de Bizkaia, 48160 Derio, Spain; jmoreno@cicbiogune.es (J.M.C.); herana@cicbiogune.es (H.E.); vvenegas@cicbiogune.es (V.V.); sgarcia@cicbiogune.es (S.G.-M.); rlopez@cicbiogune.es (R.L.-M.); egmiranda@cicbiogune.es (E.G.-M.); maperez@cicbiogune.es (M.Á.P.-C.)
[2] IKERBASQUE, Basque Foundation for Science, 48011 Bilbao, Spain
*   Correspondence: castilla@joaquincastilla.com

Received: 25 October 2017; Accepted: 12 December 2017; Published: 14 December 2017

Abstract: The misfolding of the cellular prion protein (PrP^C) into the disease-associated isoform (PrP^{Sc}) and its accumulation as amyloid fibrils in the central nervous system is one of the central events in transmissible spongiform encephalopathies (TSEs). Due to the proteinaceous nature of the causal agent the molecular mechanisms of misfolding, interspecies transmission, neurotoxicity and strain phenomenon remain mostly ill-defined or unknown. Significant advances were made using in vivo and in cellula models, but the limitations of these, primarily due to their inherent complexity and the small amounts of PrP^{Sc} that can be obtained, gave rise to the necessity of new model systems. The production of recombinant PrP using *E. coli* and subsequent induction of misfolding to the aberrant isoform using different techniques paved the way for the development of cell-free systems that complement the previous models. The generation of the first infectious recombinant prion proteins with identical properties of brain-derived PrP^{Sc} increased the value of cell-free systems for research on TSEs. The versatility and ease of implementation of these models have made them invaluable for the study of the molecular mechanisms of prion formation and propagation, and have enabled improvements in diagnosis, high-throughput screening of putative anti-prion compounds and the design of novel therapeutic strategies. Here, we provide an overview of the resultant advances in the prion field due to the development of recombinant PrP and its use in cell-free systems.

Keywords: Prion disease; TSE; recombinant PrP; in vitro propagation; PMCA; QuIC

1. Introduction

Transmissible spongiform encephalopathies (TSEs) are a group of neurodegenerative disorders which have in common the formation of amyloid plaques due to the accumulation of prion protein (PrP) which has been converted to an abnormal conformation, known as PrP^{Sc}, in the central nervous system (CNS). The misfolding of the normal cellular form of the PrP (PrP^C) to the disease–associated form, PrP^{Sc}, leads to neuronal damage, is invariably fatal but generally preceded by motor problems, such as myoclonus and ataxia, and by cognitive deficiencies. Different variants of TSE exist in many mammalian species [1]. In humans, five different prion diseases have been reported to date: Kuru [2], Gerstmann-Straüssler-Scheinker Syndrome (GSS) [3], Fatal Familial Insomnia (FFI) [4], Creutzfeldt-Jakob Disease (CJD) [5] and Variably Protease-sensitive Prionopathy (VPSPr) [6]. Each variant presents with distinct clinical signs and a different prion accumulation pattern in the brain. Besides human prion diseases, the best known examples due to the number of affected animals are: Scrapie in sheep and goat [7], Transmissible Mink Encephalopathy (TME) [8], Bovine Spongiform Encephalopathy (BSE) [9] and Chronic Wasting Disease (CWD) in cervids [10]. A slow virus was initially hypothesised to be the causal agent of these disorders. However, the lack of identification of any virus, despite extensive investigations and transmission of the disease after subjecting neural tissue to treatments known to inactivate nucleic acids, refuted this hypothesis and opened the possibility of another causal agent [11,12]. An alternative hypothesis was developed by S. Prusiner, known as the protein-only hypothesis. This controversial theory proposed that TSEs are caused solely by PrP^{Sc}, a misfolded form of the physiologically normal PrP^C which is expressed abundantly in the CNS. This misfolded form of the protein is able to induce transformation of the normal PrP^C into a pathogenic conformation, initiating an infectious process in the brain of the affected individuals [13]. Due to the long incubation periods and phenotypic variability of prion disorders, reminiscent of virus strains, the existence of a pathogen devoid of nucleic acids was not widely accepted at first. However, the weight of evidence increased inexorably during the last three decades has proven irrefutably the proteinaceous nature of this infectious agent that defied the central dogma of the molecular biology [14]. According to the source of the PrP^{Sc} seed that initiate the infectious process, TSEs can be classified as: (1) acquired, when the PrP^{Sc} comes from an exogenous source [15–17]; (2) genetic, when the PrP^C misfolds due to mutations in the PrP encoding gene [18]; (3) or sporadic, when the cause is unknown although an spontaneous misfolding of the host's wild type PrP^C is suspected [19]. The latter is the most common in humans, representing about 85% of the total cases [18].

The PrP is a glycosylphosphatidylinositol (GPI)-anchored membrane glycoprotein encoded by the *PRNP* gene, which is present in all superior animals and highly conserved in mammals. The native form of this protein is comprised of a mostly α-helical globular domain and a flexible amino terminal region [20,21]. The conformational change that results in transformation to the pathogenic isoform dramatically alters the biological and physicochemical properties of the PrP, which becomes neurotoxic, aggregation prone and partially resistant to protease digestion in most cases [22–24]. The details of this process remain largely unknown at the molecular level hindering the understanding of several aspects of TSEs. The main limitation comes from the impediment to unraveling the three-dimensional structure of the pathogenic conformer due to its amyloidogenic nature [25]. This hinders an adequate understanding of some of the most striking characteristics of prions such as the strain phenomenon, which is responsible of the existence of phenotypically distinct TSEs that share identical PrP sequences [26,27], or interspecies transmission of prions, since there is a transmission barrier between many species due to differences in their PrP amino acid sequences [7,28].

The study of TSEs and their causal agent has been limited for a long time to animal models naturally susceptible to prion diseases and started with Gajdusek and colleagues who demonstrated that both Kuru and CJD were infectious disorders by direct inoculations in the CNS of monkeys [29,30]. A similar approach was used to prove the relationship between BSE and variant CJD (vCJD) [31], and for the generation of rodent-adapted prions by inoculation of scrapie into mice [32]. The difficulties and costs associated with the maintenance, long incubation periods related to interspecies transmission barriers and the lack of ability to adapt and study certain prion strains significantly hindered progress in TSE research despite the advances achieved using naturally susceptible animal models. The emergence of the first transgenic mice expressing different PrPs [33] greatly increased the interest in animal models for research on prion diseases. These new models permitted evaluation of the transmissibility of different prion strains to transgenic animals bearing human PrP [34] and PrPs from other species [35] and also showed the effect of different *PRNP* gene mutations on the susceptibility to prion infection [36]. Moreover, models overexpressing PrP permitted shortening of the usually prolonged incubation times and facilitated obtaining large enough amounts of infectious material to study prions at the molecular level [35]. Nevertheless, generation of transgenic mice did not ameliorate all the problems related to animal models such as the high costs associated with their development and housing of high number of animals needed to reach valid conclusions.

The development of cell cultures derived from different cell lineages all susceptible to prion infection addressed some of the limitations of the animal models and their use increased rapidly in the prion field [37]. However, most of the cell lines only propagate mouse-adapted prions in a highly strain-specific manner. In fact, different clones from the same cell line can show different susceptibility to the same prion strains [38] and cell lines highly susceptible to infection by some prions can be completely resistant to others [39]. Specificity issues were recently overcome by the development of non-neuronal cell lines [40] and these in vitro models are used to study several aspects of the cellular biology of prions including the native, non-pathogenic prion protein (PrP^C). Nonetheless, developing cell models for prion infection is highly challenging and frequently unsuccessful [37].

Some of the problems associated with in vivo and in cellula models, primarily the limited quantity of PrP^{Sc} that could be obtained, were overcome in 1997 when Wüthrich and collaborators developed a novel technique to generate large amounts of recombinant PrP (rec-PrP) in *Escherichia coli* using a nickel-based purification system [41]. This technique enables the production of highly concentrated and pure rec-PrP for use in further investigations and has already proven its value in the study of TSEs. Despite the differences between brain-derived and recombinant PrP (the latter lacks glycosylations and GPI-anchoring to the cell membrane), it has enabled the atomic structure of the non-pathogenic PrP to be derived [21] and the development of several cell-free systems for prion generation in vitro [42–51]. Initially, in vitro generated misfolded protease-resistant PrPs (rec-PrP^{res}) were poorly infectious in vivo [42], limiting their use as *bona fide* prion models. However, this situation changed with the generation of a highly infectious misfolded rec-PrP able to cause prion disease in vivo and reproduce all the characteristic hallmarks of a TSE [49]. Subsequently, recombinant PrP and cell-free systems for prion propagation have become invaluable tools. Herein, we focus on the different uses of rec-PrP and infectious misfolded rec-PrP and how their development has been pivotal in the field of TSE by enabling: mechanistic and structural studies, improvements in diagnosis and high-throughput screening of anti-prion compounds and the design of new therapeutic strategies (Table 1).

2. Molecular Mechanisms

Compelling evidence has been gathered in support of the protein-only hypothesis of prion disease [13,14] resulting in the proteinaceous nature of the etiologic agent of TSEs accepted widely. Due to the novelty of a proteinaceous pathogen totally devoid of nucleic acid, the molecular mechanisms that lead to its transmission and propagation and the subsequent neurodegeneration were completely unknown. The first clues to the etiopathogeny of TSEs were from using animal models [29,52]. However, the complexity of these in vivo models limited the information that could be derived on the molecular mechanisms of the pathogen. Moreover, the long incubation times and the high costs associated with animal maintenance led to the search for a simpler, shorter term and more versatile models in order to make advances on the study of this unusual infectious agent. Cell culture models offered a simpler model that allowed unravelling several aspects of the cellular biology of the native PrP and the pathogenic isoform [53], although they are still complex models not totally suitable to study some aspects of the PrP such as tertiary structure or detailed misfolding mechanisms. The development of the first protocol to produce large amounts of pure, bacterially-expressed, recombinant PrP by Wüthrich and collaborators [41] provided a starting point for the development of cell-free systems and offered a new, reliable model that complemented the in vivo and in cellula models. The use of rec-PrP allows production of modified PrPs with chosen deletions, insertions, point mutations, distinct labeling and fusion proteins rapidly, providing versatility never seen before. Initially, the in vitro misfolded PrPs did not reproduce the hallmark characteristics of brain-derived PrP^{Sc}, being either poorly infectious in vivo or totally non-infectious [42–48], limiting their usefulness in studying the native form of PrP. It is noteworthy that the first misfolded PrP produced in cell-free systems with minimal components was with brain-derived purified PrP^C and polyA RNA which resulted in a 50–100% substrate conversion efficiency [43] and demonstrated that fully infectious prions could be produced from purified components and cofactors. The production of the first misfolded rec-PrP infective in vivo (*bona fide* prion) by Ma and collaborators [49] demonstrated that it is possible to generate recombinant prions with all the hallmarks of PrP^{Sc} in cell-free systems, although the substrate conversion efficiency was <5% and the specific infectivity <100-fold lower than the previous recombinant prion. Finally, the first recombinant prion with high specific infectivity was produced a few years later using purified phosphatidylethanolamine (PE) as a cofactor instead of the lipids used by Ma and collaborators [54]. Therefore, at present, cell-free systems are the simplest model available to study several aspects of prion disorders that will be summarized in this section.

Prior to obtaining infectious recombinant prions, rec-PrP and cell-free systems demonstrated their utility for (i) the search for molecules that could interact with the prion protein; (ii) studying the proneness to misfolding of different PrPs; and (iii) evaluating putative transmission barriers and the molecular mechanisms underlying these phenomena. The possible interactions of PrP with copper for example, have been a topic of interest as PrP has long been considered to play a central role in the homeostasis of copper in the CNS [55]. Its implication in prion propagation has also been reported [56,57], although its exact role in these process is unknown. The interaction of copper with octarepeat regions was known previously [58] but an additional domain was described using rec-PrP [59]. The effect of the interaction between copper and PrP and the misfolding process was also studied using rec-PrP-based cell-free systems [60–62]. The stability and misfolding proneness of distinct PrPs has been studied using rec-PrP and cell-free propagation systems too. Since the assessment of the influence of different domains and polymorphisms on the misfolding proneness of PrP usually requires modifying its primary structure, rec-PrP has been chosen as model for many of these studies due to the ease with which this can be accomplished. This is the case in research focused on unravelling the role of the N-terminal region of the PrP, which is intrinsically disordered [21], on misfolding and prion-associated neurotoxicity. Rec-PrPs with deletions in the N-terminal region have been used to show this domain is not necessary for PrP fibril formation, at least for mouse rec-PrP (amino acids 121-231) [63], hamster rec-PrP (90-231) [64], and human rec-PrP (90-231) [65]. Moreover, toxicity studies of the latter on cell cultures suggest that this domain is not directly involved in prion-associated neurotoxicity. The Central Lysine Cluster (CLC), encompassing amino acid residues 101-110 of the PrP, has been identified also as a critical region for the conversion to the pathogenic isoform due to the presence of several mutations associated with genetic prion disorders [66,67]. Specifically, recombinant hamster and mouse PrP were mutated to determine the effect of GSS-associated mutations in the CLC (P102L and P105L) on their proneness to misfolding. These mutations and others in the CLC were shown to play a pivotal role on the susceptibility of PrP to misfolding [68]. Similarly, the effect of different polymorphism that were suspected of influencing the susceptibility to prion infection in vivo have been confirmed in vitro using rec-PrPs. The susceptibility of humans to BSE infection is known to be influenced by a polymorphism at amino acid residue 129, where the presence of a methionine residue instead of a valine one results in increased susceptibility [69]. Experiments using human rec-PrP with both polymorphisms demonstrated that a methionine residue in position 129 makes the a α-helix more solvent-exposed, increasing its proneness to misfolding [70]. Polymorphisms in mouse PrP that define genotypes *a* (L108/T189) and *b* (F108/V189) were also known to influence significantly the pathogenesis upon infection with scrapie in vivo [71]. Recombinant prion fibrils obtained from both polymorphic forms of mouse rec-PrP determined a clear difference in the nucleation phase [72], clearly demonstrating the influence of these amino acid residues on the misfolding ability of PrP. Besides evaluating the effect of different protein regions and amino acid residues on the fibril formation process, the influence of chemical modifications can be tested easily also using rec-PrP. For instance, the role of methionine oxidation, an event that could be related to the ease of prion propagation [73,74] was proven using hamster rec-PrP, which upon methionine oxidation showed increased proneness for β-sheet structure acquisition [75] and fibril formation [76].

Despite most of the previous examples being performed using cell-free systems and giving rise to non- or poorly infectious amyloid fibrils composed of rec-PrP, their use for the investigation of possible molecular mechanism related to PrP misfolding is beyond doubt. However, the results from these methods need to be interpreted cautiously as they may not correlate completely with the molecular mechanisms that take place in vivo. For this reason, the generation of *bona fide* recombinant prions was a significant breakthrough in the field of TSE research. Nonetheless, even the methodology that allowed production of the first recombinant prions that were as infectious as those derived from brain tissue [49] show an intrinsic variability by producing different misfolded rec-PrPs with strikingly different biological properties [77]. This is most likely due to the generation of misfolded rec-PrP fibrils with distinct tertiary or quaternary structures in vitro, reminiscent of the different structures underlying different prion strains [78]. At present, this in vitro misfolding event cannot be controlled accurately thereby resulting in different misfolded rec-PrP conformations with different tertiary structures stochastically and depending on the conditions used [49,77,79]. Differences in the post-translational modifications between rec-PrP and brain-derived PrP probably hamper precise templating with brain-derived PrP^{Sc} seeds, impeding the formation of the same strains that can be found in vivo. Therefore, most of the misfolded rec-PrPs generated in vitro are non- or poorly infectious in animal models, despite showing other prion-like properties such as self-propagation or protease-resistance. In fact, prions obtained from mouse P101L recombinant PrP (the equivalent to P102L human mutation related to GSS) produced the accumulation of amyloid plaques in the CNS of TgP101L knock-in transgenic mice but no clinical signs were observed in these animals [80]. Therefore, it is important to consider the biological properties of the misfolded rec-PrPs in vivo in order to correctly interpret the significance of the results obtained with these models. Moreover, defining the factors that would allow controlling the biological properties of the misfolded rec-PrPs generated in cell-free systems is of utmost importance to obtain biologically significant and unrefutable conclusions regarding the molecular mechanisms under study.

Legname and collaborators demonstrated that obtaining infectious recombinant prions is possible using purified rec-PrP, although challenge with this misfolded rec-PrP in a wild-type mouse model resulted in unusually long incubation periods [42]. This effect was probably due to the differences between the recombinant PrP and the mammalian cellular PrP, to the variety of conformations that the rec-PrP can acquire during its in vitro misfolding or to a lower infectious titer than brain-derived prions. In subsequent passages in animals, the misfolded rec-PrP adapted to in vivo propagation through a phenomenon that has been named deformed templating, which suggests that the initial slow propagation is due to incomplete structural compatibility [81]. Moreover, the generation of structurally distinct, misfolded rec-PrPs, which showed different biological properties, was demonstrated later. However, this set of ultrastructurally distinguishable recombinant prions converged into a common conformation upon successive inoculations in vivo [82]. The relevance of an appropriate three-dimensional structure over the differences between recombinant and mammalian PrP was proved by the generation of the first recombinant mouse prion capable of infecting wild-type mice with characteristics similar to those of brain-derived brain prions [49,83]. In this case, lipids (specifically POPG) and RNA were used as cofactors in the in vitro misfolding reaction, highlighting the relevance of appropriate cofactors to drive a mammalian prion-like misfolding, as addition of different cofactors rendered structurally highly similar but non-infectious prions under the same conditions [77]. This is in contrast with the observations of Deleault and collaborators that were able to obtain consistently recombinant prions with high specific infectivity using polyethylene glycol (PE) as a cofactor. The requirement of cofactor molecules for the formation of recombinant *bona fide* prions has been studied in depth and several molecules including specific lipids and polyanions were successfully used to obtain infectious recombinant prions [64,84–88]. However, the fact that absence of these cofactors restrict the rec-PrP misfolding towards recombinant *bona fide* prions has been undermined recently. Replication of the experiment that gave rise to the first highly infectious mouse recombinant prion [49] resulted in non-infective prions in vivo [79] suggesting that a stochastic element may be present. However, the generation of infectious recombinant prions was achieved recently in the absence of cofactors using both a murine full-length rec-PrP [47] and a C-terminally deleted rec-PrP (amino acids 23-144) [89]. These results demonstrated that obtaining recombinant prions with biological characteristics similar to mammalian prions is possible, although the conditions that invariably lead to this goal are not completely understood as yet. A clear example of this is the POPG-complemented recombinant prions [49] and PE-complemented ones [54] for which differences in specific infectivity and reproducibility could be due either to different protocols or to the different cofactors used. Despite not totally controlling the process, the ability to generate highly infectious recombinant prions, besides supporting the protein-only hypothesis, provides an invaluable model to finally unravel the greatest enigma of prions, their three-dimensional structure at a molecular level.

Recombinant PrP and in vitro misfolded recombinant fibrils play a central role in several structural studies focused mainly on deciphering the changes occurring during the misfolding event. Although it is well known that changes in secondary and tertiary structure of the PrP occur during the misfolding event, the regions involved in the initial steps and the structure of the misfolded pathogenic protein are completely unknown, which is reflected clearly in the notably different structural models proposed [64,90]. Using bovine rec-PrP (amino acids 121-230) and high-resolution NMR the regional stability and structural changes occurring upon urea-induced misfolding were measured, revealing region-specific information about the initial steps of PrP misfolding [91]. Similarly, the fibrilization of mouse rec-PrP aggregated in vitro in the presence of chaotropic agents was followed by hydrogen-deuterium exchange and mass spectrometry, highlighting the significance of the C-terminal domain and the direct addition of rec-PrP monomers to the fibrils, without intermediate oligomeric states [92]. However, these studies were performed by inducing rec-PrP fibrilization with chaotropic agents, which may not reproduce the misfolding process that takes place in vivo. Cell-free systems that allow the generation of infectious *bona fide* prions contribute to solving this issue. In fact, self-propagating recombinant misfolded PrPs with distinct in vivo infectivity have been analyzed using distinct physicochemical techniques and revealed subtle structural differences in the regions 91-115 and 144-163 that could be responsible for the different infectivity [88]. Infectious amyloid fibrils composed of sheep rec-PrP were the first studied with high resolution techniques such as Atomic Force Microscopy (AFM) and solid state-NMR (ssNMR) and showed results were not in complete agreement with either of the structural models proposed for prions [93]. Apart from full-length rec-PrP, amyloid fibrils formed by C-terminal truncated recombinant prions from human, mouse and hamster were also analyzed by ssNMR, showing a parallel in-register β-core [89]. Collectively, these studies clearly show the potential of recombinant *bona fide* prions to finally obtain a high-resolution three-dimensional structure of prions.

3. Diagnosis

The similarity between the clinical signs of TSEs and other neurodegenerative diseases makes the early diagnosis of some prion disorders extremely difficult. Motor and cognitive alterations are common features in all neurodegenerative diseases, hampering definitive diagnosis. The analysis of molecular biomarkers such as the 14-3-3 protein in the cerebrospinal fluid (CSF) of suspicious cases is often used to support the diagnosis in addition to clinical signs. However, altered levels of such biomarkers is non-specific as it is a common trait in many neurodegenerative diseases [94]. Although early detection of PrP^{Sc} would be the best biomarker due to its specificity, the accumulation of detectable amounts of PrP^{Sc} is mainly restricted to the CNS and is a late event in the course of many prion disorders limiting its use as an early diagnostic tool [95,96]. Nonetheless, in the case of oral or intraperitoneal infections compelling evidence shows that minute amounts of PrP^{Sc} are present in some body fluids of the affected individuals and in some prion disease to a major extent in the lymphoreticular system. More importantly, these traits of PrP^{Sc} appear prior to its accumulation on the CNS during the pre-symptomatic stage of the disease [97,98]. The diagnosis of prion disorders in an early stage is of utmost importance to help distinguish them from other dementias [99]. This is also critical in terms of public health, since the presence of PrP^{Sc} in certain body fluids from pre-symptomatic individuals might be enough to infect others [100,101]. Thus, PrP^{Sc} detection is pivotal for biosafety in blood transfusions and surgery [102,103]. However, the exceptionally small amounts of the causal agent in easily accessible body fluids, such as blood or urine, restrict its direct detection [100,104]. In the case of the CSF, despite the specific infectivity could be much higher it is also low in the asymptomatic stage of the disease, making difficult to detect it directly [105]. In order to solve this problem two different strategies which use either PrP^{Sc} concentration or its amplification before detection have been adopted.

There are two main techniques for PrP^{Sc} concentration and detection in body fluids. Both take advantage of steel beads or wires to concentrate prions present in the blood and CSF, which for unknown reasons bind the PrP^{Sc} present in the sample [106,107]. The first one, called Direct Detection Assay (DDA), is based on direct detection of these concentrated (using a solid-state binding matrix) prions by specific antibodies [108]. The second, Standard Steel Binding Assay (SSBA) couples prion concentration with a Scrapie Cell Assay (SCA) on which PrP^{Sc} detection is based on the infection of susceptible cell lines using the concentrated prions as seeds [109,110]. However, the DDA relies on the direct detection of PrP^{Sc} after concentration which may not be present within the lower detection limit of the method depending on the disease stage of the affected individual. In the case of SSBA, the cell assay would act as reporter solving the issue of sufficient PrP^{Sc}, although it is restricted to those prions for which susceptible cell lines exist.

The development of cell-free systems using rec-PrP is the cornerstone of the detection methods based on PrP^{Sc} amplification and would overcome some of the limitations of the previous techniques. Several systems have emerged for in vitro prion propagation or amplification, all of them employing the use of a substrate that contains an excess of natively folded PrP and a small amount of PrP^{Sc} which acts as seed and promotes the misfolding of the PrP from the substrate. Subsequently, this larger amount of misfolded PrP can be detected directly using specific antibodies or amyloid-binding dyes. The development of protein misfolding cyclic amplification (PMCA) was a great step forward on this direction. Initially, PrP^{C} from brain homogenates was used as a substrate to amplify minute amounts of PrP^{Sc} from prion-infected tissue samples using serial cycles of incubation and sonication [111]. However, the use of brain homogenates as source of PrP limited its application to the existing animal models. Substitution of the PrP derived from brain homogenates for rec-PrP overcame some of these limitations [112]. Furthermore, any sequence of rec-PrP can be designed and generated including chimeric or mutant PrPs that could improve the sensitivity of the technique due to increases in proneness to misfolding. Caughey and collaborators used used hamster rec-PrP as substrate with PMCA to amplify PrP^{Sc} seeds obtained from the CSF of scrapie-infected hamsters [113]. This system permitted the detection of as little as 50 ag of PrP^{Sc}, allowing the diagnose of scrapie in hamsters using just 2 μL of CSF. However, despite the improvement by using rec-PrP, this technique was never implemented in clinical practice as it could not be used successfully with human samples and because the complexity of the incubation/sonication system and the expertise needed to interpret the results precluded its daily use in hospitals.

Another technique based on rec-PrP amplification to detect minute amounts of PrP^{Sc} is the Amyloid Seeding Assay (ASA). Instead of using natively folded rec-PrP and incubation/sonication cycles, rec-PrP is kept in a semi-denatured state through the addition of chaotropic agents, which is thought to be an intermediate state on its misfolding pathway leading to aggregation [114]. Under mild denaturing conditions (presence of 0.46 M of guanidine hydrochloride), constant temperature (37 °C) and shaking, a nucleation process takes place creating misfolded states of the rec-PrP, which is the first step in the growth of amyloid fibers that occurs later. The growth of the amyloid fibers is monitored measuring the Thioflavin T (ThT) fluorescence over time, which increases in intensity upon binding to amyloid fibers. This simplifies the detection compared to direct detection by western blotting and specific antibodies [115]. The misfolding of hamster rec-PrP into amyloid fibers in vitro that occurs under these conditions is a spontaneous and slow process with a lag phase (time needed to detect some fluorescence increase) of nearly 12 h. The addition of preformed amyloid fibers or PrP^{Sc} as seeds greatly accelerate rec-PrP fibril formation, reducing the lag phase to just 2 h. The sensitivity of the technique has been evaluated using purified fibers from phosphotungstenate (PTA)-precipitation of scrapie-infected mouse brains, hamster brains and sporadic CJD infected human brains. In all cases addition of a seed lead to accelerated ThT signal increase and estimates suggest that as little as 0.03 fg of PrP^{Sc} can be detected by ASA [115]. This technique would be more suitable as a practical diagnostic system because it does not require sonication and the rec-PrP is easily produced. Furthermore, the sensitivity is appropriate and the system requires less specialist knowledge and technical proficiency compared to PMCA. However, it has not been tested with CSF samples and the necessity of purified fibers or brain homogenate as seeds impede its use as a tool for early diagnosis of prion infections.

To date, the most successful technique and the most likely to be implemented in clinical practice is based also on the use of rec-PrP as a substrate for prion amplification. The Real Time Quaking Induced Conversion (RT-QuIC) allows the conversion of rec-PrP into the protease-resistant misfolded PrP isoform (PrPres) using shaking and controlled temperature and avoids the use of chaotropic agents [116,117]. The fibril growth is measured in real time using ThT, which is present throughout the fibril growth. It has been used with multiple strains, species and different PrPSc sources, including different classical strains of hamster, human, bovine, cervine, ovine and murine prions and some atypical strains also and all of them are detected efficiently by this technique [118–124]. Moreover, it has been further simplified because a single rec-PrP (the one from bank vole, *Myodes glareolus*) has been found to act as nearly universal substrate for the detection of prion strains coming from different species, including human samples [122]. The use of this rec-PrP as substrate for RT-QuIC allows the accurate diagnosis of humans affected by sporadic prion disorders with 100% sensitivity and 100% specificity from nasal brushing samples [99,125,126]. This sampling technique is a painless and relatively non-invasive way to obtain neuronal samples from the olfactory bulb and this, coupled to the extremely sensitive and specific RT-QuIC, may become the routine diagnostic tool for prion disease diagnosis in pre-symptomatic patients in the future.

4. Screening

The unknown molecular mechanisms underlying prion propagation and the resultant neurodegeneration are motivating researchers in the field to seek chemical compounds with anti-prion properties empiricaly, mostly through high-throughput screening of large chemical libraries [127]. Despite some compounds with putative anti-prion features have been identified based on experimental evidence [128,129] or because of their amyloid binding properties [130–132], none of them was effective in vivo [133–135]. Therefore, the search for new compounds that may become effective treatments is an area of great activity in the TSE field, as well as the development of rapid, inexpensive and reliable systems for high-throughput screening.

Although the most reliable results in the search for compounds with anti-prion properties would be by in vivo models, these are unsuitable for high-throughput screening due to the economical and ethical issues associated with the large number of animals required. Instead, cell culture models have been used to detect compounds that could impair prion propagation [136]. However, these models require expertise, have higher material costs than cell-free systems and are not suitable for human prion diseases as only some animal strains can be propagated. Moreover, despite being closer to the natural scenario, cell culture still does not result in all the characteristic hallmarks of a TSE and are slower than other in vitro systems because of the time needed to process samples and measure the decrease in PrPSc levels. Thus, the most cost-effective methods for high-throughput screening of putative anti-prion compounds relies on the use of rec-PrP. Relatively large quantities of pure rec-PrP can be produced easily and the protein can be modified or labelled in any way imaginable to facilitate the measurement of the outcome. Different in vitro techniques based on the use of rec-PrP have demonstrated their potential for the screening of putative anti-prion drugs. These techniques can be divided in two groups depending on the mechanisms sought to inhibit the progression of the disease; compounds able to bind to native PrP or those inhibiting fibril formation.

Most of the techniques in use at present seek compounds able to bind to natively folded PrP and impede its misfolding through stabilization of this conformation or blocking putative interaction sites with the disease-associated isoform. This strategy was proven valuable due to compounds such as the cyclic tetrapyrroles [137] which were identified due to their capacity to bind proteins and alter their conformational properties. Some of these compounds have been shown to interact with PrP specifically and even the binding site of some have been identified by Nuclear Magnetic Resonance (NMR) [138] which appeared to be directly related to their ability to inhibit the formation of the misfolded isoform in vitro [139]. Similarly, low-molecular-weight heparin, which binds to rec-PrP and increases its thermal stability, was shown to inhibit fibrillization of rec-PrPs from mouse and hamster by RT-QuIC [140]. Although none of these compounds reached clinical trials due to toxicity or poor blood brain barrier permeability, these examples clearly demonstrate the utility of screening for compounds with the ability to bind to rec-PrP. Several assays have been developed in order to perform high-throughput screening of chemical compounds able to bind to rec-PrP.

The fluorescence polarization-based competitive binding assay uses rec-PrP and phosphorothioate oligonucleotides (PS-ONs) which are fluorophores that bind strongly to the protein and show certain pattern of fluorescence polarization (FP) when bound. The addition of compounds that compete with the PS-ON for the binding site displaces it changing the FP of the sample. The binding site of PS-ON has already been reported to be important as when it attaches to PrP it prevents misfolding [141]. Thus, any compound that competes for this binding site could be a good candidate for treatment of prion disorders. This system can be easily adapted for high-throughput screening of compounds as 96-well plates can be used and measurement of the outcome is rapid. This technique, combining the use of hamster rec-PrP and the fluorophore Randomer-FL, has been shown to be useful in screening for anti-prion compounds [141]. A similar technique was also developed based on FP technology using rec-PrP labelled with a different fluorophore, IANBD [N-((2-(iodoacetoxy)ethyl)-N-methyl)amino-7-nitrobenz-2-oxa-1,3-diazole) ester]. In this case, the proteins need to be mutated to include a cysteine residue on its primary structure without significantly altering its secondary and tertiary structure. The use of rec-PrP allows the generation of the required mutant PrPs and the measurement of secondary and tertiary structure by standard biophysical techniques. This approach was successfully used by Collinge and collaborators using a human rec-PrP with a cysteine substitution at position 145 for the screening of a 1200-compound library [142]. They identified a compound, Chicago Sky Blue 6B, which bound strongly to rec-PrP with anti-prion activity which was demonstrated also in cell culture.

Surface Plasmon Resonance (SPR) is another technique that has demonstrated its potential for the high-throughput screening of anti-prion compounds through binding to rec-PrP. This technique is based on the measurement of changes that occur in the molecular weight of a protein upon binding of different molecules [143]. This screening system has been implemented successfully by Doh-Ura and collaborators using mouse rec-PrP (amino acids 121-231) and different compounds with known anti-prion capacity [144]. Propranolol was identified as a new anti-prion drug with the results being confirmed in cell-culture.

The other group of techniques used for high-throughput screening of compounds aims to detect those which inhibit the misfolding or fibrillization of rec-PrP in cell-free systems, regardless of the mechanism of action of the compound. The generation of amyloid fibers in vitro using rec-PrP as a substrate was a breakthrough in the prion field as it was the first cell-free system [42] able to generate a highly infectious recombinant PrP [49]. From the diverse methods developed for rec-PrP fibrillization, some can be adapted to the high-throughput screening of compounds due to their technical characteristics. This is the case for the semi-automated cell-free system [145] and RT-QuIC [117], both easily automated due to the simplicity of the equipment needed and the possibility of robotising the measurement of the fluorescence outcome. Both techniques rely on an increase in Thioflavin T (ThT) fluorescence to detect the presence of fibrils formed by rec-PrP in the sample, which occurs spontaneously in both systems but can be accelerated by the addition of seeds. Therefore, these systems are suitable for high-throughput screening because they are easy to automate, have reduced costs compared to animal or cell culture models and are faster than any other system. The semi-automated cell-free system has demonstrated its ability to screen anti-prion drugs using mouse rec-PrP and different compounds with known anti-prion effects such as curcumin, PAMAM dendrimers and TMPyP-Fe(III) [145]. Similarly, RT-QuIC has been used successfully using human rec-PrP and CJD-affected brain homogenate samples as a seed. Among others, acridine, dextran sodium sulphate and tannic acid were used to provide proof of principle, showing the suitability of this system for high-throughput screening of compounds that could inhibit prion propagation in vitro [146].

Halfway between methods measuring binding of compounds to rec-PrP and those looking for inhibition of fibrillization of rec-PrP, the Scanning for Intensely Fluorescent Targets (SIFT) is worthy of mention. In this case, mouse rec-PrP labelled with a green fluorophore is incubated in the presence of CJD-affected brain homogenate and antibodies specific for human PrP labelled with a red fluorophore. The technique is based on measuring disturbances in the interactions between rec-PrP and the PrP^{Sc} present in a CJD-affected brain homogenate after the addition of compounds that could interfere with this interaction. In the absence of anti-prion compounds, rec-PrP and the CJD fibrils form ternary complexes resulting in a mix of red and green fluorescence, while addition of compounds that inhibit this interaction shifts the fluorescence emission [147]. This method was successfully used for screening ten thousand compounds and detected eighty hits from which six were effective in cell culture models.

Together, this work demonstrate the suitability and versatility of various systems based on rec-PrP for high-throughput screening of anti-prion drugs.

5. Therapy

At present, there is no effective treatment for TSEs and they remain invariably fatal. The lack of knowledge of the atomic structure of PrP^{Sc} and the molecular mechanisms leading to protein misfolding and spongiform degeneration impede the rational design of therapeutic approaches targeting PrP^{Sc} or the cascade that results in neurodegeneration. Despite this, many research groups have focused their efforts on finding a therapy for these devastating disorders. Although none of the strategies or compounds found as yet are suitable for clinical practice, several strategies have been designed with differing levels of success. The search for chemical compounds with anti-prion properties [127], discussed above, is the most obvious strategy. Other strategies proposed include gene therapy [148], administration of rec-PrPs which impair the misfolding of endogenous PrP [149,150], and immunological therapy aimed at inducing an immune response against prions [151,152]. Recombinant PrP plays a pivotal role in the last two therapeutic strategies showing its use in this area of TSE research.

The existence of transmission barriers is a well-known phenomenon in the prion field that has led to the development of therapies involving the use of rec-PrPs able to interfere with the propagation of PrP^{Sc} at the expense of endogenous PrP. Transmission barriers were first identified when interspecies prion transmission was performed experimentally. Due to differences in the PrP amino acid sequences of prion-donor and receptor species, the transmission is hindered, as shown by prolonged incubation times and reduced attack rates (percentage of individuals that succumb to the disease) [153]. Upon subsequent challenges, with resultant brain material from individuals initially challenged, in the same receptor species, incubation times are shortened and the attack rate increased as a result of adaptation [154]. In addition to differences in PrP primary structure, differences in the three-dimensional structure of prion strains (even those with the same PrP sequence) can also result in transmission barriers [78,155]. This phenomenon, first observed in vivo, has been reproduced in vitro in cell-free systems mimicking prion misfolding. PMCA is a good example of this as a well-established method to study the transmission barriers using brain homogenates as substrate and different seeds derived from infected brain homogenates [156,157]. Using rec-PrP as substrate, it was found that differences in PrP amino acid sequence was not the only criteria imposing a transmission barrier, but that differences between brain-derived prions and rec-PrP (probably due to the absence of glycosylation and a GPI anchor) also hindered the in vitro propagation of brain-derived PrP^{Sc} [158]. Taking advantage of this phenomenon, administration of rec-PrP has been proposed as a direct treatment delaying prion misfolding and thus, the progression of the disease. Using PMCA with brain homogenates from transgenic mice overexpressing human PrP as substrate and CJD-affected patient brain homogenate as seed, it was shown that the addition of human rec-PrP to the system clearly impeded the propagation of the CJD seed in a dose-dependent manner [149]. Because of a possible adaptation of the brain-derived prions to the rec-PrP substrate, dominant negative PrPs have been sought, PrPs with lower capacity to be misfolded due to certain mutations [159], which could be the best candidates for this therapeutic approach. Although the exact molecular mechanism of the interference of rec-PrP with PrP^{Sc} is unknown, this approach was also shown to be effective in vivo. In this case, mice infected with rodent-adapted scrapie prions and treated with hamster rec-PrP administered intracerebrally showed a significant increase in the survival times in a dose-dependent manor, illustrating clearly that rec-PrP could be used as an inhibitor in the treatment of TSEs [160,161].

Another therapeutic strategy involving the use of rec-PrP is the immunological therapy. As the PrP is a ubiquitous protein abundantly expressed in the CNS and PrP^{Sc} shares the same primary structure, no immunological response arises against prions, a phenomenon known as autotolerance [162,163]. The aim of the immunological therapy is to overcome this autotolerance and induce the immune system to generate antibodies which upon binding to PrP could impede its conversion to the pathogenic isoform. For that purpose, rec-PrP is used like a vaccine antigen to induce the production of auto-antibodies. The effectiveness of the immunological therapy has been shown in different models. Using dimeric mouse rec-PrP expressed in *E. coli*, auto-antibodies were generated in mice which demonstrated their efficacy by inhibiting the formation of PrP^{Sc} in cell-culture models persistently infected with prions [164]. Mouse rec-PrP immunization was also shown to provide protection in vivo, delaying the development of the disease when used as a prophylactic therapy in mice [165–167]. Furthermore, immunization with rec-PrPs from other species, such as bovine rec-PrP, delayed the disease onset in mice inoculated with rodent-adapted prions [152]. Apart from mouse models, the ability of rec-PrPs to boost the immune system has been shown in hamsters immunized with recombinant hamster PrP [151]. Altogether these works demonstrated that the use of rec-PrP to promote an immune response against prions may be an effective prophylactic approach to prevent the development of TSEs.

Since the generation of the first recombinant misfolded PrP [42], several in vitro models have been developed based on this technology. Due to these techniques, different aspects of the molecular mechanisms responsible for prion misfolding were unravelled, important improvements in TSEs diagnosis were made, different anti-prion compounds were discovered and new therapeutic strategies were developed. The development and production of recombinant PrP has demonstrated its enormous potential for further understanding of these devastating prion disorders.

Table 1. Summary of the most relevant advances accomplished with rec-PrP in each research area.

TSE Research Area	Breakthrough	Reference
	Production of highly pure bacterially-expressed recombinant PrP	[168]
	Determination of the 3D structure of cellular PrP	[21]
	Generation of the first infectious recombinant prions	[42]
	Generation of the first recombinant prions infectious in wild type animals	[49]
	Generation of the first highly infectious recombinant prions	[54]
	Interaction of PrP with copper confirmed	[59]
Molecular mechanisms	N-terminal of PrP not necessary for misfolding	[63–65]
	Confirmation of increased misfolding proneness due to disease-associated mutations	[50,68]
	Generation of the first human infectious recombinant prions	[30]
	Confirmation of different misfolding proneness in polymorphic PrPs	[69–71]
	Description of possible mechanisms of strain generation and adaptation	[51,82]
	Description of the role of cofactors in the determination of biological properties	[47,49,51,55,78]
	Generation of models for 3D structure of recombinant misfolded PrP	[64,89,90]
Diagnosis	Development of PMCA based on rec-PrP for diagnosis from CSF	[113]
	Development of RT-QUIC for diagnosis from different body fluids and tissue samples	[99,122,126,135]

Table 1. *Cont.*

TSE Research Area	Breakthrough	Reference
Screening	Development high-throughput screening methods	[144,145,147]
Therapy	Demonstration of dominant-negative effect of exogenous rec-PrP on the propagation of prions	[149,150]
	Immunotherapy based on injection of rec-PrP	[151,152]

Author Contributions: J.C., J.M.C. and H.E. planned and designed the manuscript. J.M.C., H.E., V.V., S.G.-M., R.L.-M., E.G.-M. and M.Á.P.-C. searched and classified relevant information. J.M.C. and H.E. wrote the manuscript. J.C. revised it and gave the final approval.

Conflicts of Interest: The authors declare no conflict of interest.

References

1. Aguzzi, A.; Calella, A.M. Prions: Protein aggregation and infectious diseases. *Physiol. Rev.* **2009**, *89*, 1105–1152. [CrossRef] [PubMed]

2. Liberski, P.P.; Sikorska, B.; Brown, P. Kuru: The first prion disease. *Adv. Exp. Med. Biol.* **2012**, *724*, 143–153. [PubMed]

3. Hsiao, K.; Prusiner, S.B. Molecular genetics and transgenic model of gertsmann-straussler-scheinker disease. *Alzheimer Dis. Assoc. Disord.* **1991**, *5*, 155–162. [CrossRef] [PubMed]

4. Gallassi, R.; Morreale, A.; Montagna, P.; Cortelli, P.; Avoni, P.; Castellani, R.; Gambetti, P.; Lugaresi, E. Fatal familial insomnia: Behavioral and cognitive features. *Neurology* **1996**, *46*, 935–939. [CrossRef] [PubMed]

5. Jackson, G.S.; Collinge, J. The molecular pathology of cjd: Old and new variants. *Mol. Pathol.* **2001**, *54*, 393–399. [PubMed]

6. Zou, W.Q.; Puoti, G.; Xiao, X.; Yuan, J.; Qing, L.; Cali, I.; Shimoji, M.; Langeveld, J.P.; Castellani, R.; Notari, S.; et al. Variably protease-sensitive prionopathy: A new sporadic disease of the prion protein. *Ann. Neurol.* **2010**, *68*, 162–172. [CrossRef] [PubMed]

7. Dickinson, A.G. Scrapie in sheep and goats. *Front. Biol.* **1976**, *44*, 209–241. [PubMed]

8. Marsh, R.F.; Hadlow, W.J. Transmissible mink encephalopathy. *Rev. Sci. Tech.* **1992**, *11*, 539–550. [CrossRef] [PubMed]

9. Kimberlin, R.H. An overview of bovine spongiform encephalopathy. *Dev. Biol. Stand.* **1991**, *75*, 75–82. [PubMed]

10. Williams, E.S.; Young, S. Spongiform encephalopathies in cervidae. *Rev. Sci Tech.* **1992**, *11*, 551–567. [CrossRef] [PubMed]

11. Pattison, I.H. Resistance of the scrapie agent to formalin. *J. Comp. Pathol.* **1965**, *75*, 159–164. [CrossRef]

12. Alper, T. The nature of the scrapie agent. *J. Clin. Pathol. Suppl. (R. Coll. Pathol.)* **1972**, *6*, 154–155. [CrossRef]

13. Prusiner, S.B. Novel proteinaceous infectious particles cause scrapie. *Science* **1982**, *216*, 136–144. [CrossRef] [PubMed]

14. Soto, C. Prion hypothesis: The end of the controversy? *Trends Biochem. Sci.* **2011**, *36*, 151–158. [CrossRef] [PubMed]

15. Anderson, R.M.; Donnelly, C.A.; Ferguson, N.M.; Woolhouse, M.E.; Watt, C.J.; Udy, H.J.; MaWhinney, S.; Dunstan, S.P.; Southwood, T.R.; Wilesmith, J.W.; et al. Transmission dynamics and epidemiology of bse in british cattle. *Nature* **1996**, *382*, 779–788. [CrossRef] [PubMed]

16. Collinge, J.; Whitfield, J.; McKintosh, E.; Beck, J.; Mead, S.; Thomas, D.J.; Alpers, M.P. Kuru in the 21st century—An acquired human prion disease with very long incubation periods. *Lancet* **2006**, *367*, 2068–2074. [CrossRef]

17. Hilton, D.A. Pathogenesis and prevalence of variant creutzfeldt-jakob disease. *J. Pathol.* **2006**, *208*, 134–141. [CrossRef] [PubMed]

18. Schmitz, M.; Dittmar, K.; Llorens, F.; Gelpi, E.; Ferrer, I.; Schulz-Schaeffer, W.J.; Zerr, I. Hereditary human prion diseases: An update. *Mol. Neurobiol.* **2017**, *54*, 4138–4149. [CrossRef] [PubMed]

19. Will, R.G.; Ironside, J.W. Sporadic and infectious human prion diseases. *Cold Spring Harb. Perspect. Med.* **2017**, *7*. [CrossRef] [PubMed]

20. Westergard, L.; Christensen, H.M.; Harris, D.A. The cellular prion protein (prp(c)): Its physiological function and role in disease. *Biochim. Biophys. Acta* **2007**, *1772*, 629–644. [CrossRef] [PubMed]
21. Riek, R.; Hornemann, S.; Wider, G.; Glockshuber, R.; Wuthrich, K. NMR characterization of the full-length recombinant murine prion protein, mPrP(23-231). *FEBS Lett.* **1997**, *413*, 282–288. [CrossRef]
22. Meyer, R.K.; McKinley, M.P.; Bowman, K.A.; Braunfeld, M.B.; Barry, R.A.; Prusiner, S.B. Separation and properties of cellular and scrapie prion proteins. *Proc. Natl. Acad. Sci. USA* **1986**, *83*, 2310–2314. [CrossRef] [PubMed]
23. Cohen, F.E.; Prusiner, S.B. Pathologic conformations of prion proteins. *Annu. Rev. Biochem.* **1998**, *67*, 793–819. [CrossRef] [PubMed]
24. Safar, J.; Roller, P.P.; Gajdusek, D.C.; Gibbs, C.J., Jr. Thermal stability and conformational transitions of scrapie amyloid (prion) protein correlate with infectivity. *Protein Sci.* **1993**, *2*, 2206–2216. [CrossRef] [PubMed]
25. Erana, H.; Castilla, J. The architecture of prions: How understanding would provide new therapeutic insights. *Swiss Med. Wkly.* **2016**, *146*, w14354. [PubMed]
26. Bruce, M.E.; Fraser, H. Scrapie strain variation and its implications. *Curr. Top. Microbiol. Immunol.* **1991**, *172*, 125–138. [PubMed]
27. Bruce, M.E. Tse strain variation. *Br. Med. Bull.* **2003**, *66*, 99–108. [CrossRef] [PubMed]
28. Prusiner, S.B.; Scott, M.; Foster, D.; Pan, K.M.; Groth, D.; Mirenda, C.; Torchia, M.; Yang, S.L.; Serban, D.; Carlson, G.A.; et al. Transgenetic studies implicate interactions between homologous prp isoforms in scrapie prion replication. *Cell* **1990**, *63*, 673–686. [CrossRef]
29. Gajdusek, C.; Gibbs, C.; Alpers, M. Experimental transmission of kuru-like syndrome to chimpanzees. *Nature* **1966**, *209*, 794–796. [CrossRef] [PubMed]
30. Gibbs, C.J., Jr.; Gajdusek, D.C.; Asher, D.M.; Alpers, M.P.; Beck, E.; Daniel, P.M.; Matthews, W.B. Creutzfeldt-jakob disease (spongiform encephalopathy): Transmission to the chimpanzee. *Science* **1968**, *161*, 388–389. [CrossRef] [PubMed]
31. Hill, A.F.; Desbruslais, M.; Joiner, S.; Sidle, K.C.; Gowland, I.; Collinge, J.; Doey, L.J.; Lantos, P. The same prion strain causes vcjd and bse. *Nature* **1997**, *389*, 448–450. [CrossRef] [PubMed]
32. Chandler, R.L. Encephalopathy in mice produced by inoculation with scrapie brain material. *Lancet* **1961**, *1*, 1378–1379. [CrossRef]
33. Scott, M.; Foster, D.; Mirenda, C.; Serban, D.; Coufal, F.; Walchli, M.; Torchia, M.; Groth, D.; Carlson, G.; DeArmond, S.J.; et al. Transgenic mice expressing hamster prion protein produce species-specific scrapie infectivity and amyloid plaques. *Cell* **1989**, *59*, 847–857. [CrossRef]
34. Asante, E.A.; Linehan, J.M.; Desbruslais, M.; Joiner, S.; Gowland, I.; Wood, A.L.; Welch, J.; Hill, A.F.; Lloyd, S.E.; Wadsworth, J.D.; et al. Bse prions propagate as either variant cjd-like or sporadic cjd-like prion strains in transgenic mice expressing human prion protein. *EMBO J.* **2002**, *21*, 6358–6366. [CrossRef] [PubMed]
35. Groschup, M.H.; Buschmann, A. Rodent models for prion diseases. *Vet. Res.* **2008**, *39*, 32. [CrossRef] [PubMed]
36. Priola, S.A. Prion protein and species barriers in the transmissible spongiform encephalopathies. *Biomed. Pharmacother* **1999**, *53*, 27–33. [CrossRef]
37. Krauss, S.; Vorberg, I. Prions ex vivo: What cell culture models tell us about infectious proteins. *Int. J. Cell Biol.* **2013**, *2013*, 704546. [CrossRef] [PubMed]
38. Race, R.E.; Fadness, L.H.; Chesebro, B. Characterization of scrapie infection in mouse neuroblastoma cells. *J. Gen. Virol.* **1987**, *68 Pt 5*, 1391–1399. [CrossRef] [PubMed]
39. Mahal, S.P.; Baker, C.A.; Demczyk, C.A.; Smith, E.W.; Julius, C.; Weissmann, C. Prion strain discrimination in cell culture: The cell panel assay. *Proc. Natl. Acad. Sci. USA* **2007**, *104*, 20908–20913. [CrossRef] [PubMed]
40. Vilette, D.; Andreoletti, O.; Archer, F.; Madelaine, M.F.; Vilotte, J.L.; Lehmann, S.; Laude, H. Ex vivo propagation of infectious sheep scrapie agent in heterologous epithelial cells expressing ovine prion protein. *Proc. Natl. Acad. Sci. USA* **2001**, *98*, 4055–4059. [CrossRef] [PubMed]
41. Hornemann, S.; Korth, C.; Oesch, B.; Riek, R.; Wider, G.; Wuthrich, K.; Glockshuber, R. Recombinant full-length murine prion protein, mPrP(23231): Purification and spectroscopic characterization. *FEBS Lett.* **1997**, *413*, 277–281. [CrossRef]
42. Legname, G.; Baskakov, I.V.; Nguyen, H.O.; Riesner, D.; Cohen, F.E.; DeArmond, S.J.; Prusiner, S.B. Synthetic mammalian prions. *Science* **2004**, *305*, 673–676. [CrossRef] [PubMed]

43. Deleault, N.R.; Harris, B.T.; Rees, J.R.; Supattapone, S. Formation of native prions from minimal components in vitro. *Proc. Natl. Acad. Sci. USA* **2007**, *104*, 9741–9746. [CrossRef] [PubMed]

44. Bocharova, O.V.; Breydo, L.; Salnikov, V.V.; Gill, A.C.; Baskakov, I.V. Synthetic prions generated in vitro are similar to a newly identified subpopulation of prpsc from sporadic creutzfeldt-jakob disease. *Protein Sci.* **2005**, *14*, 1222–1232. [CrossRef] [PubMed]

45. Legname, G.; Nguyen, H.O.; Baskakov, I.V.; Cohen, F.E.; Dearmond, S.J.; Prusiner, S.B. Strain-specified characteristics of mouse synthetic prions. *Proc. Natl. Acad. Sci. USA* **2005**, *102*, 2168–2173. [CrossRef] [PubMed]

46. Chu, N.K.; Becker, C.F. Semisynthesis of membrane-attached prion proteins. *Methods Enzymol.* **2009**, *462*, 177–193. [PubMed]

47. Kim, J.I.; Cali, I.; Surewicz, K.; Kong, Q.; Raymond, G.J.; Atarashi, R.; Race, B.; Qing, L.; Gambetti, P.; Caughey, B.; et al. Mammalian prions generated from bacterially expressed prion protein in the absence of any mammalian cofactors. *J. Biol. Chem.* **2010**, *285*, 14083–14087. [CrossRef] [PubMed]

48. Colby, D.W.; Wain, R.; Baskakov, I.V.; Legname, G.; Palmer, C.G.; Nguyen, H.O.; Lemus, A.; Cohen, F.E.; DeArmond, S.J.; Prusiner, S.B. Protease-sensitive synthetic prions. *PLoS Pathog.* **2010**, *6*, e1000736. [CrossRef] [PubMed]

49. Wang, F.; Wang, X.; Yuan, C.G.; Ma, J. Generating a prion with bacterially expressed recombinant prion protein. *Science* **2010**, *327*, 1132–1135. [CrossRef] [PubMed]

50. Elezgarai, S.R.; Fernández-Borges, N.; Erana, H.; Sevillano, A.; Moreno, J.; Harrathi, C.; Saá, P.; Gil, D.; Kong, Q.; Requena, J.R.; et al. Generation of a new infectious recombinant prion: A model to understand gerstmann–sträussler–scheinker syndrome. *Sci Rep.* **2017**, *7*, 9584. [CrossRef] [PubMed]

51. Fernández-Borges, N.; Di Bari, M.A.; Eraña, H.; Sánchez-Martín, M.A.; Pirisinu, L.; Parra, B.; Elezgarai, S.R.; Vanni, I.; López-Moreno, R.; Vaccari, G.; et al. De novo generation of a variety of different infectious recombinant prion strains. *Acta Neuropathol.* **2017**. Submitted.

52. Gibbs, C.J., Jr.; Gajdusek, D.C. Transmission of scrapie to the cynomolgus monkey (*Macaca fascicularis*). *Nature* **1972**, *236*, 73–74. [CrossRef] [PubMed]

53. Bedecs, K. Cell culture models to unravel prion protein function and aberrancies in prion diseases. *Methods Mol. Biol.* **2008**, *459*, 1–20. [PubMed]

54. Deleault, N.R.; Piro, J.R.; Walsh, D.J.; Wang, F.; Ma, J.; Geoghegan, J.C.; Supattapone, S. Isolation of phosphatidylethanolamine as a solitary cofactor for prion formation in the absence of nucleic acids. *Proc. Natl. Acad. Sci. USA* **2012**, *109*, 8546–8551. [CrossRef] [PubMed]

55. Brown, D.R.; Qin, K.; Herms, J.W.; Madlung, A.; Manson, J.; Strome, R.; Fraser, P.E.; Kruck, T.; von Bohlen, A.; Schulz-Schaeffer, W.; et al. The cellular prion protein binds copper in vivo. *Nature* **1997**, *390*, 684–687. [CrossRef] [PubMed]

56. Hijazi, N.; Shaked, Y.; Rosenmann, H.; Ben-Hur, T.; Gabizon, R. Copper binding to prpc may inhibit prion disease propagation. *Brain Res.* **2003**, *993*, 192–200. [CrossRef] [PubMed]

57. Sigurdsson, E.M.; Brown, D.R.; Alim, M.A.; Scholtzova, H.; Carp, R.; Meeker, H.C.; Prelli, F.; Frangione, B.; Wisniewski, T. Copper chelation delays the onset of prion disease. *J. Biol. Chem.* **2003**, *278*, 46199–46202. [CrossRef] [PubMed]

58. Hornshaw, M.P.; McDermott, J.R.; Candy, J.M.; Lakey, J.H. Copper binding to the N-terminal tandem repeat region of mammalian and avian prion protein: Structural studies using synthetic peptides. *Biochem. Biophys. Res. Commun.* **1995**, *214*, 993–999. [CrossRef] [PubMed]

59. Burns, C.S.; Aronoff-Spencer, E.; Legname, G.; Prusiner, S.B.; Antholine, W.E.; Gerfen, G.J.; Peisach, J.; Millhauser, G.L. Copper coordination in the full-length, recombinant prion protein. *Biochemistry* **2003**, *42*, 6794–6803. [CrossRef] [PubMed]

60. Wong, E.; Thackray, A.M.; Bujdoso, R. Copper induces increased beta-sheet content in the scrapie-susceptible ovine prion protein prpvrq compared with the resistant allelic variant prparr. *Biochem. J.* **2004**, *380*, 273–282. [CrossRef] [PubMed]

61. Bocharova, O.V.; Breydo, L.; Salnikov, V.V.; Baskakov, I.V. Copper(ii) inhibits in vitro conversion of prion protein into amyloid fibrils. *Biochemistry* **2005**, *44*, 6776–6787. [CrossRef] [PubMed]

62. Yu, S.; Yin, S.; Pham, N.; Wong, P.; Kang, S.C.; Petersen, R.B.; Li, C.; Sy, M.S. Ligand binding promotes prion protein aggregation—Role of the octapeptide repeats. *FEBS J.* **2008**, *275*, 5564–5575. [CrossRef] [PubMed]

63. Martins, S.M.; Frosoni, D.J.; Martinez, A.M.; De Felice, F.G.; Ferreira, S.T. Formation of soluble oligomers and amyloid fibrils with physical properties of the scrapie isoform of the prion protein from the c-terminal domain of recombinant murine prion protein mPrP-(121-231). *J. Biol. Chem.* **2006**, *281*, 26121–26128. [CrossRef] [PubMed]

64. Groveman, B.R.; Dolan, M.A.; Taubner, L.M.; Kraus, A.; Wickner, R.B.; Caughey, B. Parallel in-register intermolecular beta-sheet architectures for prion-seeded prion protein (PrP) amyloids. *J. Biol. Chem.* **2014**, *289*, 24129–24142. [CrossRef] [PubMed]

65. Corsaro, A.; Thellung, S.; Villa, V.; Nizzari, M.; Aceto, A.; Florio, T. Recombinant human prion protein fragment 90–231, a useful model to study prion neurotoxicity. *OMICS* **2012**, *16*, 50–59. [CrossRef] [PubMed]

66. Ironside, J.W.; Head, M.W. Biology and neuropathology of prion diseases. *Handb. Clin. Neurol.* **2008**, *89*, 779–797. [PubMed]

67. Ghetti, B.; Tagliavini, F.; Takao, M.; Bugiani, O.; Piccardo, P. Hereditary prion protein amyloidoses. *Clin. Lab. Med.* **2003**, *23*, 65–85. [CrossRef]

68. Kraus, A.; Anson, K.J.; Raymond, L.D.; Martens, C.; Groveman, B.R.; Dorward, D.W.; Caughey, B. Prion protein prolines 102 and 105 and the surrounding lysine cluster impede amyloid formation. *J. Biol. Chem.* **2015**, *290*, 21510–21522. [CrossRef] [PubMed]

69. Wadsworth, J.D.; Asante, E.A.; Desbruslais, M.; Linehan, J.M.; Joiner, S.; Gowland, I.; Welch, J.; Stone, L.; Lloyd, S.E.; Hill, A.F.; et al. Human prion protein with valine 129 prevents expression of variant cjd phenotype. *Science* **2004**, *306*, 1793–1796. [CrossRef] [PubMed]

70. Pham, N.; Yin, S.; Yu, S.; Wong, P.; Kang, S.C.; Li, C.; Sy, M.S. Normal cellular prion protein with a methionine at position 129 has a more exposed helix 1 and is more prone to aggregate. *Biochem. Biophys. Res. Commun.* **2008**, *368*, 875–881. [CrossRef] [PubMed]

71. Moore, R.C.; Hope, J.; McBride, P.A.; McConnell, I.; Selfridge, J.; Melton, D.W.; Manson, J.C. Mice with gene targetted prion protein alterations show that prnp, sinc and prni are congruent. *Nat. Genet.* **1998**, *18*, 118–125. [CrossRef] [PubMed]

72. Cortez, L.M.; Kumar, J.; Renault, L.; Young, H.S.; Sim, V.L. Mouse prion protein polymorphism phe-108/val-189 affects the kinetics of fibril formation and the response to seeding: Evidence for a two-step nucleation polymerization mechanism. *J. Biol. Chem.* **2013**, *288*, 4772–4781. [CrossRef] [PubMed]

73. Canello, T.; Engelstein, R.; Moshel, O.; Xanthopoulos, K.; Juanes, M.E.; Langeveld, J.; Sklaviadis, T.; Gasset, M.; Gabizon, R. Methionine sulfoxides on prpsc: A prion-specific covalent signature. *Biochemistry* **2008**, *47*, 8866–8873. [CrossRef] [PubMed]

74. Canello, T.; Frid, K.; Gabizon, R.; Lisa, S.; Friedler, A.; Moskovitz, J.; Gasset, M.; Gabizon, R. Oxidation of helix-3 methionines precedes the formation of pk resistant prp. *PLoS Pathog.* **2010**, *6*, e1000977. [CrossRef] [PubMed]

75. Requena, J.R.; Dimitrova, M.N.; Legname, G.; Teijeira, S.; Prusiner, S.B.; Levine, R.L. Oxidation of methionine residues in the prion protein by hydrogen peroxide. *Arch. Biochem. Biophys.* **2004**, *432*, 188–195. [CrossRef] [PubMed]

76. Breydo, L.; Bocharova, O.V.; Makarava, N.; Salnikov, V.V.; Anderson, M.; Baskakov, I.V. Methionine oxidation interferes with conversion of the prion protein into the fibrillar proteinase k-resistant conformation. *Biochemistry* **2005**, *44*, 15534–15543. [CrossRef] [PubMed]

77. Wang, F.; Wang, X.; Orru, C.D.; Groveman, B.R.; Surewicz, K.; Abskharon, R.; Imamura, M.; Yokoyama, T.; Kim, Y.S.; Vander Stel, K.J.; et al. Self-propagating, protease-resistant, recombinant prion protein conformers with or without in vivo pathogenicity. *PLoS Pathog.* **2017**, *13*, e1006491. [CrossRef] [PubMed]

78. Safar, J.; Wille, H.; Itri, V.; Groth, D.; Serban, H.; Torchia, M.; Cohen, F.E.; Prusiner, S.B. Eight prion strains have prp(sc) molecules with different conformations. *Nat. Med.* **1998**, *4*, 1157–1165. [CrossRef] [PubMed]

79. Timmes, A.G.; Moore, R.A.; Fischer, E.R.; Priola, S.A. Recombinant prion protein refolded with lipid and rna has the biochemical hallmarks of a prion but lacks in vivo infectivity. *PLoS ONE* **2013**, *8*, e71081. [CrossRef] [PubMed]

80. Barron, R.M.; King, D.; Jeffrey, M.; McGovern, G.; Agarwal, S.; Gill, A.C.; Piccardo, P. Prp aggregation can be seeded by pre-formed recombinant prp amyloid fibrils without the replication of infectious prions. *Acta Neuropathol.* **2016**, *132*, 611–624. [CrossRef] [PubMed]

81. Makarava, N.; Kovacs, G.G.; Savtchenko, R.; Alexeeva, I.; Budka, H.; Rohwer, R.G.; Baskakov, I.V. Genesis of mammalian prions: From non-infectious amyloid fibrils to a transmissible prion disease. *PLoS Pathog.* **2011**, *7*, e1002419. [CrossRef] [PubMed]

82. Ghaemmaghami, S.; Colby, D.W.; Nguyen, H.O.; Hayashi, S.; Oehler, A.; DeArmond, S.J.; Prusiner, S.B. Convergent replication of mouse synthetic prion strains. *Am. J. Pathol.* **2013**, *182*, 866–874. [CrossRef] [PubMed]

83. Zhang, Y.; Wang, F.; Wang, X.; Zhang, Z.; Xu, Y.; Yu, G.; Yuan, C.; Ma, J. Comparison of 2 synthetically generated recombinant prions. *Prion* **2014**, *8*, 215–220. [CrossRef]

84. Deleault, N.R.; Walsh, D.J.; Piro, J.R.; Wang, F.; Wang, X.; Ma, J.; Rees, J.R.; Supattapone, S. Cofactor molecules maintain infectious conformation and restrict strain properties in purified prions. *Proc. Natl. Acad. Sci. USA* **2012**, *109*, E1938–E1946. [CrossRef] [PubMed]

85. Miller, M.B.; Wang, D.W.; Wang, F.; Noble, G.P.; Ma, J.; Woods, V.L., Jr.; Li, S.; Supattapone, S. Cofactor molecules induce structural transformation during infectious prion formation. *Structure* **2013**, *21*, 2061–2068. [CrossRef] [PubMed]

86. Supattapone, S. Elucidating the role of cofactors in mammalian prion propagation. *Prion* **2014**, *8*, 100–105. [CrossRef] [PubMed]

87. Supattapone, S. Synthesis of high titer infectious prions with cofactor molecules. *J. Biol. Chem.* **2014**, *289*, 19850–19854. [CrossRef] [PubMed]

88. Noble, G.P.; Wang, D.W.; Walsh, D.J.; Barone, J.R.; Miller, M.B.; Nishina, K.A.; Li, S.; Supattapone, S. A structural and functional comparison between infectious and non-infectious autocatalytic recombinant prp conformers. *PLoS Pathog.* **2015**, *11*, e1005017. [CrossRef] [PubMed]

89. Choi, J.K.; Cali, I.; Surewicz, K.; Kong, Q.; Gambetti, P.; Surewicz, W.K. Amyloid fibrils from the n-terminal prion protein fragment are infectious. *Proc. Natl. Acad. Sci. USA* **2016**, *113*, 13851–13856. [CrossRef] [PubMed]

90. Vazquez-Fernandez, E.; Vos, M.R.; Afanasyev, P.; Cebey, L.; Sevillano, A.M.; Vidal, E.; Rosa, I.; Renault, L.; Ramos, A.; Peters, P.J.; et al. The structural architecture of an infectious mammalian prion using electron cryomicroscopy. *PLoS Pathog.* **2016**, *12*, e1005835. [CrossRef] [PubMed]

91. Julien, O.; Chatterjee, S.; Thiessen, A.; Graether, S.P.; Sykes, B.D. Differential stability of the bovine prion protein upon urea unfolding. *Protein Sci.* **2009**, *18*, 2172–2182. [CrossRef] [PubMed]

92. Singh, J.; Udgaonkar, J.B. Dissection of conformational conversion events during prion amyloid fibril formation using hydrogen exchange and mass spectrometry. *J. Mol. Biol.* **2013**, *425*, 3510–3521. [CrossRef] [PubMed]

93. Muller, H.; Brener, O.; Andreoletti, O.; Piechatzek, T.; Willbold, D.; Legname, G.; Heise, H. Progress towards structural understanding of infectious sheep prp-amyloid. *Prion* **2014**, *8*, 344–358. [CrossRef] [PubMed]

94. Collins, S.; Boyd, A.; Fletcher, A.; Gonzales, M.F.; McLean, C.A.; Masters, C.L. Recent advances in the pre-mortem diagnosis of creutzfeldt-jakob disease. *J. Clin. Neurosci.* **2000**, *7*, 195–202. [CrossRef] [PubMed]

95. Farquhar, C.F.; Dornan, J.; Somerville, R.A.; Tunstall, A.M.; Hope, J. Effect of sinc genotype, agent isolate and route of infection on the accumulation of protease-resistant prp in non-central nervous system tissues during the development of murine scrapie. *J. Gen. Virol.* **1994**, *75 Pt 3*, 495–504. [CrossRef] [PubMed]

96. Farquhar, C.F.; Dornan, J.; Moore, R.C.; Somerville, R.A.; Tunstall, A.M.; Hope, J. Protease-resistant prp deposition in brain and non-central nervous system tissues of a murine model of bovine spongiform encephalopathy. *J. Gen. Virol.* **1996**, *77 Pt 8*, 1941–1946. [CrossRef] [PubMed]

97. Fraser, H.; Dickinson, A.G. Pathogenesis of scrapie in the mouse: The role of the spleen. *Nature* **1970**, *226*, 462–463. [CrossRef] [PubMed]

98. Fraser, H.; Dickinson, A.G. Studies of the lymphoreticular system in the pathogenesis of scrapie: The role of spleen and thymus. *J. Comp. Pathol.* **1978**, *88*, 563–573. [CrossRef]

99. Zanusso, G.; Monaco, S.; Pocchiari, M.; Caughey, B. Advanced tests for early and accurate diagnosis of creutzfeldt-jakob disease. *Nat. Rev. Neurol.* **2016**, *12*, 325–333. [CrossRef] [PubMed]

100. Brown, P.; Rohwer, R.G.; Dunstan, B.C.; MacAuley, C.; Gajdusek, D.C.; Drohan, W.N. The distribution of infectivity in blood components and plasma derivatives in experimental models of transmissible spongiform encephalopathy. *Transfusion* **1998**, *38*, 810–816. [CrossRef] [PubMed]

101. Reichl, H.; Balen, A.; Jansen, C.A. Prion transmission in blood and urine: What are the implications for recombinant and urinary-derived gonadotrophins? *Hum. Reprod.* **2002**, *17*, 2501–2508. [CrossRef] [PubMed]

102. Ward, H.J.; MacKenzie, J.M.; Llewelyn, C.A.; Knight, R.S.; Hewitt, P.E.; Connor, N.; Molesworth, A.; Will, R.G. Variant creutzfeldt-jakob disease and exposure to fractionated plasma products. *Vox Sang.* **2009**, *97*, 207–210. [CrossRef] [PubMed]

103. Bonda, D.J.; Manjila, S.; Mehndiratta, P.; Khan, F.; Miller, B.R.; Onwuzulike, K.; Puoti, G.; Cohen, M.L.; Schonberger, L.B.; Cali, I. Human prion diseases: Surgical lessons learned from iatrogenic prion transmission. *Neurosurg. Focus* **2016**, *41*, E10. [CrossRef] [PubMed]

104. Moda, F.; Gambetti, P.; Notari, S.; Concha-Marambio, L.; Catania, M.; Park, K.W.; Maderna, E.; Suardi, S.; Haik, S.; Brandel, J.P.; et al. Prions in the urine of patients with variant creutzfeldt-jakob disease. *N. Engl. J. Med.* **2014**, *371*, 530–539. [CrossRef] [PubMed]

105. Foutz, A.; Appleby, B.S.; Hamlin, C.; Liu, X.; Yang, S.; Cohen, Y.; Chen, W.; Blevins, J.; Fausett, C.; Wang, H.; et al. Diagnostic and prognostic value of human prion detection in cerebrospinal fluid. *Ann. Neurol.* **2017**, *81*, 79–92. [CrossRef] [PubMed]

106. Zobeley, E.; Flechsig, E.; Cozzio, A.; Enari, M.; Weissmann, C. Infectivity of scrapie prions bound to a stainless steel surface. *Mol. Med.* **1999**, *5*, 240–243. [PubMed]

107. Flechsig, E.; Hegyi, I.; Enari, M.; Schwarz, P.; Collinge, J.; Weissmann, C. Transmission of scrapie by steel-surface-bound prions. *Mol. Med.* **2001**, *7*, 679–684. [PubMed]

108. Edgeworth, J.A.; Farmer, M.; Sicilia, A.; Tavares, P.; Beck, J.; Campbell, T.; Lowe, J.; Mead, S.; Rudge, P.; Collinge, J.; et al. Detection of prion infection in variant creutzfeldt-jakob disease: A blood-based assay. *Lancet* **2011**, *377*, 487–493. [CrossRef]

109. Edgeworth, J.A.; Jackson, G.S.; Clarke, A.R.; Weissmann, C.; Collinge, J. Highly sensitive, quantitative cell-based assay for prions adsorbed to solid surfaces. *Proc. Natl. Acad. Sci. USA* **2009**, *106*, 3479–3483. [CrossRef] [PubMed]

110. Sawyer, E.B.; Edgeworth, J.A.; Thomas, C.; Collinge, J.; Jackson, G.S. Preclinical detection of infectivity and disease-specific prp in blood throughout the incubation period of prion disease. *Sci. Rep.* **2015**, *5*, 17742. [CrossRef] [PubMed]

111. Saa, P.; Castilla, J.; Soto, C. Ultra-efficient replication of infectious prions by automated protein misfolding cyclic amplification. *J. Biol. Chem.* **2006**, *281*, 35245–35252. [CrossRef] [PubMed]

112. Fernández-Borges, N.; Erana, H.; Elezgarai, S.R.; Harrathi, C.; Venegas, V.; Castilla, J. A quick method to evaluate the effect of the amino acid sequence in the misfolding proneness of the prion protein. In *Prions: Methods and Protocols*; Lawson, V.A., Ed.; Springer: New York, NY, USA, 2017.

113. Atarashi, R.; Moore, R.A.; Sim, V.L.; Hughson, A.G.; Dorward, D.W.; Onwubiko, H.A.; Priola, S.A.; Caughey, B. Ultrasensitive detection of scrapie prion protein using seeded conversion of recombinant prion protein. *Nat. Methods* **2007**, *4*, 645–650. [CrossRef] [PubMed]

114. Zaman, M.H. Misfolding dynamics of human prion protein. *Mol. Cell. Biomech.* **2005**, *2*, 179–190. [PubMed]

115. Colby, D.W.; Zhang, Q.; Wang, S.; Groth, D.; Legname, G.; Riesner, D.; Prusiner, S.B. Prion detection by an amyloid seeding assay. *Proc. Natl. Acad. Sci. USA* **2007**, *104*, 20914–20919. [CrossRef] [PubMed]

116. Atarashi, R.; Wilham, J.M.; Christensen, L.; Hughson, A.G.; Moore, R.A.; Johnson, L.M.; Onwubiko, H.A.; Priola, S.A.; Caughey, B. Simplified ultrasensitive prion detection by recombinant prp conversion with shaking. *Nat. Methods* **2008**, *5*, 211–212. [CrossRef] [PubMed]

117. Wilham, J.M.; Orru, C.D.; Bessen, R.A.; Atarashi, R.; Sano, K.; Race, B.; Meade-White, K.D.; Taubner, L.M.; Timmes, A.; Caughey, B. Rapid end-point quantitation of prion seeding activity with sensitivity comparable to bioassays. *PLoS Pathog.* **2010**, *6*, e1001217. [CrossRef] [PubMed]

118. McGuire, L.I.; Peden, A.H.; Orru, C.D.; Wilham, J.M.; Appleford, N.E.; Mallinson, G.; Andrews, M.; Head, M.W.; Caughey, B.; Will, R.G.; et al. Real time quaking-induced conversion analysis of cerebrospinal fluid in sporadic creutzfeldt-jakob disease. *Ann. Neurol.* **2012**, *72*, 278–285. [CrossRef] [PubMed]

119. Peden, A.H.; McGuire, L.I.; Appleford, N.E.; Mallinson, G.; Wilham, J.M.; Orru, C.D.; Caughey, B.; Ironside, J.W.; Knight, R.S.; Will, R.G.; et al. Sensitive and specific detection of sporadic creutzfeldt-jakob disease brain prion protein using real-time quaking-induced conversion. *J. Gen. Virol.* **2012**, *93*, 438–449. [CrossRef] [PubMed]

120. Sano, K.; Satoh, K.; Atarashi, R.; Takashima, H.; Iwasaki, Y.; Yoshida, M.; Sanjo, N.; Murai, H.; Mizusawa, H.; Schmitz, M.; et al. Early detection of abnormal prion protein in genetic human prion diseases now possible using real-time quic assay. *PLoS ONE* **2013**, *8*, e54915. [CrossRef] [PubMed]

121. Orru, C.D.; Hughson, A.G.; Race, B.; Raymond, G.J.; Caughey, B. Time course of prion seeding activity in cerebrospinal fluid of scrapie-infected hamsters after intratongue and intracerebral inoculations. *J. Clin. Microbiol.* **2012**, *50*, 1464–1466. [CrossRef] [PubMed]

122. Orru, C.D.; Groveman, B.R.; Raymond, L.D.; Hughson, A.G.; Nonno, R.; Zou, W.; Ghetti, B.; Gambetti, P.; Caughey, B. Bank vole prion protein as an apparently universal substrate for rt-quic-based detection and discrimination of prion strains. *PLoS Pathog.* **2015**, *11*, e1004983.

123. Dassanayake, R.P.; Orru, C.D.; Hughson, A.G.; Caughey, B.; Graca, T.; Zhuang, D.; Madsen-Bouterse, S.A.; Knowles, D.P.; Schneider, D.A. Sensitive and specific detection of classical scrapie prions in the brains of goats by real-time quaking-induced conversion. *J. Gen. Virol.* **2016**, *97*, 803–812. [CrossRef] [PubMed]

124. Haley, N.J.; Siepker, C.; Walter, W.D.; Thomsen, B.V.; Greenlee, J.J.; Lehmkuhl, A.D.; Richt, J.A. Antemortem detection of chronic wasting disease prions in nasal brush collections and rectal biopsy specimens from white-tailed deer by real-time quaking-induced conversion. *J. Clin. Microbiol.* **2016**, *54*, 1108–1116. [CrossRef] [PubMed]

125. Bessen, R.A.; Shearin, H.; Martinka, S.; Boharski, R.; Lowe, D.; Wilham, J.M.; Caughey, B.; Wiley, J.A. Prion shedding from olfactory neurons into nasal secretions. *PLoS Pathog.* **2010**, *6*, e1000837. [CrossRef] [PubMed]

126. Orru, C.D.; Bongianni, M.; Tonoli, G.; Ferrari, S.; Hughson, A.G.; Groveman, B.R.; Fiorini, M.; Pocchiari, M.; Monaco, S.; Caughey, B.; et al. A test for creutzfeldt-jakob disease using nasal brushings. *N. Engl. J. Med.* **2014**, *371*, 519–529. [CrossRef] [PubMed]

127. Giles, K.; Olson, S.H.; Prusiner, S.B. Developing therapeutics for prp prion diseases. *Cold Spring Harb. Perspect. Med.* **2017**, *7*. [CrossRef] [PubMed]

128. Caughey, B.; Race, R.E. Potent inhibition of scrapie-associated prp accumulation by congo red. *J. Neurochem.* **1992**, *59*, 768–771. [CrossRef] [PubMed]

129. Margalith, I.; Suter, C.; Ballmer, B.; Schwarz, P.; Tiberi, C.; Sonati, T.; Falsig, J.; Nystrom, S.; Hammarstrom, P.; Aslund, A.; et al. Polythiophenes inhibit prion propagation by stabilizing prion protein (prp) aggregates. *J. Biol. Chem.* **2012**, *287*, 18872–18887. [CrossRef] [PubMed]

130. Korth, C.; May, B.C.; Cohen, F.E.; Prusiner, S.B. Acridine and phenothiazine derivatives as pharmacotherapeutics for prion disease. *Proc. Natl. Acad. Sci. USA* **2001**, *98*, 9836–9841. [CrossRef] [PubMed]

131. Cortez, L.M.; Campeau, J.; Norman, G.; Kalayil, M.; Van der Merwe, J.; McKenzie, D.; Sim, V.L. Bile acids reduce prion conversion, reduce neuronal loss, and prolong male survival in models of prion disease. *J. Virol.* **2015**, *89*, 7660–7672. [CrossRef] [PubMed]

132. Dinkel, K.D.; Stanton, J.B.; Boykin, D.W.; Stephens, C.E.; Madsen-Bouterse, S.A.; Schneider, D.A. Antiprion activity of db772 and related monothiophene- and furan-based analogs in a persistently infected ovine microglia culture system. *Antimicrob. Agents Chemother.* **2016**, *60*, 5467–5482. [CrossRef] [PubMed]

133. Barret, A.; Tagliavini, F.; Forloni, G.; Bate, C.; Salmona, M.; Colombo, L.; De Luigi, A.; Limido, L.; Suardi, S.; Rossi, G.; et al. Evaluation of quinacrine treatment for prion diseases. *J. Virol.* **2003**, *77*, 8462–8469. [CrossRef] [PubMed]

134. Haik, S.; Brandel, J.P.; Salomon, D.; Sazdovitch, V.; Delasnerie-Laupretre, N.; Laplanche, J.L.; Faucheux, B.A.; Soubrie, C.; Boher, E.; Belorgey, C.; et al. Compassionate use of quinacrine in creutzfeldt-jakob disease fails to show significant effects. *Neurology* **2004**, *63*, 2413–2415. [CrossRef] [PubMed]

135. Collinge, J.; Gorham, M.; Hudson, F.; Kennedy, A.; Keogh, G.; Pal, S.; Rossor, M.; Rudge, P.; Siddique, D.; Spyer, M.; et al. Safety and efficacy of quinacrine in human prion disease (prion-1 study): A patient-preference trial. *Lancet Neurol.* **2009**, *8*, 334–344. [CrossRef]

136. Pollera, C.; Carcassola, G.; Ponti, W.; Poli, G. Development of in vitro cell cultures for the evaluation of molecules with antiprionic activity. *Vet. Res. Commun* **2003**, *27* (Suppl. S1), 719–721. [CrossRef] [PubMed]

137. Priola, S.A.; Raines, A.; Caughey, W.S. Porphyrin and phthalocyanine antiscrapie compounds. *Science* **2000**, *287*, 1503–1506. [CrossRef] [PubMed]

138. Nicoll, A.J.; Trevitt, C.R.; Tattum, M.H.; Risse, E.; Quarterman, E.; Ibarra, A.A.; Wright, C.; Jackson, G.S.; Sessions, R.B.; Farrow, M.; et al. Pharmacological chaperone for the structured domain of human prion protein. *Proc. Natl. Acad. Sci. USA* **2010**, *107*, 17610–17615. [CrossRef] [PubMed]

139. Massignan, T.; Cimini, S.; Stincardini, C.; Cerovic, M.; Vanni, I.; Elezgarai, S.R.; Moreno, J.; Stravalaci, M.; Negro, A.; Sangiovanni, V.; et al. A cationic tetrapyrrole inhibits toxic activities of the cellular prion protein. *Sci. Rep.* **2016**, *6*, 23180. [CrossRef] [PubMed]

140. Vieira, T.C.; Cordeiro, Y.; Caughey, B.; Silva, J.L. Heparin binding confers prion stability and impairs its aggregation. *FASEB J.* **2014**, *28*, 2667–2676. [CrossRef] [PubMed]
141. Kocisko, D.A.; Bertholet, N.; Moore, R.A.; Caughey, B.; Vaillant, A. Identification of prion inhibitors by a fluorescence-polarization-based competitive binding assay. *Anal. Biochem.* **2007**, *363*, 154–156. [CrossRef] [PubMed]
142. Risse, E.; Nicoll, A.J.; Taylor, W.A.; Wright, D.; Badoni, M.; Yang, X.; Farrow, M.A.; Collinge, J. Identification of a compound that disrupts binding of amyloid-beta to the prion protein using a novel fluorescence-based assay. *J. Biol. Chem.* **2015**, *290*, 17020–17028. [CrossRef] [PubMed]
143. Frostell-Karlsson, A.; Reameus, A.; Roos, H.; Andersson, K.; Borg, P.; Hämäläinen, M.; Karlsson, R. Biosensor analysis of the interaction between immobilized human serum albumin and drug compounds for prediction of human serum albumin binding levels. *J. Med. Chem.* **2000**, *43*, 1986–1992. [CrossRef] [PubMed]
144. Kawatake, S.; Nishimura, Y.; Sakaguchi, S.; Iwaki, T.; Doh-ura, K. Surface plasmon resonance analysis for the screening of anti-prion compounds. *Biol. Pharm. Bull.* **2006**, *29*, 927–932. [CrossRef] [PubMed]
145. Breydo, L.; Bocharova, O.V.; Baskakov, I.V. Semiautomated cell-free conversion of prion protein: Applications for high-throughput screening of potential antiprion drugs. *Anal. Biochem.* **2005**, *339*, 165–173. [CrossRef] [PubMed]
146. Hyeon, J.W.; Kim, S.Y.; Lee, S.M.; Lee, J.; An, S.S.; Lee, M.K.; Lee, Y.S. Anti-prion screening for acridine, dextran, and tannic acid using real time-quaking induced conversion: A comparison with prpsc-infected cell screening. *PLoS ONE* **2017**, *12*, e0170266. [CrossRef] [PubMed]
147. Bertsch, U.; Winklhofer, K.F.; Hirschberger, T.; Bieschke, J.; Weber, P.; Hartl, F.U.; Tavan, P.; Tatzelt, J.; Kretzschmar, H.A.; Giese, A. Systematic identification of antiprion drugs by high-throughput screening based on scanning for intensely fluorescent targets. *J. Virol.* **2005**, *79*, 7785–7791. [CrossRef] [PubMed]
148. White, M.D.; Farmer, M.; Mirabile, I.; Brandner, S.; Collinge, J.; Mallucci, G.R. Single treatment with rnai against prion protein rescues early neuronal dysfunction and prolongs survival in mice with prion disease. *Proc. Natl. Acad. Sci. USA* **2008**, *105*, 10238–10243. [CrossRef] [PubMed]
149. Yuan, J.; Zhan, Y.A.; Abskharon, R.; Xiao, X.; Martinez, M.C.; Zhou, X.; Kneale, G.; Mikol, J.; Lehmann, S.; Surewicz, W.K.; et al. Recombinant human prion protein inhibits prion propagation in vitro. *Sci. Rep.* **2013**, *3*, 2911. [CrossRef] [PubMed]
150. Skinner, P.J.; Kim, H.O.; Bryant, D.; Kinzel, N.J.; Reilly, C.; Priola, S.A.; Ward, A.E.; Goodman, P.A.; Olson, K.; Seelig, D.M. Treatment of prion disease with heterologous prion proteins. *PLoS ONE* **2015**, *10*, e0131993. [CrossRef] [PubMed]
151. Xiao, X.L.; Jiang, H.Y.; Zhang, J.; Han, J.; Nie, K.; Zhou, X.B.; Huang, Y.X.; Chen, L.; Zhou, W.; Zhang, B.Y.; et al. Preparation of monoclonal antibodies against prion proteins with full-length hamster prp. *Biomed. Environ. Sci.* **2005**, *18*, 273–280. [PubMed]
152. Ishibashi, D.; Yamanaka, H.; Yamaguchi, N.; Yoshikawa, D.; Nakamura, R.; Okimura, N.; Yamaguchi, Y.; Shigematsu, K.; Katamine, S.; Sakaguchi, S. Immunization with recombinant bovine but not mouse prion protein delays the onset of disease in mice inoculated with a mouse-adapted prion. *Vaccine* **2007**, *25*, 985–992. [CrossRef] [PubMed]
153. Bruce, M.; Chree, A.; McConnell, I.; Foster, J.; Pearson, G.; Fraser, H. Transmission of bovine spongiform encephalopathy and scrapie to mice: Strain variation and the species barrier. *Philos. Trans. R. Soc. Lond. B Biol. Sci.* **1994**, *343*, 405–411. [CrossRef] [PubMed]
154. Bartz, J.C.; McKenzie, D.I.; Bessen, R.A.; Marsh, R.F.; Aiken, J.M. Transmissible mink encephalopathy species barrier effect between ferret and mink: Prp gene and protein analysis. *J. Gen. Virol.* **1994**, *75 Pt 11*, 2947–2953. [CrossRef] [PubMed]
155. Caughey, B.; Raymond, G.J.; Bessen, R.A. Strain-dependent differences in beta-sheet conformations of abnormal prion protein. *J. Biol. Chem.* **1998**, *273*, 32230–32235. [CrossRef] [PubMed]
156. Fernandez-Borges, N.; de Castro, J.; Castilla, J. In vitro studies of the transmission barrier. *Prion* **2009**, *3*, 220–223. [CrossRef] [PubMed]
157. Castilla, J.; Morales, R.; Saa, P.; Barria, M.; Gambetti, P.; Soto, C. Cell-free propagation of prion strains. *EMBO J.* **2008**, *27*, 2557–2566. [CrossRef] [PubMed]
158. Kim, J.I.; Surewicz, K.; Gambetti, P.; Surewicz, W.K. The role of glycophosphatidylinositol anchor in the amplification of the scrapie isoform of prion protein in vitro. *FEBS Lett.* **2009**, *583*, 3671–3675. [CrossRef] [PubMed]

159. Lee, C.I.; Yang, Q.; Perrier, V.; Baskakov, I.V. The dominant-negative effect of the q218k variant of the prion protein does not require protein X. *Protein Sci.* **2007**, *16*, 2166–2173. [CrossRef] [PubMed]
160. Seelig, D.M.; Goodman, P.A.; Skinner, P.J. Potential approaches for heterologous prion protein treatment of prion diseases. *Prion* **2016**, *10*, 18–24. [CrossRef] [PubMed]
161. Diaz-Espinoza, R.; Morales, R.; Concha-Marambio, L.; Moreno-Gonzalez, I.; Moda, F.; Soto, C. Treatment with a non-toxic, self-replicating anti-prion delays or prevents prion disease in vivo. *Mol. Psychiatry* **2017**. [CrossRef] [PubMed]
162. Williamson, R.A.; Peretz, D.; Smorodinsky, N.; Bastidas, R.; Serban, H.; Mehlhorn, I.; DeArmond, S.J.; Prusiner, S.B.; Burton, D.R. Circumventing tolerance to generate autologous monoclonal antibodies to the prion protein. *Proc. Natl. Acad. Sci. USA* **1996**, *93*, 7279–7282. [CrossRef] [PubMed]
163. Prusiner, S.B.; Groth, D.; Serban, A.; Koehler, R.; Foster, D.; Torchia, M.; Burton, D.; Yang, S.L.; DeArmond, S.J. Ablation of the prion protein (prp) gene in mice prevents scrapie and facilitates production of anti-prp antibodies. *Proc. Natl. Acad. Sci. USA* **1993**, *90*, 10608–10612. [CrossRef] [PubMed]
164. Gilch, S.; Wopfner, F.; Renner-Muller, I.; Kremmer, E.; Bauer, C.; Wolf, E.; Brem, G.; Groschup, M.H.; Schatzl, H.M. Polyclonal anti-prp auto-antibodies induced with dimeric prp interfere efficiently with prpsc propagation in prion-infected cells. *J. Biol. Chem.* **2003**, *278*, 18524–18531. [CrossRef] [PubMed]
165. Sigurdsson, E.M.; Brown, D.R.; Daniels, M.; Kascsak, R.J.; Kascsak, R.; Carp, R.; Meeker, H.C.; Frangione, B.; Wisniewski, T. Immunization delays the onset of prion disease in mice. *Am. J. Pathol.* **2002**, *161*, 13–17. [CrossRef]
166. White, A.R.; Enever, P.; Tayebi, M.; Mushens, R.; Linehan, J.; Brandner, S.; Anstee, D.; Collinge, J.; Hawke, S. Monoclonal antibodies inhibit prion replication and delay the development of prion disease. *Nature* **2003**, *422*, 80–83. [CrossRef] [PubMed]
167. Xanthopoulos, K.; Lagoudaki, R.; Kontana, A.; Kyratsous, C.; Panagiotidis, C.; Grigoriadis, N.; Yiangou, M.; Sklaviadis, T. Immunization with recombinant prion protein leads to partial protection in a murine model of tses through a novel mechanism. *PLoS ONE* **2013**, *8*, e59143. [CrossRef] [PubMed]
168. Zahn, R.; von Schroetter, C.; Wuthrich, K. Human prion proteins expressed in escherichia coli and purified by high-affinity column refolding. *FEBS Lett.* **1997**, *417*, 400–404. [CrossRef]

pathogens

MDPI

Review

How do PrP^{Sc} Prions Spread between Host Species, and within Hosts?

Neil A. Mabbott

The Roslin Institute & Royal (Dick) School of Veterinary Sciences, University of Edinburgh, Easter Bush, Midlothian EH25 9RG, UK; neil.mabbott@roslin.ed.ac.uk; Tel.: +44-131-651-9100

Received: 2 November 2017; Accepted: 21 November 2017; Published: 24 November 2017

Abstract: Prion diseases are sub-acute neurodegenerative diseases that affect humans and some domestic and free-ranging animals. Infectious prion agents are considered to comprise solely of abnormally folded isoforms of the cellular prion protein known as PrP^{Sc}. Pathology during prion disease is restricted to the central nervous system where it causes extensive neurodegeneration and ultimately leads to the death of the host. The first half of this review provides a thorough account of our understanding of the various ways in which PrP^{Sc} prions may spread between individuals within a population, both horizontally and vertically. Many natural prion diseases are acquired peripherally, such as by oral exposure, lesions to skin or mucous membranes, and possibly also via the nasal cavity. Following peripheral exposure, some prions accumulate to high levels within the secondary lymphoid organs as they make their journey from the site of infection to the brain, a process termed neuroinvasion. The replication of PrP^{Sc} prions within secondary lymphoid organs is important for their efficient spread to the brain. The second half of this review describes the key tissues, cells and molecules which are involved in the propagation of PrP^{Sc} prions from peripheral sites of exposure (such as the lumen of the intestine) to the brain. This section also considers how additional factors such as inflammation and aging might influence prion disease susceptibility.

Keywords: prions; prion protein; PrP^{Sc}; horizontal transmission; vertical transmission; secondary lymphoid organs; intestine; central nervous system

---

1. Introduction

The prion diseases are a unique group of sub-acute neurodegenerative diseases that affect humans and certain domestic and free-ranging animals. The infectious prion agent is considered to comprise solely of abnormally folded isoforms of the host-encoded cellular prion protein PrP^{C}, termed prion disease-specific PrP^{Sc}. Prion infectivity co-purifies with PrP^{Sc} and appears to constitute the major component of the infectious agent [1,2]. The pathology caused during prion disease is considered to be restricted almost entirely to the central nervous system (CNS), where it causes extensive neurodegeneration which ultimately leads to death. The characteristic histopathological features of CNS prion diseases include vacuolation in the brain (spongiform pathology), neurodegeneraton, microgliosis, astrocytosis, and abnormal accumulations of PrP^{Sc}.

Several different forms of prion diseases have been described: spontaneous, genetic, or acquired through various routes of exposure (Table 1). Many prion diseases such as natural sheep scrapie, chronic wasting disease (CWD) in cervid species, and bovine spongiform encephalopathy (BSE) in cattle are considered to be orally-acquired; for example through the consumption of prion-contaminated food or pasture. Examples of vertical prion transmission between an infected mother to the developing fetus or offspring have also been reported. In addition to their important health and economic impacts to livestock industries, some prion disease also have zoonotic potential. The consumption of BSE-contaminated food during the UK BSE epidemic was responsible for the occurrence of a novel human prion disease, variant Creutzfeldt-Jakob (vCJD), which was predominantly described in young adults. Whether other animal prion diseases also have zoonotic potential is an important human health concern. Sporadic CJD (sCJD), in contrast, typically affects elderly individuals and has an unknown etiology and incidence of approximately 1/million population per annum throughout the world. Whether this prion disease is also acquired is uncertain, but a study using experimental mice has proposed that sheep scrapie prions may also have zoonotic potential and cause a disease in the recipients with characteristics identical to sCJD [3]. Instances of accidental iatrogenic prion transmission between humans have also been documented following the transplantation of sCJD-contaminated tissues (dura mater grafts) or tissue products (pituitary-derived human growth hormones), transfusion of blood or blood products from vCJD-infected donors, or use of prion-contaminated surgical instruments or medical devices.

Table 1. PrP^{Sc} prion diseases of humans and animals.

Prion Disease	Affected Species	Transmission Route
Iatrogenic Creutzfeldt-Jakob disease (CJD)	Human	Accidental medical exposure to CJD-contaminated tissues or tissue products
Sporadic Creutzfeldt-Jakob disease	Human	Unknown. Theories include somatic mutation or spontaneous conversion of PrP^c to PrP^{Sc}
Variant Creutzfeldt-Jakob disease	Human	Ingestion of BSE-contaminated food or transfusion of blood or blood products from variant CJD-infected blood donor
Familial Creutzfeldt-Jakob disease	Human	Germ-line mutations of the *PRNP* gene
Gerstmann-Straussler-Scheinker syndrome	Human	Germ-line mutations of the *PRNP* gene
Kuru	Human	Ritualistic cannibalism
Fatal familial insomnia	Human	Germ-line mutations of the *PRNP* gene
Bovine spongiform encephalopathy	Cattle	Ingestion of contaminated food
Scrapie	Sheep, goats, mouflon	Acquired. Ingestion, horizontal transmission, vertical transmission unclear
Chronic wasting disease	Elk, deer, moose	Acquired, ingestion, horizontal transmission, vertical transmission unclear
Transmissible mink encephalopathy	Mink	Acquired (ingestion) source unknown
Feline spongiform encephalopathy	Domestic and zoological cats	Ingestion of BSE-contaminated food
Exotic ungulate encephalopathy	Nyala, kudu	Ingestion of BSE-contaminated food

The first half of this review provides a thorough account of our current understanding of the various ways in which PrP^{Sc} prions may spread between individuals within a population, both horizontally and vertically. Following peripheral exposure, the prions often accumulate to high levels within the secondary lymphoid organs (SLO). This initial replication prion replication phase within the secondary lymphoid tissues is essential for the efficient spread of the prions from the site of exposure to the CNS, a process termed neuroinvasion. The second half of this review goes on to discuss the key tissues, cells, and molecules that facilitate the propagation of PrP^{Sc} prions from peripheral sites of exposure (such as the lumen of the intestine) to the brain. This section also describes how additional factors such as inflammation and aging can influence prion disease susceptibility.

2. Transmission of PrP^{Sc} Prions between Host Species

The precise route by which many of the natural prion diseases are acquired or transmitted between host species is uncertain. Prion diseases in animals have the potential to be transmitted by a variety of routes depending on factors such as the stage of host development and the husbandry conditions within which they are maintained. The sections below describe our understanding on contribution of a range of exposure routes summarized in Figure 1 and the factors which can affect disease susceptibility.

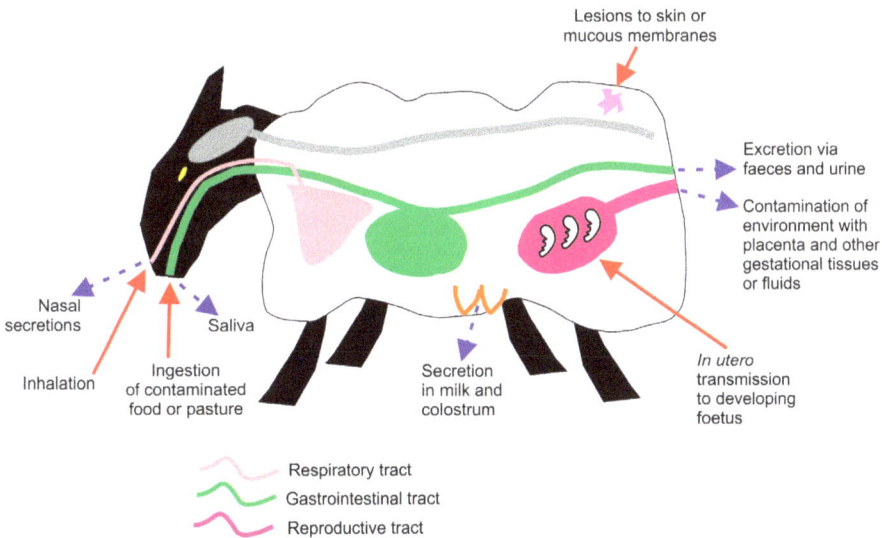

Figure 1. Cartoon summarizing the potential routes of prion exposure in animals such as sheep, and mechanisms in which prions may be disseminated between animals. Red arrows, routes of prion exposure; Broken blue arrows, routes of prion shedding or secretion from an infected animal.

2.1. Horizontal Transmission

Many natural prion diseases are horizontally transmitted between host species. Indeed, analysis of mathematical models derived from flocks of sheep affected with natural scrapie, and captive mule deer with CWD, have revealed that horizontal transmission is remarkably efficient and can play an important role in sustaining prion disease epidemics within affected populations [4,5]. The horizontal transmission of CWD between reindeer has also been demonstrated in an experimental setting [6].

2.1.1. Oral/Ingestion

Studies using experimental rodents and domestic animals (sheep, deer and cattle) indicate that the many natural prion diseases such as natural scrapie, CWD, and BSE are most likely to be orally acquired. For example, the oral consumption of meat and bone meal contaminated with BSE prions was responsible for the efficient transmission of BSE prions amongst the UK cattle herd in the late 1980s [7,8]. Indeed, the introduction of control measures to remove ruminant materials from feed was instrumental in controlling the UK cattle BSE epidemic. Unfortunately BSE has since been shown to have zoonotic potential and serious human health concerns, as the consumption of BSE prion contaminated beef products was similarly responsible for the occurrence of vCJD in humans [9,10].

Humans were not the only non-bovine species shown to be susceptible to BSE, as domestic and exotic cats [11] as well as other exotic species such as Arabian oryx, greater kudu, and nyala [12,13] also developed a BSE-related prion disease after consumption of BSE prion-contaminated food.

2.1.2. Prions Can Be Shed into the Environment and Can Remain Infectious

On the farm and amongst free-ranging animals the horizontal transmission prions most likely occurs via the ingestion of contaminated pasture. Studies of sheep with natural scrapie, cervids with CWD, and experimentally affected rodents show that prion-affected individuals can shed infectious prions into the environment via the excretion of urine [14–18], feces [19,20] and oral (saliva) [15,21–23] and nasal [24,25] secretions. Analysis of CWD-infected cervids shows that low levels of prions may be excreted in urine and feces throughout the asymptomatic phase [26]. Furthermore, factors such as inflammation within the kidney (nephritis) may enhance the amount of infectious prions excreted into the urine [27]. The chronic excretion or secretion of prions from affected animals provides the opportunity for significant environmental contamination to occur throughout the disease course. Studies show that on farms with an incidence of prion disease, the prions may persist within the environment for long periods [28]. Prions may persist in soil for at least 18 months [29,30] and can retain their infectious properties, even when bound to plants [31]. Although the mechanism of action is not known, the binding of prions to soil may also enhance their ability to infect the host after oral exposure [32]. The persistence of the prions within the farm environment also introduces practical issues for disease control, as the removal and culling of infected animals from farms on its own may be insufficient to prevent further cases of prion disease occurring when disease-free stock are reintroduced.

2.1.3. Nasal Cavity Is a Potential Portal for Prion Entry

Natural prion disease susceptible animal host species such as sheep, goats and cervids have highly developed olfactory systems which they use to detect food, select mates and sense predators. Although no natural cases of prion transmission via inhalation or the nasal cavity have been reported, experimental studies using mice, hamsters and sheep have shown that prion infection can be established by this route [33–38]. These studies imply that small amounts of soil-bound prions might be inhaled and infect the host as the animal forages for food amongst pasture or bedding.

2.1.4. Lesions to Skin and Mucous Membranes

Intact skin normally acts as a barrier against prion transmission [39], but experimental studies in mice and sheep show that skin lesions also represent an efficient route by which prion infection can be established [39,40]. Lesions to the oral and nasal mucosa similarly enhance prion disease susceptibility, most likely by increasing the efficiency of prion uptake across these epithelial surfaces [41–44].

2.1.5. Accidental Iatrogenic Transmission in Humans

Many instances of accidental iatrogenic CJD transmission have been recorded where disease was transmitted through the use of prion contaminated neurosurgical instruments and stereotactic electroencephalography electrodes, or transplantation of tissues (cornea, dura mater), or preparations (pituitary-derived growth hormone, follicular stimulating hormone) from sCJD-affected cadavers [45]. However, there is no evidence to suggest that these iatrogenic CJD patients are a risk of horizontal sCJD transmission to family members or close contacts [46]. Studies in rodents have also shown that PrP^{Sc} can accumulate in the dental tissues, suggesting the potential for iatrogenic prion transmission during invasive dental procedures such as tooth extractions or root canal treatment [47]. However, no convincing evidence has been found to suggest an increased risk of vCJD transmission due to dental treatment [48].

Data from mice experimentally infected with the mouse-adapted Fukuoka-1 strain of GSS disease have suggested that blood during the clinical phase of disease contains 100 infectious units (IU) of prion infectivity/mL of buffy coat, and approximately 10 IU/mL of plasma [49–51]. Much lower levels were detected in buffy coat during the pre-clinical phase, with infectivity undetectable in plasma. A similar level and distribution of infectivity had been shown in mice infected with vCJD [52]. Fractionation of individual blood components from prion infected mice detected low levels of prion infectivity in buffy coat, plasma, cryoprecipitate, and fraction I + II + III [49]. None was detected in association with fractions IV or V [49] or highly purified platelets [53].

Two experimental studies in sheep have shown prions can be transmitted to recipients by transfusion of blood from donor sheep which were infected with either natural scrapie or BSE [54,55]. Whole blood or buffy coat drawn during the pre-clinical and clinical phases of disease transmitted disease to at least 10% of the transfusion recipients [54,55]. These studies provided strong evidence that there may be sufficient levels of infectious prions present in the peripheral blood of some pre-clinically affected humans to transmit disease to recipients by transfusion of blood or blood products.

There has been little evidence to suggest that sCJD may be horizontally transmitted between humans by blood or blood products [56]. Several epidemiological case control, look-back, and surveillance studies on sentinel populations such as hemophiliacs have failed to demonstrate an increased risk of sCJD infection due to blood transfusion or exposure to plasma products [56–60]. However a recent study has reported prion infectivity in the plasma of two of four individuals who were infected with sCJD [61].

Several vCJD patients are known to have been blood donors. In the UK four cases of vCJD have been reported in recipients of blood or blood products derived from vCJD-infected donors [62–65]. In the first of these cases [62], a blood donation was made in 1996 and the donor, who was well at the time, went on to develop clinical vCJD, confirmed in 2001. Non-leucodepleted red cell concentrate from this donation was administered to a patient who subsequently developed vCJD. In the second case [63], the donation was made in 1999 and the donor subsequently developed vCJD and succumbed to the disease in 2001. A single unit of non-leucodepleted red cell concentrate derived from this donor was administered to a patient who died of unrelated causes in 2004. Post mortem examination revealed evidence of PrP^{Sc} accumulation only in the patient's spleen and one of the cervical lymph nodes. To reduce the risk of potential transmission of variant CJD by blood transfusion the UK implemented universal leucodepletion in 1999. This rationale was based on observations that PrP^{Sc} could be detected in lymphoid tissues of vCJD patients [66,67] implying that cells such as lymphocytes might potentially contaminate the blood-stream with prions. Estimates suggest that although leucodepletion can potentially remove approximately 42% of the prion infectivity in blood, a significant fraction remains [68]. This is consistent with the detection of both cell-associated and soluble prion infectivity in peripheral blood [49–51]. No new cases of transfusion-associated vCJD have been reported in the UK since 2007 [60].

2.2. Vertical Transmission

Depending on the host species and prion isolate, infected mothers have been shown to have the potential to transmit prion infection to their offspring. As the examples below show, maternal transmission may play a significant role in sustaining the prion disease prevalence in affected populations. For example, a study of natural scrapie-affected ewes revealed that the incidence of scrapie was increased in their offspring [69]. The risk of developing scrapie was not influenced in the offspring of scrapie-affected sires [69], consistent with the absence of detectable prion infectivity in semen from infected rams [70].

Experimental studies using Reeve's muntjac deer have also demonstrated the potential for CWD to be maternally transmitted in cervid species [71]. This was further supported in a separate postmortem study of maternal and fetal tissues collected from free-ranging Rocky Mountain elk. This study similarly concluded that mother to offspring prion transmission may contribute to the efficient transmission of CWD amongst naturally affected cervids [72].

A study of embryos collected from BSE-affected dams suggested that cattle embryos were unlikely to be infected with BSE prions, even when collected from clinically affected mothers, when the risk of maternal transmission may considered the greatest [73]. However, other studies have estimated that the maternal transmission of BSE prions may occur in approximately 10% of calves born to BSE-affected dams [74]. Detailed analyses of UK maternal cohorts has suggested that risk of BSE transmission is increased in calves born to infected dams, especially those born up to two years before the onset of clinical signs of BSE in the dam [75,76].

BSE can efficiently transmit to other host species including sheep, goats and mice. Although studies in mice have suggested that maternal BSE prion transmission was possible [74], no evidence of maternal transmission to goat embryos was also reported [77]. If sheep in the UK had been infected with BSE prions during the cattle epidemic, the possibility that maternal transmission might help sustain this disease within the UK sheep flock was an important concern. Maternal BSE transmission in sheep was shown to be possible in an experimental study, but the low frequency at which it occurred was considered to be unlikely to maintain this disease within a population [78].

Whether prion disease-affected dams transmit prion disease to their offspring during gestation, or around the time of parturition (birth) has been the subject of much investigation. No cases of scrapie were recorded in offspring derived from the embryos of scrapie-infected dams when they were transferred into scrapie-free recipient sheep [79–81]. This implied that embryo transfer could be used as a method to prevent the maternal transmission of natural sheep scrapie, even in offspring with high risk *PRNP* genotypes [82]. These studies also suggested that rather than transmitting disease to the developing fetus in utero, infection from dam to offspring most likely occurred during birth or the post-natal period [83].

Placental tissues derived from infected dams may be contaminated with prions [83], and the placenta of goats infected with classical scrapie was able to transmit disease to susceptible goat kids and lambs via oral route [84]. Thus the contamination of pasture or the farm environment with prion-infected placenta or other birth-associated tissues and fluids may contribute to the post-natal transmission of disease between mother and offspring, as well as the horizontal transmission between other animals within the same population [85].

However, other studies in scrapie-affected sheep have reported the in utero transmission of prions to the developing fetus [86–90]. The in utero transmission of CWD in free ranging Rocky Mountain elk has also been reported, as PrP^{Sc} was detected in approximately 80% ($n = 12/15$) of the fetuses analyzed from infected dams in one study, regardless of the gestational stage of the fetus [72]. In animals such as sheep where births from multiple fetuses may occur, the sharing of blood components between developing the fetuses in the same uterine horn may aid the dissemination of prions to the cotyledons of fetuses with scrapie-resistant genotypes [91].

The possibility that human prion diseases may be maternally transmitted has obviously raised concern, especially as a number of children have been born to CJD-affected parents. However, current analyses have found no evidence that human prion diseases are maternally transmitted. For example, one study in 2011 analyzed 125 children born to parents who were diagnosed with vCJD [92]. None of these children developed vCJD during the study period or were classified as suffering from a progressive neurodegenerative disorder. The mothers of nine of these children were symptomatic at conception, birth or within a year of clinical onset, and one child was known to have been breast fed. A study in primates also found no evidence of maternal transmission of kuru, sCJD or scrapie [93], consistent with the absence of prion disease-specific PrP^{Sc} in the uterus and gestational tissues, including the placenta and amniotic fluid, of a pregnant woman with sCJD [94]. Of course, obvious caveats to these studies are the small numbers of cases analyzed, and the potentially long duration of the preclinical phase of the disease in the children. However, despite these concerns, the available data do not support the conclusion that human prion diseases can be maternally transmitted.

Milk and Colostrum

The presence of PrP^{Sc} within the mammary glands of scrapie-affected sheep has been reported [95], and the abundance may be enhanced in tissues with chronic inflammation or mastitis [96–98]. Colostrum and milk from scrapie-affected ewes have similarly been shown to contain infectious prions [99,100], which may also be enhanced in milk derived from animals with scrapie and mastitis [97]. These studies demonstrate the potential risk of prion spread between sheep and other species through the consumption of sheep milk or milk products. These studies raised important concerns that prions BSE may also be transmitted to humans through consumption of cattle milk or milk products. However, the risk is considered to be extremely low as abnormal PrP was undetectable in the milk from BSE-infected cattle [101]. As mentioned above, no evidence to support a role for vCJD transmission from an infected mother to child in humans has been reported [92].

3. Transmission of PrP^{Sc} Prions within Host Species

3.1. Prions and the Prion Protein

Expression of the cellular prion protein, PrP^{C}, is obligatory for cells to be able to replicate prions [102,103]. The prion hypothesis proposes that abnormally folded prion disease-specific PrP^{Sc} proteins are able to self-propagate by recruiting cellular PrP^{C}, which is then transformed into the disease-causing PrP^{Sc} isoform [104]. In support of this hypothesis, independent studies have shown that when recombinant mouse PrP is misfolded into the disease-specific form in vitro, the de novo-generated misfolded protein can transmit a prion disease to recipient mice [2,105]. Cellular PrP^{C} is a 30–35 kDa glycoprotein which is encoded by the *PRNP* gene. This protein is expressed on the outer leaflet of the cell membrane via its glycosylphosphatidylinositol (GPI) anchor [106]. The N-terminal portion of the prion protein is mostly unstructured, comprising a long, flexible tail. The secondary structure of the globular C-terminal domain of PrP^{C} contains three α-helices and a short, two-strand β-pleated sheet [107,108]. During prion disease changes occur in the secondary, tertiary and quaternary structures of the PrP molecule, increasing the amount of β-pleated sheet [107,109]. These changes have profound effects on the physico-chemical and biological characteristics of PrP, as the disease-specific PrP^{Sc} isoform is neurotoxic, relatively resistant to proteinase digestion, and accumulates in affected tissues in insoluble aggregates. The precise mechanisms by which these conformational and biological changes occur are unknown, but the requirement for additional chaperone molecules such as RNA, proteoglycans and lipids has been demonstrated [110,111]. Within the CNS it has been revealed that PrP^{C} plays an important role in promoting myelin homeostasis through interactions with the G-protein-coupled receptor Gpr126 (also known as Adgrg6) on Schwann cells [112]. Many other neuronal functions have also been reported, including regulation of circadian rhythms [113], synaptic transmission [114], cognition [115], seizure sensitivity [116], signal transduction [117,118], regulation of apoptosis [119], and protection from oxidative stress [120].

Once peripherally acquired PrP^{Sc} prions infect the host via one of the routes described above, many of them first accumulation within the secondary lymphoid organs (SLO) and persist within them at high levels for the duration of the disease. For example, after oral exposure of mice to scrapie prions, the agent accumulates first in the gut-associated lymphoid tissues (GALT) such as the Peyer's patches [121–123]. Similarly, PrP^{Sc} is first detected in the GALT following experimental oral infection of mule deer fawns (*Odocoileus hemionus*) with CWD [124], or sheep with some strains of natural scrapie [125–127].

PrP^{C} is also expressed in many cell populations within the immune system and SLO, including lymphocytes, leukocytes, granulocytes, mononuclear phagocytes, and stromal cells. Although PrP^{C}-deficient mice appear to show no obvious immune deficits [128], PrP^{C} may play a role in cell activation [129–131], T cell differentiation [132] and intercellular interactions [133], and phagocytosis [134,135]. The ubiquitous cellular expression of PrP^{C} has important immunological consequences as PrP is tolerated by the host's immune system. This prevents the development of specific immune responses against PrP^C and PrP^{Sc} prions. Thus, the accumulation of prions within SLO does not lead to their eradication from the host, and little evidence of PrP^{Sc}/prion-specific immunity has been reported in affected animals [136–138].

Cellular Sites of PrP^C to PrP^{Sc} Conversion

Many studies have attempted to identify the sites of PrP^C to PrP^{Sc} (prion) conversion, as such knowledge may reveal novel targets to block the de novo generation of infectious prions within the host. After synthesis, PrP^C is first processed in the Golgi before it is expressed upon the plasma membrane [139]. Following subsequent internalization, PrP^C traffics to early endosomes. From these the PrP^C is sorted either into recycling endosomes and returned to the plasma membrane, or alternatively, sorted into late endosomes for degradation within lysosomes [140]. Several intracellular prion conversion sites have been suggested including the endocytic pathway [141], lysosomes [142], the endosomal recycling compartment [143,144], and the trans-Golgi network [145], implying that PrP^{Sc} traffics along the same endocytic route as PrP^C. Despite these advances, it is possible that the intracellular location of the prion conversion site varies according to cell type, and host and prion species. For example, high resolution electron microscopy and cryo-electron microscopy studies have not observed any intracellular PrP^{Sc} accumulations within prion-infected stromal follicular dendritic cells in SLO (see FDC below, Section 3.3.6), suggesting that prion conversion in these cell populations occurs upon the cell membrane [146,147]. Cell surface prion conversion may be a more widespread occurrence. Evidence from in vitro studies using neuroblastoma cells has proposed that PrP^{Sc} conversion takes place on the cell within minutes of exposure to prions [148]. The de novo generated PrP^{Sc} is then rapidly endocytosed with some recycled back to the plasma membrane in association with recycling endosome, whereas the remainder predominantly undergoes lysosomal degradation [149].

3.2. The Accumulation of PrP^{Sc} Prions in SLO is Essential for Their Efficient Spread to the CNS

Studies using immunodeficient mice undertaken across the past four decades have been instrumental in determining the contribution of the host's SLO and immune cell populations to prion disease pathogenesis. Original experiments using asplenic mice revealed, contrary to expectations, that the early accumulation of prions within the SLO may actually help facilitate efficient CNS infection. These revealed that prion disease survival times after intraperitoneal infection were extended in mice which lacked a spleen [150,151]. Other studies have shown that disease pathogenesis is similarly delayed in the absence of the SLO draining the exposure site, such as the Peyer's patches in the small intestine after oral prion exposure [122,123,152,153], or the skin draining lymph nodes after infection via skin lesions [154].

As well as providing useful biomedical insight into the dissemination and pathogenesis of the acquired PrP^{Sc} prion disease within the host, the information from these studies has proven to have important practical applications. The detection of PrP^{Sc} within the GALT and other SLO soon after exposure has provided a means to identify some prion-infected individuals during the pre-clinical phase [155–160]. In a human vCJD-infected patient, PrP^{Sc} was detected within the GALT before the onset of clinical signs [160]. The retrospective analysis of archived appendix and tonsil tissues has since been used in the United Kingdom to provide an estimate of the prevalence of vCJD prions in the human population [161–163].

3.3. The Cellular Dissemination of PrP^{Sc} Prions within the Host

Peripherally-acquired PrP^{Sc} prions appear to exploit an elegant cellular relay to ensure their efficient propagation from the site of exposure to the SLO, where they accumulate before establishing infection within the nervous system (Figure 2). Much of our understanding of this early phase of the disease process has been gained from the study of experimental prion transmissions to a large range of transgenic and immunodeficiency mice, especially those using mouse-passaged scrapie prion isolates. The conclusions from many of these experimental mouse scrapie prions transmissions are discussed below, but comparisons with data from studies of natural prion disease-infected hosts are included where possible. Since many natural prion infections are considered to be orally acquired, these descriptions are mainly focused on the cells and tissues involved in the propagation of prions from the intestine. However, data from other routes of exposure are also discussed at the appropriate places.

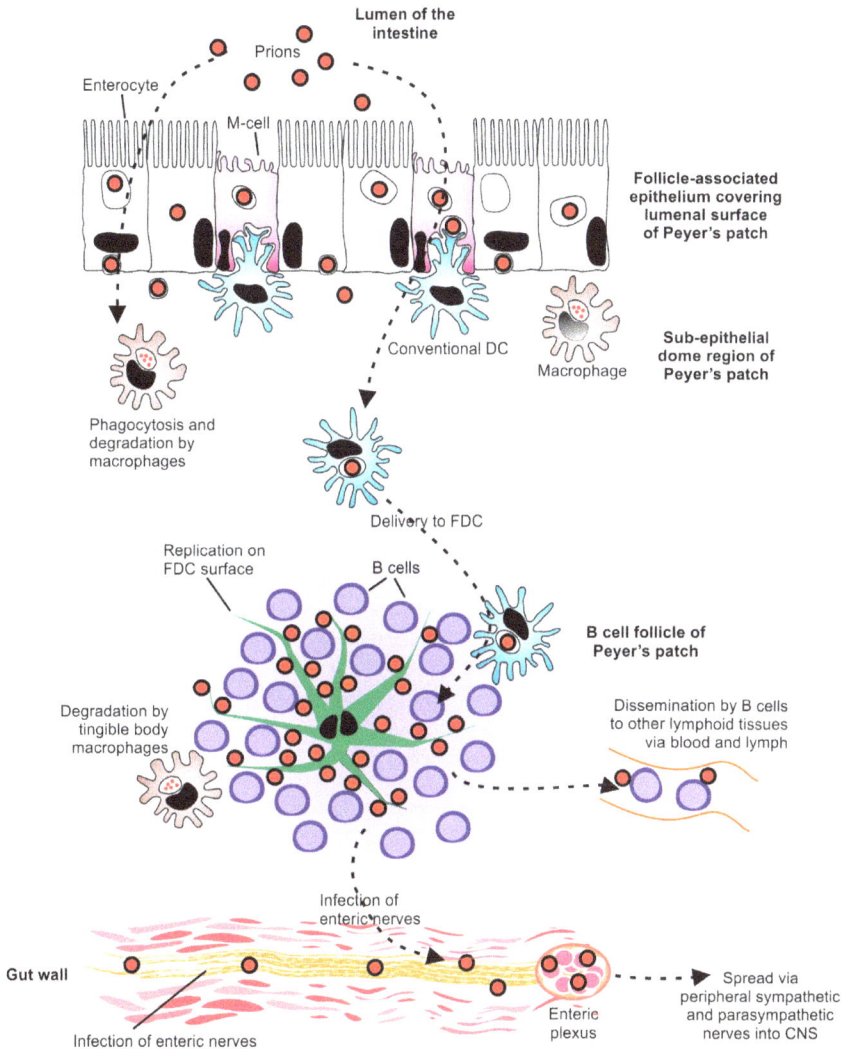

Figure 2. The cells involved in the spread of prions from the intestine to the central nervous system (CNS). After oral exposure the replication of prions upon follicular dendritic cells (FDC) in the Peyer's patches in the intestine is essential to establish host infection. With the Peyer's patches, the prions exploit an elegant cellular relay to make their way from the lumen of intestine to the nervous system.

3.3.1. Prions Cross the Gut Epithelium via M Cells

After oral infection, the prions must first cross the gut epithelium, but this single layer of tightly bound epithelial cells acts as an impermeable barrier to macromolecules, the commensal gut microflora, and many orally-acquired pathogenic microorganisms. However, in order for orally-acquired prions to establish infection within the GALT, they first have to cross the gut epithelium in sufficient quantities. The specialized follicle-associated epithelium (FAE) which covers the GALT contains a unique population of highly phagocytic epithelial cells, termed M cells. These cells are specialised for the transcytosis of particulate antigens and microorganisms from the gut lumen into the GALT [164]. The sampling of antigens and pathogens by M cells is important for the initiation of efficient mucosal immune responses against some pathogenic bacteria [165] and the commensal microflora [166]. However, some pathogenic bacteria and viruses have evolved to exploit the transcytotic properties of M cells to cross the gut epithelium and establish host infection [134,167–170].

Prions also appear to exploit M cells to cross the gut epithelium and establish infection with the GALT [171–175]. The accumulation of mouse-passaged ME7 scrapie prions in the GALT and disease susceptibility after oral exposure were both reduced in mice that lacked M cells, or in mice which M cells were transiently depleted before infection [174,175]. Other studies using mouse-passaged RML scrapie prions [171], Fukuoka-1 prions [173], BSE prions [172], and 263 K hamster prions [176] indicate that M cells also play an important role in the transfer of other orally-acquired PrP^{Sc} prion isolates across the gut epithelium.

Certain pathogenic bacteria [167,177] or inflammatory stimuli such as cholera toxin [178] can increase the density of M cells in the gut epithelium. The differentiation of M cells is dependent on stimulation from underlying stromal cells via their production of the cytokine RANKL [166,179,180]. When mice were treated with RANKL to increase the density of M cells within in the gut epithelium, the uptake of prions from the gut lumen was similarly enhanced, and disease susceptibility was increased by approximately 10-fold [175]. This study indicates that factors that increase the density of M cells in the gut epithelium, such as concurrent pathogen infection, may profoundly affect host susceptibility to orally-acquired prion infections. The binding of CWD prions to soil particles such as smectite clay montmorillonite have also been shown to increase the efficiency of prion uptake from the intestine [30]. Whether the binding of prions to certain types of soil particles enhances their ability to be transferred across the gut epithelium by M cells remains to be determined.

Data on the role of M cells in the initial uptake of prions from the gut lumen have occasionally been conflicting. Some studies in rodents in which prions were immunohistologically traced after oral exposure have suggested that M cells are the initial sites of prion uptake in the gut epithelium [173,181]. However, other studies in lambs [182] and mice [146] have described M cell-independent uptake pathways. The reasons for these discrepancies are uncertain, and it is plausible that both M cell-dependent and M cell-independent routes may contribute to differing degrees in some circumstances. In vitro-based studies have shown that undifferentiated gut epithelial cell lines (Caco-2 cells) act as a barrier to prion uptake [171], but transcytosis of PrP^{Sc} was evident when it was complexed with ferritin [183]. However, if enterocyte-mediated transfer plays a major role, one would not expect oral prion disease susceptibility to be blocked in mice which specifically lack M cells [174,175].

Antigen sampling by M cells has also been demonstrated within the epithelia covering the nasal associated lymphoid tissue (NALT) in the nasal passages [184,185]. After intra-nasal exposure of hamsters to 263 K scrapie prions, transient PrP^{Sc} uptake was detected with NALT-associated M cells. However, a greater abundance of paracellular transport across the epithelia within the nasal cavity was evident [186], indicating that cells involved in the transepithelial transport of PrP^{Sc} prions may vary depending on the exposure route.

Highly sensitive PrP^{Sc}-based detection assays have detected low/trace levels of prions in the blood-stream almost immediately after oral exposure [187,188]. The route through which the prions initially contaminated the blood-stream after oral exposure was not determined, but it was suggested that the amount of PrP^{Sc} that was initially present within the blood-stream was sufficient to establish CNS infection [188]. This conclusion appears to contradict data from many other independent studies which shown that prion replication in the GALT after oral exposure is essential for the subsequent transmission of disease to the CNS [121–123,152,153,189]. Thus, although low levels of PrP^{Sc} may be detected in the blood-stream within minutes of oral infection [187,188], the levels within it are insufficient to directly establish infection within the nervous system.

M cells express a diverse array of receptors on their apical surfaces which specifically bind to certain pathogenic microorganisms [164]. Whether the uptake of prions by M cells is also mediated via by a specific receptor is not known. M cells express PrP^{C} highly [134,190], but PrP^{C}-deficiency in the gut epithelium does not affect the uptake of PrP^{Sc} from the intestine [146,173].

3.3.2. Conventional Dendritic Cells Aid the Delivery of Prions to SLO

Particles that have been transported across the gut epithelium by M cells are released into the basolateral pocket where they are sampled by mononuclear phagocytes [191]. Mononuclear phagocytes differentiate from bone marrow precursor cells and comprise a heterogeneous population of monocytes, conventional dendritic cells (DC) and tissue macrophages. Conventional DC are strategically positioned to sample their local environment for pathogens and their antigens. After antigen uptake, these cells undergo maturation and migrate towards the draining lymphoid tissue to initiate a specific immune response. Conventional DC possess both degradative and non-degradative antigen uptake pathways to enable them to present processed (partially digested) antigens to T cells or native (intact) antigens to B cells [192,193]. These cells are also centrally involved in the transport of antigens within Peyer's patches, and on towards the mesenteric lymph nodes [194–196]. The migratory characteristics of conventional DC have been exploited by some pathogens to mediate their delivery to SLO [197–200]. The ability of conventional DC to capture and retain unprocessed antigens [201,202] and migrate into B cell follicles [203–205] suggested that classical DC were plausible candidates for the propagation of prions to and within SLO. This hypothesis was supported by the observation some migrating intestinal DC in the afferent mesenteric lymph had acquired PrP^{Sc} after its injection into the gut lumen [206]. Subsequent studies showed that the early replication of prions within the draining SLO was impeded when conventional DC where transiently depleted at the time of exposure [189,207–209]. Thus, like certain other pathogens, prions may also exploit conventional DC to establish host infection after peripheral exposure, perhaps using them as "Trojan horses".

Whether specific subsets of conventional DC are able to propagate prions to and within SLO is uncertain. However, CD8^{+} conventional DC are unlikely to play a role, as these cells are rarely encountered within the subepithelial dome region immediately beneath the M cell-containing FAE [210], and the specific depletion of CD8^{+}CD11c^{+} cells does not influence oral prion disease pathogenesis [208]. Similarly, although prion pathogenesis following infection via skin lesions was impaired in the specific absence of CD11c^{+}langerin^{-} dermal DC, the absence of epidermal Langerhan's cells or langerin^{+} dermal DC had no effect on disease pathogenesis [209]. High levels of infectious prions have also been detected within splenic plasmacytoid DC [211], but these cells are also unlikely to contribute to prion propagation as they do not migrate in the lymphatics [212].

Mononuclear phagocytes express cellular PrP^{C} [213–215], but the propagation of prions to SLO is not affected by the lack of PrP^{C} expression in hematopoietic cells [216–221]. This demonstrates that prions are acquired by conventional DC in a PrP^{C}-independent manner, and also that they are not important sites of prion replication. Conventional DC can acquire prions after their opsonization by complement components such as C1q and C3 [221,222]. Depending on their location, phenotype and activation status, conventional DC can express a variety of complement-binding receptors including CR1 (CD35), CR2 (CD21), CR4 (CD11c/CD18), calreticulin, CD93 and SIGN-R1 (CD209b), but whether these also mediate the uptake of prions by conventional DC is uncertain [221,222]. However, prion disease pathogenesis is unaffected in the absence of SIGN-R1 expression at the time of exposure [223].

Conventional DC may simply acquire prions non-specifically by fluid phase micropinocytosis, as they constitutively sample their microenvironments.

Tunneling nanotubes (TNT) are thin membrane-bound cylinders of cytoplasm which can connect cells to enable cell-to-cell communication and the intercellular transfer of plasma membrane or cytoplasmic components. TNT structures are exploited by HIV-1 as a means of intercellular transfer between T cells [224], and to shuttle virus-encoded immunosuppressive factors from infected macrophages to B cells to suppress host antibody responses [225]. A study has also suggested that intracellular transfer between M cells and neighboring cells can also occur via TNT [226]. In vitro co-culture studies show infectious prions can also transfer between conventional DC and neurones via endolysosomal vesicles within TNT [227–230]. Whether TNT mediate the intercellular transfer of prions in vivo remains to be determined. Infectious prions can also be released from infected cells in the form of small endosomal-derived vesicles termed exosomes [231] which have the potential to infect neighboring cells [211]. However, the relative contribution of exosomes in this process may vary depending on the prion strain [232].

3.3.3. Macrophages Can Phagocytose and Destroy Prions

While some mononuclear phagocytes such as conventional DC can propagate infectious prions to and within SLO, some mononuclear phagocyte populations may sequester and destroy them [233,234]. Tingible body macrophages, for example, are specifically located within the germinal centers of B cell follicles. These macrophages are characteristically loaded with the remnants of phagocytosed apoptotic lymphocytes (tingible bodies), and during prion disease also contain heavy accumulations of PrP^{Sc} within their endosomal compartments [146,147,220]. Transient macrophage depletion prior to peripheral prion exposure has been shown to enhance the accumulation of PrP^{Sc} within SLO [235,236]. These data suggest that macrophages typically scavenge and degrade prions in an attempt to protect the host from infection.

The burden of infectious prions within the SLO rapidly reaches a plateau level within a few weeks of exposure which is maintained for the duration of the infection [216,237]. How this plateau is maintained is uncertain. It is plausible that a competitive state is reached within these tissues whereby the rate of prion amplification matches the rate of degradation by macrophages [234,235].

3.3.4. Cell Free

Although data from many of the studies described suggest that prions are propagated to and within SLO in a cell-associated manner, the possibility that a fraction of the prions are also conveyed in a cell-free manner cannot also be excluded [40,186,209,221].

3.3.5. B Cells Indirectly Support Prion Replication in SLO

In stark contrast to their susceptibility to infection with most other pathogenic microorganisms, mice which lack mature B and T cells, such as severe combined immunodeficient (SCID) mice, $Rag-1^{-/-}$, $Rag-2^{-/-}$, and $Agr^{-/-}$ mice [238–241], are refractory to peripheral prion infection. However, the T cells themselves do not influence prion disease pathogenesis as prion accumulation in SLO and neuroinvasion are not affected in T cell deficient thymectomised mice [151,242], or in transgenic mice with specific T-cell deficiencies (CD4$^{-/-}$, CD8$^{-/-}$, $\beta2$-$\mu^{-/-}$, TCR$\alpha^{-/-}$ or $Perforin^{-/-}$ mice) [217,240]. T cells appear to lack the cellular factors required to sustain prion infection as even the artificial ectopic expression of high levels of PrP^{C} in T cells (20 × $Prnp$ copies) is insufficient to sustain prion infection within them [243].

In the SLO of some prion-infected hosts, heavy accumulations of disease-specific PrP are detectable within the B cell follicles [124,125,160,238,244]. In the specific absence of B cells, the accumulation of prions in the spleen and subsequent neuroinvasion are both significantly reduced [240]. However, B cells also do not replicate prions as transgenic mice which express high levels of PrP^{C} only on B cells were also unable to directly replicate prions [245]. This indicated that B cells most likely played an indirect role in prion pathogenesis, perhaps through the provision of homeostatic support to other cell populations.

3.3.6. Follicular Dendritic Cells Retain and Replicate Prions

SCID mice and other B cell-deficient mouse lines are also indirectly deficient in follicular dendritic cells (FDC), as these cells require constitutive stimulation from B cells to maintain them in their differentiated state [246–249]. FDC are an important stromal cell subset that resides within the B cell follicles and germinal centers of SLO. FDC differentiate from ubiquitous perivascular precursor cells (pericytes) and are a distinct lineage from the bone-marrow-derived classical DC described above (Section 3.3.2) [250–252]. Immunohistochemical analysis shows that prions accumulate upon FDC in the SLO of experimentally-infected mice, some sheep with natural scrapie, cervids with CWD and patients with vCJD [124,125,160,238,244]. As discussed below, the accumulation and replication of certain prion strains upon FDC is essential to establish host infection and neuroinvasion.

The transfusion of SCID mice with wild-type (immunocompetent) bone marrow restores the B and T cells in these mice, and by doing so, indirectly induces the maturation of FDC in their SLO [246]. Coincident with the induction of FDC maturation, this treatment also renders the mice susceptible to peripheral prion infection [216–218,241].

B cells express the cytokines tumor necrosis factor (TNF)α and lymphotoxins (LT), which are the essential stimuli that maintain FDC differentiation [248]. In the absence of these cytokines, the FDC rapidly de-differentiate [248,253–257]. Prion accumulation in SLO and neuroinvasion are significantly impaired in mice deficient in TNFα or LT stimulation, demonstrating the requirement for FDC in the establishment of prion infections [123,237,258]. Prion accumulation in SLO and disease susceptibility are also reduced when the FDC are temporarily de-differentiated by administration of soluble receptors which block the LT- or TNFα-mediated signaling between the B cells and FDC [121,259–263]. Although FDC are important for maintaining germinal center responses, an absence of germinal centers or germinal center B cells alone does not influence prion disease pathogenesis or susceptibility [237,264].

FDC characteristically trap and retain native antigens on their cell surfaces, which they display to B cells within the follicle and germinal center. The long-term retention of these antigens by FDC helps to promote immunoglobulin (antibody)-isotype class switching, affinity maturation of naïve B cells and the maintenance of immunological memory [265–271]. By secreting the factor MFGE-8 (which specifically binds to phosphatidylserine on the surface of apoptotic cells) the FDC also mediate the phagocytosis of apoptotic B cells by tingible body macrophages [272]. The ability of FDC to retain native antigens for long periods raised the possibility that they might also simply trap and retain prions produced by other infected cell populations such as neurones. Several studies have exploited the non-haematopoietic-origin of FDC [246,250] to help address this issue. Mismatches were created in *Prnp* gene expression between the FDC-containing stromal and lymphocyte/leukocyte-containing compartments of the SLO by transfusing hematopoietic cells from PrP^C-deficient mice into PrP^C-expressing (wild-type) mice, and vice versa [216–218]. In these studies, the FDC were derived from the recipient, whereas all the hematopoietic cell populations were derived from the donor bone marrow. When these mice were infected with prions, prion accumulation upon FDC was only detected in the spleens of mice which had a PrP^C-expressing stromal compartment. These studies provided strong evidence that FDC were important sites of prion accumulation in SLO. However, roles for other stromal cells could not be entirely excluded as in these studies it was not possible to dissociate the PrP^C expression status of the FDC from that of the nervous system, or other stromal cell populations within the SLO.

FDC and mature B cells express high levels of *Cr2* which encodes the complement receptors (CR) CR2/CR1 (CD21/35) [273,274]. CD21-cre mice [273] have been used to specifically control *Prnp* expression only in FDC, enabling PrP^C expression to be "switched on" or "switched off" only in FDC [220]. The expression of PrP^C only on FDC was sufficient on its own to sustain high levels of prion replication in the spleen. Conversely, prion replication in the spleen was blocked when PrP^C expression was specifically ablated only on the FDC. Data from all the above studies together definitively show that FDC are the essential sites of prion replication in SLO.

After their replication upon FDC within the SLO, the prions subsequently infect both the sympathetic and parasympathetic nervous systems and spread along the nerves within them to the CNS, where they ultimately cause neurodegeneration [176,275,276]. The role of FDC in prion disease pathogenesis appears to be to amplify the prions above the threshold required for neuroinvasion. How the prions subsequently infect the peripheral nervous system is uncertain, as no significant direct physical contacts or synapses between FDC and nerves has been described. However, the rate of neuroinvasion from SLO is influenced by the distance between FDC and the peripheral sympathetic nerves [262].

3.3.7. FDC Acquire Prions as Complement-Opsonized Complexes

FDC express high levels of PrP^C [216,217,220,277], but many other cell lineages also express PrP^C highly but do not play an essential role in prion disease pathogenesis. Clearly, other FDC characteristics in addition to PrP^C expression enable them to replicate prions. FDC have many slender dendritic processes which extend throughout the B cell follicle. These dendrites enable the FDC to trap and retain large amounts of native antigen upon their surfaces in the form of immune complexes, consisting of antigen-antibody and/or opsonizing complement components. Complement components C1q and the regulatory protein factor H can bind to PrP^{Sc} [278,279], and the specific absence of opsonizing complement components (C1q, C2, C3, C4 and factor H) or CR expression on FDC impedes prion accumulation in the spleen and delays neuroinvasion [274,280–284]. Comparison of the relative contributions of the CR1 and CR2 receptors has revealed a more prominent role for CR2 in prion disease pathogenesis [285]. Activation of the terminal complement activation pathway leads to formation of the membrane attack complex which can lyse the target cells. Deficiency in complement component C5 (an important component of the membrane attack complex) in contrast, has no influence on prion disease pathogenesis [286]. This suggests that soon after the prions infect the host, they are bound by soluble complement components and are acquired by FDC in the SLO as complement-opsonized complexes in a CR-dependent manner [274,280–282,286].

The immune complexes that are retained by FDC are initially internalized before undergoing cyclical rounds of display on the FDC surface [287]. This cyclical mode of immune complex expression on the FDC surface helps to protect the antigens from degradation, enabling them to be retained for much longer periods [271]. Despite this, high resolution immunohistochemical analyses indicate that prion replication occurs on the FDC surface, as PrP^{Sc} has not been detected within them [146,147,288–290]. This appears to be in contrast to prion infection in nerves where PrP^{Sc} conversion has been demonstrated within the endosomal recycling compartment [143].

3.3.8. Conventional DC Can Shuttle Prions towards FDC

Chemokines help to attract lymphocytes and leukocytes to SLO and control their positioning within them. The chemokine CXCL13 is expressed by FDC and other stromal cells in the B cell-follicles of lymphoid tissues and recruits CXCR5-expressing cells towards them [291,292]. The migration of certain populations of conventional DC towards the FDC-containing B cell-follicles is also mediated by CXCL13-CXCR5 signaling [203,205,293]. In the specific absence of CXCR5-expressing conventional DC, the early accumulation of prions upon FDC in Peyer's patches was impeded [294]. This suggests that once the prions have been transported across the gut epithelium by M cells, they are subsequently acquired by conventional DC [146,189] and propagated by them in a CXCL13-CXCR5-dependent manner towards the FDC within the B cell-follicles of Peyer's patches [294]. The prions are then acquired by FDC and amplified upon their surfaces above the threshold required to achieve neuroinvasion [121,123,146,153,220].

The positioning of conventional DC within the inter-follicular T-cell regions of the Peyer's patches, and their steady-state migration from Peyer's patches to the mesenteric lymph nodes are both dependent upon CCR7-CCL19/CCL21-signaling [295]. Consistent with the demonstrations that T cells do not influence prion disease pathogenesis [240], an absence of CCR7-CCL19/CCL21-signaling does not influence oral prion disease pathogenesis [296].

Although the transient depletion of CD11c$^+$ cells or deficiency in CXCR5-expressing conventional DC impedes the accumulation of prions in Peyer's patches and reduces disease susceptibility, a small number of mice in these studies did develop clinical prion disease [189,294]. This indicates that conventional DC provide an efficient route by which prions are initially conveyed to FDC. However, in the absence of conventional DC at the time of oral exposure, small quantities of prions are able to avoid clearance by cells such as tissue macrophages [235,236], and establish infection upon FDC via less efficient routes [40,209,221].

Once the prions have been transported towards the B cell follicles by conventional DC, it remains to be determined how they are subsequently transferred to the FDC. Follicular B cells in the sub-capsular sinus region of the lymph nodes can acquire lymph-borne immune complexes via their CR and deliver them to FDC [297–299]. In the spleen, marginal zone B cells play a similar role in the shuttling of blood stream-borne immune complexes to FDC [300]. Once the B cells are within the follicle, the higher immune complex-binding affinities of the FDC enables them to strip the immune-complexes from the surfaces of the B cells. Conventional DC can retain PrP^{Sc} on both the cell surface and in intracellular compartments [221]. The higher expression levels of CR on the cell membranes of FDC may similarly enable them to strip complement-opsonized prions from conventional DC.

3.3.9. FDC-Independent Prion Accumulation and Neuroinvasion

Although many prion isolates in different host species may replicate first upon FDC in the SLO, examples have been described where prion accumulation in SLO and/or neuroinvasion occur independently of FDC. The factors which determine the requirement for FDC in prion disease pathogenesis are uncertain, but this may be influenced by prion agent strain, host species, *PRNP* genotype, and exposure route. The dose of prions used to infect the host can also influence disease susceptibility, as high doses can bypass the requirement for amplification within SLO prior to neuroinvasion [258]. The infecting prion strain and host *PRNP* genotype can influence disease pathogenesis in sheep [125,301]. Some prion diseases, such as sporadic CJD in humans are not associated with early prion accumulation in the periphery tissues including the SLO [302], and the susceptibility mice to the mouse-adapted FU agent strain of CJD prions was unaffected by an absence of mature B cells and FDC [303]. BSE prions in cattle are considered to have little SLO involvement during the preclinical phase [304], but PrP^{Sc} and/or infectious prions have been detected in the small intestines of most cattle after experimental (oral) infection and some cattle after natural exposure [305–307]. However, when BSE prions are transmitted to other species such as humans (in the form of vCJD), sheep and mice, their accumulation in the lymphoid tissues is a characteristic feature [66,308,309]. Whereas mouse-adapted ME7 prions are unable to accumulate in the spleens of TNFα-deficient mice due to an absence of FDC [237], this is not true for RML prions, which can also accumulate within the high endothelial venules of lymph nodes [258,310]. Inflammatory stromal cells that are a distinct lineage from FDC have also been shown to have the potential to replicate prions under certain circumstances [311].

Despite the presence of FDC within the NALT, prion neuroinvasion after exposure via the nasal cavity can occur independently of the NALT and SLO, implying direct infection of the nervous system [35,37].

3.4. Prion Infections Cause Limited Pathology in SLO

The accumulation of PrP^{Sc} within the CNS ultimately leads to the development of neuropathology. Despite the detection of high levels PrP^{Sc} upon FDC in the SLO throughout the duration of the disease, no gross immunological deficiencies have been reported [312–315]. However, ultrastructural analysis of prion-affected SLO has revealed evidence of morphological changes to the FDC. These include adversely affected maturation cycles, abnormal dendritic folding and exacerbated accumulation of immune complexes between the FDC dendrites [147,289]. The immunological consequences of these pathological disturbances to FDC and germinal centers are uncertain, as antibody production appears to be unaffected in prion-infected animals [312].

3.5. Orally-Acquired Prions Replicate First in the GALT of the upper Gastrointestinal Tract

The FDC within the Peyer's patches in the small intestine are the essential early sites of prion replication and neuroinvasion in mice, as oral prion disease susceptibility is blocked in their absence [122,123,152,153,175]. Many natural prion diseases also accumulate upon FDC in the large intestinal GALT after oral exposure, such as the recto-anal mucosa-associated lymphoid tissues (RAMALT) of scrapie-affected sheep and goats, and CWD-affected deer and elk [155–158,316–318]. The detection of PrP^{Sc} within these tissues has helped to detect prion-infected animal and individuals during the pre-clinical phase [155–160], and has been used in the UK to estimate the prevalence of vCJD in the human population [161–163]. However, studies using mice have shown that the large intestinal GALT are not important early sites of prion accumulation or neuroinvasion from the intestine [153].

Detailed analyses of sheep with natural scrapie has revealed a similar mode of pathogenesis. Scrapie prions are first detected in the GALT of the upper gastrointestinal tract before spreading to the draining lymph nodes and onwards to other lymphoid tissues. The large intestinal GALT, such as the caecal patches, were not early sites of prion accumulation [319–322]. A similar temporal distribution was observed after oral exposure of sheep to BSE [323] and in humans with vCJD, where the distribution of PrP^{Sc} in lymphoid tissues is restricted during the pre-clinical phase, and more wide-spread at the clinical stage [63,67]. Likewise, RAMALT biopsy studies in scrapie-infected goats [316] and CWD-infected white-tailed deer [317,324] show lower incidences of prion accumulation in the RAMALT during the earlier stages of disease, supporting the notion that these tissues are not early sites of prion accumulation. Indeed, an analysis of CWD prevalence in elk showed that PrP^{Sc} was not reliably detected in the RAMALT until it was also detectable in the CNS [318], consistent with the conclusion that the large intestinal GALT become infected with prions until much later in the disease process [157,325]. Experimental transmissions of CWD to white-tailed deer have identified the oropharynx as the initial site of prion entry after oronasal exposure [326]. The possibility therefore cannot be excluded that biopsy specimens of large intestinal GALT may miss individuals if sampled during the early stages of oral prion infection, and significantly underestimate the disease prevalence.

Antigens from the intestinal lumen may be delivered directly to the mesenteric lymph nodes [327,328]. Prions are similarly detected upon FDC in the mesenteric lymph nodes and the spleen soon after oral exposure (most likely after their dissemination from the Peyer's patches). The absence of the mesenteric lymph nodes does not affect prion neuroinvasion or disease susceptibility [123,153,175]. This implies that the levels of prions initially delivered to the mesenteric lymph nodes immediately after oral exposure are insufficient to establish accumulation and replication upon the FDC within them. Consistent with the conclusion that the GALT of the upper gastrointestinal tract are the essential sites of prion accumulation and neuroinvasion after oral prion exposure, the absence of the spleen also does not influence oral prion disease pathogenesis or susceptibility in mice [294,329].

3.6. B Cells Aid the Spread of Prions between SLO

After residing within the Peyer's patches, B cells migrate to the mesenteric lymph nodes and then return to the circulation [330]. B cells can recirculate between lymphoid tissues for several weeks [331] and often acquire antigens from FDC as they migrate through the germinal centers within them [332]. Studies in mice have suggested that recirculating B cells also appear to mediate the initial propagation of prions from the draining lymphoid tissue to other SLO. When the migration of B cells between SLO was specifically blocked, the dissemination of prions from the draining SLO to other SLO was also blocked [333]. Whether the same occurs in natural prion infections is uncertain, but prions have been detected in association with lymphocytes [334,335] and B cells in the blood of sheep with scrapie [336] and deer with CWD [337].

3.7. Prion Infection of the CNS Occurs via the Sympathetic and Parasympathetic Nervous Systems

The circumventricular organs in the bran are sites of molecular exchange between the blood-stream and the CNS. The detection of prion disease-specific PrP within these brain regions has been described in some scrapie-affected sheep, suggesting the potential for the hematogenous spread of prions into the CNS [338]. However, monocytic infiltration into the circumventricular organs is not observed in prion disease-affected hosts, arguing against the cell-associated hematogenous spread of prions into the CNS. Furthermore, studies in mice show that an absence of recruitment of circulating monocytes into the CNS does not influence prion disease pathogenesis within the CNS [339]. Although infectious prions are present in the blood-stream of vCJD-infected individuals, the spatial distribution of the PrP^{Sc} deposits in the brain in relation to the blood vessels also does not support a major role for the heematogenous spread of vCJD prions into the CNS [340].

The heematogenous spread of prions directly into the CNS cannot be entirely excluded, as low/trace levels of PrP^{Sc} may be initially detected with the CNS soon after exposure, using highly sensitive assays [188]. However, data from many studies suggests that the prions initially establish infection within the CNS after their spread from the SLO along peripheral nerves. The SLO and the GALT are highly innervated with sympathetic neurones [341]. Detailed immunohistochemical tracing studies show that after replication upon FDC in the GALT, orally acquired prions subsequently infect the enteric nervous system and spread along sympathetic (e.g., splanchnic nerve) and parasympathetic (e.g., vagus nerve) efferent nerves fibers to infect the CNS [146,176,275,342–345]. Furthermore, the specific depletion of sympathetic nerves, impairs prion neuroinvasion from SLO [276]. Conversely, prion disease pathogenesis is exacerbated in mice in which the density of sympathetic nerves in SLO is increased [276], or in those in which the distance between FDC and sympathetic nerves is reduced [262].

Within the intestine, mononuclear phagocytes are abundant in the muscular layer where they interact with enteric nerves to regulate gastrointestinal motility [346,347]. Data from in vivo and in vitro studies have proposed that prion-infected conventional DC or other mononuclear phagocytes may also play a role in the transfer prions to peripheral nerves [228–230,348].

3.8. Effect of Inflammation and Pathogen Co-Infection on Prion Disease Pathogenesis

The majority of the studies above describe prion disease pathogenesis during the steady state. However, inflammation can have a significant impact on prion disease pathogenesis, either by enhancing the uptake of prions from the exposure site, or by expanding their tissue distribution. For example, pathology to the gut mucosa such as that caused by bacterial colitis [349], enhanced M cell-density [175], or lesions to the mucosal surfaces of the oral [43] or nasal [44] cavities can each enhance disease susceptibility by increasing the prion uptake. Mitogen stimulation or repetitive immunization to non-PrP antigens around the time of peripheral prion exposure can also similarly increase disease susceptibility [350,351]. Chronic inflammation, by inducing the formation of FDC-containing ectopic tertiary lymphoid tissues, can expand the distribution of prions within the infected host [27,123,153,352,353]. The effects of chronic inflammation on prion disease pathogenesis could have important human and animal health consequences and aid their vertical and horizontal transmission, for example, by enhancing the burden of prions secreted into milk (in animals with mastitis) or urine (in animals with nephritis).

Despite the apparent widespread exposure of the UK human population to BSE prions during the BSE epidemic (approximately 500,000 infected cattle are estimated to have entered the food chain [354]) the numbers of clinical cases of vCJD in humans have fortunately been rare (Reference [355]; 178 definite or probable cases, as of 2 October 2017; www.cjd.ed.ac.uk). This does however, raise the possibility that oral prion susceptibility may differ between individuals. Studies using transgenic mice which expressed human PrP^C proposed that a significant species barrier restricts BSE transmission to humans [356]. After interspecies prion exposure, the processing and amplification of prions upon FDC in SLO is important for their adaptation to the new host, and to achieve neuroinvasion [309,357]. How this host adaptation occurs is not known, but may be influenced by the sialylation status of the PrP^{Sc} [358]. It is plausible that inflammation [175] or enteritis [349] may enable a greater burden of prions to be acquired from the gut lumen, increasing the probability that more will be able to avoid clearance by cells such as macrophages [123,235]. This may help to reduce the transmission barrier to some orally acquired prion strains by providing a greater opportunity for the prion quasi-species with zoonotic potential to be selected and to undergo adaptation and amplification upon FDC [282].

3.9. Effects of Host Age on Prion Disease Pathogenesis and Susceptibility

Clinical sporadic CJD cases have predominantly occurred in the elderly (median age at onset of disease = 67 years), whereas the majority of the clinical vCJD cases have almost exclusively occurred in young adults (median age at onset of disease = 26 years). The age-related incidence of vCJD is not simply due to the exposure of young adults to greater levels of BSE prions through dietary preference [359]. Aging has a profound effect on immune function, termed immunosenescence [360–362]. As a consequence of this immunosenescence, the elderly respond less effectively to vaccination, have increased susceptibility to viral and bacterial infections and increased incidence of cancer and autoimmune diseases. Immunosenescence can also influence the pathogenesis of peripherally acquired prion infections, by impairing prion accumulation/replication in SLO and reducing disease susceptibility. In sheep, cattle, cervids, and humans, susceptibility to peripheral prion infections is associated with GALT development [363]. For example, in Cheviot sheep, a marked fall in the size of their ileal Peyer's patches and lymphoid follicle density is apparent from the onset of puberty [364].

In aged (≥600 days old) mice the development and function of FDC [365] and M cells [366] which play a central role in prion neuroinvasion is adversely affected. The expression of PrP^C by aged FDC is also reduced [365]. In the spleen, the marginal zone forms a barrier around the lymphocyte-containing white pulp, and is important for the capture of blood-borne antigens and immune complexes. The immune complexes are then captured by marginal zone B cells and delivered to the FDC in the B cell follicles [300,367]. Due to gross disturbances to the splenic marginal zone, the FDC in aged mice are also unable to efficiently trap immune complexes and prions on their surfaces [277,309,368] (see author's online video https://media.ed.ac.uk/media/1_4wmqldeh). These ageing-related impairments to M cells and FDC impede the accumulation of prions in the SLO of aged mice, reducing disease susceptibility [277,309,365].

Although peripherally-exposed aged mice did not develop clinical prion disease, PrP^{Sc} was detected in some of their brains but was undetectable in their spleens [365]. This has important implications for the reliability of preclinical diagnostic tests based on the detection of PrP^{Sc} in blood or SLO, as these may be much less sensitive when used on elderly individuals. For example, low numbers of lymphoid follicles were shown to be present in RAMALT biopsy specimens from older elk (*Cervus elaphus nelson*; >8.5 years old) [369] and mule deer (*Odocoileus hemionus*; ≥3 years old) [370] reducing the reliability of the histopathological detection of PrP^{Sc} within the RAMALT of CWD-affected animals.

In pre-weaning animals, the developmental status of the GALT or intestine can also have a profound influence oral prion disease susceptibility. As a consequence of the underdeveloped status of FDC in neonatal mice, prion accumulation in their SLO after peripheral exposure is reduced and neuroinvasion is delayed [371,372]. The GALT tissues are more developed in neonatal sheep when compared to mice. However, the oral susceptibility of lambs to BSE prions is greater during the pre-weaning stage than during the post-weaning stage [373]. Multiple factors are likely to contribute to this age-related difference in susceptibility, but differences in gut development, mucosal permeability, and possibly also the presence of maternal immunoglobulin may each enhance the transfer of PrP^{Sc} across the gut epithelium [374].

4. Opportunities for Prophylactic and Therapeutic Intervention

At the time of writing, no safe or effective treatments had been developed for clinical use to block or prevent further spread of prion diseases in humans or domestic animals. However, many varied approaches have been reported. Unfortunately, some potentially useful anti-prion drugs identified from in vitro studies have had quite differing efficacies when translated in vivo into animal models or prion disease-affected patients.

4.1. PrP^{Sc} as a Therapeutic Target

A large variety of experimental in vitro studies have attempted to identify potential drugs or compounds which may block prion conversion or accumulation within cells. From these, many have been translated into animal models, and a small number have made it into clinical trials in human prion disease patients.

4.1.1. Quinacrine

Quinacrine is an antimalarial drug which can also inhibit prion accumulation in infected cells [375]. However, subsequent studies showed this drug had no effect on survival times in mice experimentally infected with prions [376]. Two independent clinical studies in humans also failed to demonstrate any beneficial effect of quinacrine treatment on the clinical course of prion disease in affected patients [377,378].

4.1.2. Pentosan Polysulphate

Pentosan polysulphate is a polyanion and heparin analogue which can significantly extend survival times, or reduce the susceptibility of mice to peripherally administered prions when administered around the time of infection [379], or by direct intraventricular infusion into the brain [380]. This compound has also been administered to some human CJD patients in the UK and Japan. In these limited trials, pentosan polysulphate was administered continuously to the brain by intraventricular infusion. Treatment may have extended survival times in some patients, but no apparent improvement in clinical signs were reported [381].

4.1.3. Tetracyclic Antibiotics

The tetracyclic antibiotics doxycycline and tetracycline can reduce prion infectivity and inhibit the neurotoxicity of PrP-peptides in vitro [382,383]. Unfortunately, the results of clinical trials using doxycycline in human patients with clinical prion disease were negative [384].

4.2. Prions as Anti-Prions

A series of studies undertaken in the 1970's showed that the infection of mice with a prion agent strain with a long disease duration could block subsequent infection with a prion agent strain with a shorter disease duration [385,386]. This competition between prion agents raised the hypothesis of whether such blocking could be achieved against prion diseases in domestic animals or humans through administration of prion agent strains with incubation periods which exceed the longevity of the host, or the use of synthetic molecules. Later independent studies showed that an attenuated "slow" strain of mouse-passaged CJD prions could similarly impede the pathogenesis of subsequent a infection with a more virulent or "faster" prion agent strain [387]. On a similar theme, a recent novel approach generated non-pathogenic, self-replicating PrP^{Sc}-like forms, termed "anti-prions". A single injection with these apparently innocuous PrP^{Sc}-like forms into hamsters infected with 263 K prions was shown to be sufficient to compete with prion replication and delay survival times [388]. Substantial safety trials and risk assessments will need to be undertaken before the efficacy of such approaches are tested in domestic animal species and human prion disease patients.

4.3. Targeting Prion-Induced Neurodegeneration

Detailed analyses of the molecular mechanisms by which prions cause neurodegeneration have begun to identify potential molecular targets for intervention during prion disease. For example, prion replication in the brain leads to sustained over-activity of the unfolded protein response that controls the initiation of protein synthesis. This causes a persistent repression of protein translation which ultimately leads to synaptic failure and neuronal death [389]. Oral treatment of mice with a specific inhibitor of the kinase PERK (protein kinase RNA-like endoplasmic reticulum kinase), a key mediator of the unfolded protein response pathway, can prevent the development of neurodegeneration and clinical prion disease [390].

4.4. Immunization

There have been many elegant experimental attempts to develop novel immunotherapeutic approaches to induce host immunity to prions [391,392]. A major barrier to the effectiveness of many of these is T cell-tolerance towards PrP due to the widespread expression of cellular PrP^C throughout the mammalian body. This has made the development of effective PrP^{Sc} prion-specific vaccines extremely challenging. However, studies predominantly performed in mice have shown that immunization against PrP or the passive administration of PrP-specific monoclonal antibodies are potential approaches to block prion disease transmission [393,394]. At the time of writing, the MRC Prion Unit in the UK planned to undergo a clinical trial to passively administer a small number of sCJD patients with a human anti-PrP antibody to attempt to block further neurodegeneration.

Mucosal vaccination appears to be the most appropriate method for prophylactic protection against orally acquired prion infections [395,396]. However, a mucosal vaccine may offer little protection against accidental iatrogenic CJD transmissions where infection occurs via intravenous transfusion of contaminated blood or blood products, transplantation of tissues, or use of contaminated surgical instruments. Therefore, a useful anti-prion vaccine should be able to induce both strong mucosal and systemic anti-PrP antibody responses.

Since the cellular prion protein is almost ubiquitously expressed, the potential for an anti-prion vaccine to recognize host PrP^C and cause autoimmunity must not be ignored. However, detailed comparisons of several anti-prion antibodies have identified those that target the α1 and α3 helices of PrP^C can induce rapid neurotoxicity [397].

5. Conclusions

Substantial progress have been made in our understanding of how PrP^{Sc} prions disseminate between and within individuals. However, many important issues with implications for animal and human health remain unresolved. The transmission of prions from one host species to other hosts of the same species is typically efficient, and causes disease in the recipients with highly reproducible disease characteristics. However, inter-species prion transmissions upon first passage are typically characterized by their low efficiency and extended disease durations. This effect on prion transmission is termed the 'species barrier' effect. Many factors are known to have an important influence on the inter-species transmission of prions such as polymorphisms and mutations in the *PRNP* gene, and biophysical aspects of templating events are clearly important (see above). Unfortunately the precise molecular mechanism/s responsible for the species barrier effect is uncertain. An ability to predict the potential for a novel prion isolate to have the potential to transmit to other species, especially humans, is crucial to restrict and control future prion disease outbreaks.

Despite the widespread exposure of the UK population to BSE-contaminated food in the 1980s, there have fortunately been much fewer human clinical vCJD cases than anticipated. However, retrospective analyses of human tonsils and appendix indicate a higher incidence of pre-clinically infected individuals (~1/2000) than the clinical data suggest [161–163]. This implies that many individuals may harbor detectable levels of prions in their tissues in the absence of clinical signs, and the potential existence of a subclinical carrier state. The factors which can influence the progression of CNS prion disease in preclinically-affected individuals are not known. A thorough understanding the factors which can influence the progression of the preclinical phase will identify those which enhance the risk of developing clinical prion disease.

Once the prions have been amplified on the surfaces of FDC above the threshold required for neuroinvasion, they infect the enteric nerves within the intestine [146,176,276]. Although the relative positioning of the FDC and sympathetic nerves within the SLO appears to influence the rate of neuroinvasion [262], but how the infection is propagated between FDC and enteric nerves is unknown. By identifying how the prions initially infect the nervous system may identify a novel methods to block the spread of prions to the CNS.

Unfortunately there are no safe or effective treatments which can be used to cure or block further spread of these devastating neurodegenerative diseases in humans. Some exciting experimental advances have been made, but trials in larger animal species including humans will be required to determine their efficacy and safety. Currently, advances in gene editing techniques are rapidly enhancing the ability to repair the genomes of human cell populations to treat certain previously incurable diseases associated with specific gene mutations. Whether a similar gene-editing approach can be used to block prion infection in host tissues by inserting protective mutations within the *PRNP* gene [398] is an exciting prospect for future research.

Acknowledgments: This work was supported by project (grants BB/F019726/1; BB/J014672/1; BB/K021257/1) and Institute Strategic Programme Grant funding (grant numbers BB/J004332/1 and BB/P013740/1) from the Biotechnology and Biological Sciences Research Council. Funds to support Open Access Funding were obtained from the RCUK UK Open Access Fund provided to the University of Edinburgh.

Author Contributions: N.A.M. wrote the paper.

Conflicts of Interest: The author declares no conflict of interest. The sponsors had no role in the writing of the manuscript, and in the decision to publish the results.

References

1. Bolton, D.C.; McKinley, M.P.; Prusiner, S.B. Identification of a protein that purifies with the scrapie prion. *Science* **1982**, *218*, 1309–1311. [CrossRef] [PubMed]
2. Legname, G.; Baskakov, I.V.; Nguyen, H.-O.B.; Riesner, D.; Cohen, F.E.; DeArmond, S.J.; Prusiner, S.B. Synthetic mammalian prions. *Science* **2004**, *305*, 673–676. [CrossRef] [PubMed]
3. Cassard, H.; Torres, J.M.; Lacroux, C.; Douet, J.Y.; Benestad, S.L.; Lantier, F.; Lugan, S.; Lantier, I.; Costes, P.; Aron, N.; et al. Evidence for zoonotic potential of ovine scrapie prions. *Nat. Commun.* **2014**, *5*, 5821. [CrossRef] [PubMed]
4. Woolhouse, M.E.; Mathews, L.; Coen, P.; Stringer, S.M.; Foster, J.D.; Hunter, N. Population dynamics of scrapie in a flock. *Philos. Trans. R. Soc. Lond. B Biol. Sci.* **1999**, *354*, 751–756. [CrossRef] [PubMed]
5. Miller, M.W.; Williams, E.S. Prion disease: Horizontal prion transmission in mule deer. *Nature* **2003**, *425*, 35–36. [CrossRef] [PubMed]
6. Moore, S.J.; Kunkle, R.; Greenlee, M.H.; Nicholson, E.; Richt, J.; Hamir, A.; Waters, W.R.; Greenlee, J. Horizontal transmission of chronic wasting disease. *Emerg. Infect. Dis.* **2016**, *22*, 2142–2145. [CrossRef] [PubMed]
7. Ferguson, N.M.; Donnelly, C.A.; Woolhouse, M.E.; Anderson, R.M. The epidemiology of BSE in cattle herds in Great Britain. II. Model construction and analysis of transmission dynamics. *Philos. Trans. R. Soc. Lond. B Biol. Sci.* **1997**, *352*, 803–838. [CrossRef] [PubMed]

8. Donnelly, C.A.; Ferguson, N.M.; Ghani, A.C.; Woolhouse, M.E.; Watt, C.J.; Anderson, R.M. The epidemiology of BSE in cattle herds in Great Britain. I. Epidemiological processes, demography of cattle and approaches to control by culling. *Philos. Trans. R. Soc. Lond. B Biol. Sci.* **1997**, *352*, 781–801. [CrossRef] [PubMed]

9. Bruce, M.E.; Will, R.G.; Ironside, J.W.; McConnell, I.; Drummond, D.; Suttie, A.; McCardle, L.; Chree, A.; Hope, J.; Birkett, C.; et al. Transmissions to mice indicate that 'new variant' CJD is caused by the BSE agent. *Nature* **1997**, *389*, 498–501. [CrossRef] [PubMed]

10. Hill, A.F.; Desbruslais, M.; Joiner, S.; Sidle, K.C.L.; Gowland, I.; Collinge, J. The same prion strain causes vCJD and BSE. *Nature* **1997**, *389*, 448–450. [CrossRef] [PubMed]

11. Fraser, H.; Pearson, G.R.; McConnell, I.; Bruce, M.E.; Wyatt, M.E.; Gruffydd-Jones, T.J. Transmission of feline spongiform encephalopathy to mice. *Vet. Rec.* **1994**, *134*, 449. [CrossRef] [PubMed]

12. Kirkwood, J.K.; Wells, G.A.; Wilesmith, J.W.; Cunningham, A.A.; Jackson, S.I. Spongiform encephalopathy in an arabian oryx (*Oryx leucoryx*) and a greater kudu (*Tragelaphus strepsiceros*). *Vet. Rec.* **1990**, *127*, 418–420. [PubMed]

13. Jeffrey, M.; Scott, J.R.; Williams, A.; Fraser, H. Ultrastructural features of spongiform encephalopathy transmitted to mice from three species of bovidae. *Acta Neuropathol.* **1992**, *84*, 559–569. [CrossRef] [PubMed]

14. John, T.R.; Schatzl, H.M.; Gilch, S. Early detection of chronic wasting disease prions in urine of pre-symptomatic deer by real-time quaking-induced coversion assay. *Prion* **2013**, *7*, 253–258. [CrossRef] [PubMed]

15. Haley, N.J.; Mathiason, C.K.; Carver, S.; Zabel, M.; Telling, G.C.; Hoover, E.A. Detection of chronic wasting disease prions in salivary, urinary, and intestinal tissues of deer: Potential mechanisms of prion shedding and transmission. *J. Virol.* **2011**, *85*, 6309–6318. [CrossRef] [PubMed]

16. Gregori, L.; Kovacs, G.G.; Alexeeva, I.; Budka, H.; Rohwer, R.G. Excretion of transmissible spongiform encephalopathy infectivity in urine. *Emerg. Infect. Dis.* **2008**, *14*, 1406–1412. [CrossRef] [PubMed]

17. Murayama, Y.; Yoshioka, M.; Okada, H.; Takata, M.; Yokoyama, T.; Mohri, S. Urinary excretion and blood level of prions in scrapie-infected hamsters. *J. Gen. Virol.* **2007**, *88*, 2890–2898. [CrossRef] [PubMed]

18. Ligios, C.; Cancedda, G.M.; Margalith, I.; Santuccii, C.; Madau, L.; Maestrale, C.; Basagni, M.; Saba, M.; Heikenwalder, M. Intraepithelial and interstitial deposition of pathological prion protein in kidneys of scrapie-affected sheep. *PLoS ONE* **2007**, *2*, e859. [CrossRef] [PubMed]

19. Tamguney, G.; Miller, M.W.; Wolfe, L.L.; Sirochman, T.M.; Glidden, D.V.; Palmer, C.; Lemus, A.; DeArmond, S.J.; Prusiner, S.B. Asymptomatic deer excrete infectious prions in faeces. *Nature* **2009**, *461*, 529–532. [CrossRef] [PubMed]

20. Safar, J.G.; Lessar, P.; Tamguney, G.; Freyman, Y.; Deering, C.; Letessier, F.; DeArmond, S.J.; Prusiner, S.B. Transmission and detection of prions in feces. *J. Infect. Dis.* **2008**, *198*, 81–89. [CrossRef] [PubMed]

21. Tamguney, G.; Richt, J.A.; Hamir, A.N.; Greenlee, J.J.; Miller, M.W.; Wolfe, L.L.; Sirochman, T.M.; Young, A.J.; Gidden, D.V.; Johnson, N.L.; et al. Salivary prions in sheep and deer. *Prion* **2012**, *6*, 52–61. [CrossRef] [PubMed]

22. Gough, K.C.; Baker, C.A.; Rees, H.C.; Terry, L.A.; Spiropoulos, J.; Thorne, L.; Maddison, B.C. The oral secretion of infectious scrapie prions occurs in preclinical sheep with a range of *PRNP* genotypes. *J. Virol.* **2012**, *86*, 566–571. [CrossRef] [PubMed]

23. Vascellari, M.; Nonno, R.; Mutinelli, F.; Bigolaro, M.; Di Bari, M.A.; Melchiotti, E.; Marcon, S.; D'Agostino, C.; Vaccari, G.; Conte, M.; et al. PrP^{Sc} in salivary glands of scrapie-affected sheep. *J. Virol.* **2007**, *81*, 4872–4876. [CrossRef] [PubMed]

24. Bessen, R.A.; Shearin, H.; Martinka, S.; Boharski, R.; Lowe, D.; Wilham, J.M.; Caughey, B.; Wiley, J.A. Prion shedding from olfactory neurons into nasal secretions. *PLoS Pathog.* **2010**, *6*, e1000837. [CrossRef] [PubMed]

25. Bessen, R.A.; Wilham, J.M.; Lowe, D.; Watschke, C.P.; Shearin, H.; Martinka, S.; Caughey, B.; Wiley, J.A. Accelerated shedding of prions following damage to the olfactory epithelium. *J. Virol.* **2012**, *86*, 1777–1788. [CrossRef] [PubMed]

26. Plummer, I.H.; Wright, S.D.; Johnson, C.J.; Pedersen, J.A.; Samuel, M.D. Temporal patterns of chronic wasting disease prion excretion in three cervid species. *J. Gen. Virol.* **2017**, *98*, 1932–1942. [CrossRef] [PubMed]

27. Seeger, H.; Heikenwalder, M.; Zeller, N.; Kranich, J.; Schwarz, P.; Gaspert, A.; Seifert, B.; Miele, G.; Aguzzi, A. Coincident scrapie infection and nephritis lead to urinary prion excretion. *Science* **2005**, *310*, 324–326. [CrossRef] [PubMed]

28. Maddison, B.C.; Baker, C.A.; Terry, L.A.; Bellworthy, S.J.; Thorne, L.; Rees, H.C.; Gough, K.C. Environmental sources of scrapie prions. *J. Virol.* **2010**, *84*, 11560–11562. [CrossRef] [PubMed]

29. Maddison, B.C.; Owen, J.P.; Bishop, K.; Shaw, G.; Rees, H.C.; Gough, K.C. The interaction of ruminant PrP(Sc) with soils is influenced by prion source and soil type. *Environ. Sci. Technol.* **2010**, *44*, 8503–8508. [CrossRef] [PubMed]

30. Wyckoff, A.C.; Kane, S.; Lockwood, K.; Seligman, J.; Michel, B.; Hill, D.; Ortega, A.; Mangalea, M.R.; Telling, G.C.; Miller, M.W.; et al. Clay components in soil dictate environmental stability and bioavailability of cervid prions in mice. *Front. Microb.* **2016**, *7*, 1885. [CrossRef] [PubMed]

31. Pritzkow, S.; Morales, R.; Moda, F.; Khan, U.; Telling, G.C.; Hoover, E.; Soto, C. Grass plants bind, retain, uptake, and transport infectious prions. *Cell Rep.* **2015**, *11*, 1168–1175. [CrossRef] [PubMed]

32. Johnson, C.J.; Pedersen, J.A.; Chappell, R.J.; McKenzie, D.; Aiken, J.M. Oral transmissibility of prion disease is enhanced by binding soil particles. *PLoS Pathog.* **2007**, *3*, e93. [CrossRef] [PubMed]

33. Kincaid, A.E.; Bartz, J.C. The nasal cavity is a route for prion infection in hamsters. *J. Virol.* **2007**, *81*, 4482–4491. [CrossRef] [PubMed]

34. Hamir, A.N.; Kunkle, R.A.; Richt, J.A.; MIller, J.M.; Greenlee, J.J. Experimental transmission of US scrapie agent by nasal, peritoneal, and conjuctival routes to genetically susceptible sheep. *Vet. Pathol.* **2008**, *45*, 7–11. [CrossRef] [PubMed]

35. Bessen, R.A.; Martinka, S.; Kelly, J.; Gonzales, D. Role of the lymphoreticular system in prion neuroinvasion from the oral and nasal mucosa. *J. Virol.* **2009**, *83*, 6435–6445. [CrossRef] [PubMed]

36. Denkers, N.D.; Seelig, D.M.; Telling, G.C.; Hoover, E.A. Aerosol and nasal transmission of chronic wasting disease in cervidized mice. *J. Gen. Virol.* **2010**, *91*, 1651–1658. [CrossRef] [PubMed]

37. Haybaeck, J.; Heikenwalder, M.; Klevenz, B.; Schwarz, P.; Margalith, I.; Bridel, C.; Mertz, K.; Zirdum, E.; Petsch, B.; Fuchs, T.J.; et al. Aerosols transmit prions to immunocompetent and immunodeficient mice. *PLoS Pathog.* **2011**, *7*, e1001257. [CrossRef] [PubMed]

38. Kincaid, A.E.; Ayers, J.I.; Bartz, J.C. Specificity, size and frequency of spaces that characterize the mechanism of bulk transepithelial transport of prions in the nasal cavities of hamsters and mice. *J. Virol.* **2016**, *90*, 8293–8301. [CrossRef] [PubMed]

39. Taylor, D.M.; McConnell, I.; Fraser, H. Scrapie infection can be established readily through skin scarification in immunocompetent but not immunodeficient mice. *J. Gen. Virol.* **1996**, *77*, 1595–1599. [CrossRef] [PubMed]

40. Gossner, A.; Hunter, N.; Hopkins, J. Role of lymph-borne cells in the early stages of scrapie agent dissemination from the skin. *Vet. Immunol. Immunopathol.* **2005**, *109*, 267–278. [CrossRef] [PubMed]

41. Bartz, J.C.; Kincaid, A.E.; Bessen, R.A. Rapid prion neuroinvasion following tongue infection. *J. Virol.* **2003**, *77*, 583–591. [CrossRef] [PubMed]

42. Carp, R. Transmission of scrapie by oral route: Effect of gingival scarification. *Lancet* **1982**, *1*, 170–171. [CrossRef]

43. Denkers, N.D.; Telling, G.C.; Hoover, E.A. Minor oral lesions facilitate transmission of chronic wasting disease. *J. Virol.* **2011**, *85*, 1396–1399. [CrossRef] [PubMed]

44. Crowell, J.; Wiley, J.A.; Bessen, R.A. Lesion of the alfactory epithelium accelerates prion neuroinvasion and disease onset when prion replication is restricted to neurons. *PLoS ONE* **2015**, *10*, e0119863. [CrossRef] [PubMed]

45. Brown, P.; Preece, M.A.; Will, R.G. "Friendly fire" in medicine: Hormones, homografts and Creutzfeldt-Jakob disease. *Lancet* **1992**, *340*, 24–27. [CrossRef]

46. Frontzek, K.; Moos, R.; Schaper, E.; Jann, L.; Herfs, G.; ZImmermann, D.R.; Aguzzi, A.; Budka, H. Iatrogenic and sporadic Creutzfeldt-Jakob disease in 2 sisters without mutation in the prion protein gene. *Prion* **2015**, *9*, 444–448. [CrossRef] [PubMed]

47. Okada, H.; Sakurai, M.; Yokoyama, T.; Mohri, S. Disease-associated prion protein in the dental tissue of mice infected with scrapie. *J. Comp. Pathol.* **2010**, *143*, 218–222. [CrossRef] [PubMed]

48. Everington, D.; Smith, A.J.; Ward, H.J.T.; Letters, S.; Will, R.G.; Bagg, J. Dental treatment and risk of variant CJD—A case control study. *Br. Dent. J.* **2007**, *202*, E19. [CrossRef] [PubMed]

49. Brown, P.; Rohwer, R.G.; Dunstan, B.C.; MacAuley, C.; Gajdusek, D.C.; Drohan, W.N. The distribution of infectivity in blood components and plasma derivatives in experimental models of transmissible spongiform encephalopathy. *Transfusion* **1998**, *38*, 810–816. [CrossRef] [PubMed]

50. Brown, P.; Cervenakova, L.; McShane, L.M.; Barber, P.; Rubenstein, R.; Drohan, W.N. Further studies of blood infectivity in an experimental model of transmissible spongiform encephalopathy, with an explanation of why blood components do not transmit Creutzfeldt-Jakob disease in humans. *Transfusion* **1999**, *39*, 1169–1178. [CrossRef] [PubMed]

51. Brown, P. Creutzfeldt-Jakob disease: Blood infectivity and screeing tests. *Semin. Hematol.* **2001**, *38* (Suppl. 9), 2–6. [CrossRef]

52. Cervenakova, L.; Yakovleva, O.; McKenzie, C.; Kolchinsky, S.; McShane, L.M.; Drohan, W.N.; Brown, P. Similar levels of infectivity in the blood of mice infected with human-derived vCJD and GSS strains of transmissible spongiform encephalopathy. *Transfusion* **2003**, *43*, 1687–1694. [CrossRef] [PubMed]

53. Holada, K.; Vostal, J.G.; Theisen, P.W.; MacAuley, C.; Gregori, L.; Rohwer, R.G. Scrapie infectivity in hamster blood is not associated with platelets. *J. Virol.* **2002**, *76*, 4649–4650. [CrossRef] [PubMed]

54. Houston, F.; Foster, J.D.; Chong, A.; Hunter, N.; Bostock, C.J. Transmission of BSE by blood transfusion in sheep. *Lancet* **2000**, *356*, 999. [CrossRef]

55. Hunter, N.; Foster, J.; Chong, A.; McCutcheon, S.; Parnham, D.; Eaton, S.; MacKenzie, C.; Houston, F. Transmission of prion diseases by blood transfusion. *J. Gen. Virol.* **2002**, *83*, 2897–2905. [CrossRef] [PubMed]

56. Brown, P. The risk of blood-borne Creutzfeldt-Jakob Disease. In *Transmissible Subacute Spongiform Encephalopathies: Prion Diseases*; Court, L., Dodet, B., Eds.; Elsevier: Paris, France, 1996; pp. 447–450.

57. Wietjens, D.P.W.M.; Davanipour, Z.; Hofman, A.; Kondo, K.; Martthews, W.B.; Will, R.G.; van Duijn, C.M. Risk factors for Creutzfeldt-Jakob disease: A reanalysis of case-control studies. *Neurology* **1996**, *47*, 1287–1291. [CrossRef]

58. Van Duijn, C.M.; Delasnerie-Lauprêtre, N.; Masullo, C.; Zerr, I.; De Silva, R.; Wietjens, D.P.W.M.; Brandel, J.-P.; Weber, T.; Bonavita, V.; Zeilder, M.; et al. Case-control study of risk factors of Creutzfeldt-Jakob disease in Europe during 1993–1995. *Lancet* **1998**, *351*, 1081–1085. [CrossRef]

59. Collins, S.; Law, M.G.; Fletcher, A.; Boyd, A.; Kaldor, J.; Masters, C.L. Surgical treatment and risk of sporadic Creutzfeldt-Jakob disease: A case-control study. *Lancet* **1999**, *353*, 693–697. [CrossRef]

60. Unwin, P.J.; Mackenzie, J.M.; Llewelyn, C.A.; Will, R.G.; Hewitt, P.E. Creutzfeldt-Jakob disease and blood transfusion: Updated results of the UK transfusion medicine epidemiology review study. *Vox Sang.* **2016**, *110*, 310–316.

61. Douet, J.Y.; Zafar, S.; Perret-Liaudet, A.; Lacroux, C.; Lugan, S.; Aron, N.; Cassard, H.; Ponto, C.; Corbiere, F.; Torres, J.M.; et al. Detection of infectivity in blood of persons with variant and sporadic Creutzfeldt-Jakob disease. *Emerg. Infect. Dis.* **2014**, *20*, 114–117. [CrossRef] [PubMed]

62. Llewelyn, C.A.; Hewitt, P.E.; Knight, R.S.G.; Amar, K.; Cousens, S.; Mackenzie, J.; Will, R.G. Possible transmission of variant Creutzfeldt-Jakob disease by blood transfusion. *Lancet* **2004**, *363*, 417–421. [CrossRef]

63. Peden, A.H.; Head, M.W.; Ritchie, D.L.; Bell, J.E.; Ironside, J.W. Preclinical vCJD after blood transfusion in a *PRNP* codon 129 heterozygous patient. *Lancet* **2004**, *354*, 527–529. [CrossRef]

64. Wroe, S.J.; Pal, S.; Siddique, D.; Hyare, H.; Macfarlane, R.; Joiner, S.; Lineham, J.M.; Brandner, S.; Wadsworth, J.D.F.; Hewitt, P.; et al. Clinical presentation and pre-mortem diagnosis of variant Creutzfeldt-Jakob disease associated with blood transfusion: A case report. *Lancet* **2006**, *368*, 2061–2067. [CrossRef]

65. Health Protection Agency. vCJD Abnormal Protein Found in a Patient with Haemophilia at Post Mortem. Available online: http://webarchive.nationalarchives.gov.uk/20140714094822tf_/http://www.hpa.org.uk/NewsCentre/NationalPressReleases/2009PressReleases/090217vCJDABNORMALPRIONPROTEINFOUNDINAPATIENTWITH/ (accessed on 23 November 2017).

66. Hill, A.F.; Zeidler, M.; Ironside, J.; Collinge, J. Diagnosis of new variant Creutzfeldt-Jakob disease by tonsil biopsy. *Lancet* **1997**, *349*, 99–100. [CrossRef]

67. Hill, A.F.; Butterworth, R.J.; Joiner, S.; Jackson, G.; Rossor, M.N.; Thomas, D.J.; Frosh, A.; Tolley, N.; Bell, J.E.; Spencer, M.; et al. Investigation of variant Creutzfeldt-Jakob disease and other prion diseases with tonsil biopsy samples. *Lancet* **1999**, *353*, 183–189. [CrossRef]

68. Gregori, L.; McCombie, N.; Palmer, D.; Birch, P.; Sowemimo-Coker, S.O.; Giulivi, A.; Rohwer, R.G. Effectiveness of leucoreduction for removal of infectivity of transmissible spongiform encephalopathy. *Lancet* **2004**, *364*, 529–531. [CrossRef]

69. Hoinville, L.J.; Tongue, S.C.; Wilesmith, J.W. Evidence for maternal transmission of scrapie in naturally affected flocks. *Prev. Vet. Med.* **2010**, *93*, 121–128. [CrossRef] [PubMed]

70. Sarradin, P.; Melo, S.; Barc, C.; Lecomte, C.; Andréoletti, O.; Lantier, F.; Dacheux, J.L.; Gatti, J.L. Semen from scrapie-infected rams does not transmit prion infection to transgenic mice. *Reproduction* **2008**, *135*, 415–418. [CrossRef] [PubMed]

71. Nalls, A.V.; McNulty, E.; Powers, J.; Seelig, D.M.; Hoover, C.; Haley, N.J.; Hayes-Klug, J.; Anderson, K.; Stewart, P.; Goldmann, W.; et al. Mother to offspring transmission of chronic wasting disease in reeves' muntjac deer. *PLoS ONE* **2013**, *8*, e71844. [CrossRef] [PubMed]

72. Selariu, A.; Powers, J.G.; Nalls, A.; Brandhuber, M.; Mayfield, A.; Fullaway, S.; Wyckoff, C.A.; Goldmann, W.; Zabel, M.M.; Wild, M.A.; et al. In utero transmission and tissue distribution of chronic wasting disease-associated prions in free-ranging Rocky Mountain elk. *J. Gen. Virol.* **2015**, *96*, 3444–3455. [CrossRef] [PubMed]

73. Wrathall, A.E.; Brown, K.F.; Sayers, A.R.; WElls, G.A.; Simmons, M.M.; Farrelly, S.S.; Bellerby, P.; Squirrell, J.; Spencer, Y.I.; Wells, M.; et al. Studies of embryo transfer from cattle clinically affected by bovine spongiform encephalopathy (BSE). *Vet. Rec.* **2002**, *150*, 365–378. [CrossRef] [PubMed]

74. Castilla, J.; Brun, A.; Díaz-San Segundo, F.; Gutiérrez-Adán, A.; Pintado, B.; Ramírez, M.A.; del Riego, L.; Torres, J.M. Vertical transmission of bovine spongiform encephalopathy prions evaluated in a transgenic mouse model. *J. Virol.* **2005**, *79*, 8665–8668. [CrossRef] [PubMed]

75. Wilesmith, J.W.; Wells, G.A.; Ryan, J.B.; Gavier-Widen, D.; Simmons, M.M. A cohort study to examine maternally-associated risk factors for bovine spongiform encephalopathy. *Vet. Rec.* **1997**, *141*, 239–243. [CrossRef] [PubMed]

76. Donnelly, C.A.; Ferguson, N.M.; Ghani, A.C.; Wilesmith, J.W.; Anderson, R.M. Analysis of dam-calf pairs of BSE cases: Confirmation of maternal risk enhancement. *Proc. R. Soc. Lond. B* **1997**, *264*, 1647–1656. [CrossRef] [PubMed]

77. Foster, J.; McKelvey, W.; Fraser, H.; Chong, A.; Ross, A.; Parnham, D.; Goldmann, W.; Hunter, N. Experimentally induced bovine spongiform encephalopathy did not transmit via goat embryos. *J. Gen. Virol.* **1999**, *80*, 517–524. [CrossRef] [PubMed]

78. Foster, J.D.; Goldmann, W.; McKenzie, C.; Smith, A.; Parnham, D.; Hunter, N. Maternal transmission studies of BSE in sheep. *J. Gen. Virol.* **2004**, *85*, 3159–3163. [CrossRef] [PubMed]

79. Foote, W.C.; Clarke, W.; Maciulis, A.; Call, J.W.; Hourrigan, J.; Evans, R.C.; Marshall, M.R.; de Camp, M. Prevention of scrapie transmission in sheep, using embryo transfer. *Am. J. Vet. Res.* **1993**, *54*, 1863–1868. [PubMed]

80. Wang, S.; Foote, W.C.; Sutton, D.L.; Maciulis, A.; Miller, J.M.; Evans, R.C.; Holyoak, G.R.; Call, J.W.; Bunch, T.D.; Taylor, W.D.; et al. Preventing experimental vertical transmission of scrapie by embryo transfer. *Theriogenology* **2001**, *56*, 315–327. [CrossRef]

81. Low, J.C.; Chambers, J.; McKelvey, W.A.; McKendrick, I.J.; Jeffrey, M. Failure to transmit scrapie infection by transferring preimplantation embryos from naturally infected donor sheep. *Theriogenology* **2009**, *72*, 809–816. [CrossRef] [PubMed]

82. Wang, S.; Cockett, N.E.; MIller, J.M.; Shay, T.L.; Maciulis, A.; Sutton, D.L.; Foote, W.C.; Holyoak, G.R.; Evans, R.C.; Bunc, T.D.; et al. Polymorphic distribution of the ovine prion protein (PrP) gene in scrapie-infected sheep flocks in which embryo transfer was used to circumvent the transmission of scrapie. *Theriogenology* **2002**, *57*, 1865–1875. [CrossRef]

83. Andréoletti, O.; Lacroux, C.; Chabert, A.; Monnereau, L.; Tabouret, G.; Lantier, F.; Berthon, P.; Eychenne, P.; Lafond-Benestad, S.; Elsen, J.M.; et al. PrP(Sc) accumulation in placentas of ewes exposed to natural scrapie: Influence of foetal PrP genotype and effect on ewe-to-lamb transmission. *J. Gen. Virol.* **2002**, *83*, 2607–2616. [CrossRef] [PubMed]

84. Schneider, D.A.; Madsen-Bouterse, S.A.; Zhuang, D.; Truscott, T.C.; Dassayake, R.P.; O'Rourke, K.I. The placenta shed from goats with classical scrapie is infectious to goat kids and lambs. *J. Gen. Virol.* **2015**, *96*, 2464–2469. [CrossRef] [PubMed]

85. Gonzalez, L.; Dalgleish, M.P.; Martin, S.; Finlayson, J.; Siso, S.; Eaton, S.L.; Goldmann, W.; Witz, J.; Hamilton, S.; Stewart, P.; et al. Factors influencing temporal variation of scrapie incidence within a clsoed Suffolk sheep flock. *J. Gen. Virol.* **2012**, *93*, 203–211. [CrossRef] [PubMed]

86. Foster, J.D.; Hunter, N.; Williams, A.; Mylne, M.J.; McKelvey, W.A.; Hope, J.; Fraser, H.; Bostock, C. Observations on the transmission of scrapie in experiments using embryo transfer. *Vet. Rec.* **1996**, *138*, 559–562. [CrossRef] [PubMed]

87. Foster, J.D.; McKelvey, W.A.; Mylne, M.J.; Williams, A.; Hunter, N.; Hope, J.; Fraser, H. Studies on maternal transmission of scarpie in sheep by embryo transfer. *Vet. Rec.* **1992**, *130*, 341–343. [CrossRef] [PubMed]
88. Foster, J.D.; Goldmann, W.; Hunter, N. Evidence in sheep for pre-natal transmission of scrapie to lambs from infected mothers. *PLoS ONE* **2013**, *8*, e79433. [CrossRef] [PubMed]
89. Garza, M.C.; Fernandez-Borges, N.; Bolea, R.; Badiola, J.J.; Castilla, J.; Monleon, E. Detection of PrPres in genetically susceptible fetuses from sheep with natural scrapie. *PLoS ONE* **2011**, *6*, e27525. [CrossRef] [PubMed]
90. Spiropoulos, J.; Hawkins, S.A.; Simmons, M.M.; Bellworthy, S.J. Evidence of in utero transmission of classical scrapie in sheep. *J. Virol.* **2014**, *88*, 4591–4594. [CrossRef] [PubMed]
91. Alverson, J.; O'Rourke, K.I.; Baszler, T.V. PrP^{Sc} accumulation in fetal cotyledons of scrapie-resistant lambs is influenced by fetus location in the uterus. *J. Gen. Virol.* **2006**, *87*, 1035–1041. [CrossRef] [PubMed]
92. Murray, K.; Peters, J.; Stellitano, L.; Winstone, A.M.; Verity, C.; Will, R.G. Is there evidence of vertical transmission of variant Creutzfeldt-Jakob disease. *J. Neurol. Neurosurg. Psychiatry* **2011**, *82*, 729–731. [CrossRef] [PubMed]
93. Amyx, H.L.; Gibbs, C.J., Jr.; Gadjusek, D.C.; Greer, W.E. Absence of vertical transmission of subacute spongiform viral encephalopathies in experimental primates. *Proc. Soc. Exp. Biol. Med.* **1981**, *166*, 469–471. [CrossRef]
94. Xiao, X.; Miravalle, L.; Yuan, J.; McGheehan, J.; Dong, Z.; Wyza, R.; MacLennan, G.T.; Golichowski, A.M.; Kneale, G.; King, N.; et al. Failure to detect the presence of prions in the uterine and gestational tissues from a Gravida with Creutzfeldt-Jakob disease. *Am. J. Pathol.* **2009**, *174*, 1602–1608. [CrossRef] [PubMed]
95. Garza, M.C.; Monzon, M.; Marin, B.; Badiola, J.J.; Monleon, E. Distribution of peripheral PrP(Sc) in sheep with naturally acquired scrapie. *PLoS ONE* **2014**, *9*, e97768. [CrossRef] [PubMed]
96. Maestrale, C.; Di Guardo, G.; Cancedda, M.G.; Marruchella, G.; Masia, M.; Sechi, S.; Macciocu, S.; Santucciu, C.; Petruzzi, M.; Ligios, C. A lympho-follicular microenvironment is required for pathological prion protein deposition in chronically inflamed tissues from scrapie-affected sheep. *PLoS ONE* **2013**, *8*, e62830. [CrossRef] [PubMed]
97. Ligios, C.; Cancedda, M.G.; Carta, A.; Santucciu, C.; Maestrale, C.; Demontis, F.; Saba, M.; Patta, C.; DeMartini, J.C.; Aguzzi, A.; et al. Sheep with scrapie and mastitis transmit infectious prions through the milk. *J. Virol.* **2011**, *85*, 1136–1139. [CrossRef] [PubMed]
98. Salazar, E.; Monleon, E.; Bolea, R.; Acin, C.; Perez, M.; Alvarez, N.; Leginagoikoa, I.; Juste, R.; Minguijon, E.; Reina, R.; et al. Detection of PrP^{Sc} in lung and mammary gland is favoured by the presence of Visna/maedi virus lesions in naturally coinfected sheep. *Vet. Res.* **2010**, *41*, 58. [CrossRef] [PubMed]
99. Konold, T.; Moore, S.J.; Bellworthy, S.J.; Terry, L.A.; Thorne, L.; Ramsay, A.; Salguerro, F.J.; Simmons, M.M.; Simmons, H.A. Evidence of effective scrapie transmission via colostrum and milk in sheep. *BMC Vet. Res.* **2013**, *9*, 99. [CrossRef] [PubMed]
100. Lacroux, C.; Simon, S.; Benenstad, S.L.; Maillet, S.; Mathey, J.; Lugan, S.; Corbiere, F.; Cassard, H.; Costes, P.; Bergonier, D.; et al. Prions in milk from ewes incubating natural scrapie. *PLoS Pathog.* **2008**, *4*, e1000238. [CrossRef] [PubMed]
101. Everest, S.J.; Thorne, L.T.; Hawthorne, J.A.; Jenkins, R.; Hammersley, C.; Ramsay, A.M.; Hawkins, S.A.; Venables, L.; Flynn, L.; Sayers, R.; et al. No abnormal prion protein detected in the milk of cattle infected with the bovine spongiform encephalopathy agent. *J. Gen. Virol.* **2006**, *87*, 2433–2441. [CrossRef] [PubMed]
102. Manson, J.C.; Clarke, A.R.; McBride, P.A.; McConnell, I.; Hope, J. PrP gene dosage determines the timing but not the final intensity or distribution of lesions in scrapie pathology. *Neurodegeneration* **1994**, *3*, 331–340. [PubMed]
103. Prusiner, S.B.; Groth, D.; Serban, A.; Koehler, R.; Foster, D.; Torchia, M.; Burton, D.; Yang, S.-L.; DeArmond, S.J. Ablation of the Prion Protein (PrP) Gene in Mice Prevents Scrapie and Facilitates Production of Anti-PrP Antibodies. *Proc. Natl. Acad. Sci. USA* **1993**, *90*, 10608–10612. [CrossRef] [PubMed]
104. Prusiner, S.B. Novel proteinaceous infectious particles cause scrapie. *Science* **1982**, *216*, 136–144. [CrossRef] [PubMed]
105. Wang, F.; Wang, X.; Yuan, C.G.; Ma, J. Generating a prion with bacterially expressed recombinant prion protein. *Science* **2010**, *327*, 1132–1135. [CrossRef] [PubMed]
106. Stahl, N.; Borchelt, D.R.; Hsiao, K.; Prusiner, S.B. Scrapie prion protein contains a phosphatidylinositol gylcolipid. *Cell* **1987**, *51*, 229–240. [CrossRef]

107. Pan, K.-M.; Baldwin, M.; Nguyen, J.; Gasset, M.; Serban, A.; Groth, D.; Mehlhorn, I.; Huang, Z.; Fletterick, R.J.; Cohen, F.E.; et al. Conversion of alpha-helices into beta-sheets features in the formation of the scrapie prion protein. *Proc. Natl. Acad. Sci. USA* **1993**, *90*, 10962–10966. [CrossRef] [PubMed]

108. Riek, R.; Hornemann, S.; Wider, G.; Glockshuber, R.; Wuthrich, K. NMR characterization of the full-length recombinant murine prion protein, *m*PrP. *FEBS Lett.* **1997**, *413*, 282–288. [CrossRef]

109. Caughey, B.W.; Dong, A.; Bhat, K.S.; Ernst, D.; Hayes, S.F.; Caughey, W.S. Secondary structure analysis of the scrapie-associated protein PrP 27–30 in water by infrared spectroscopy. *Biochemistry* **1991**, *30*, 7672–7680. [CrossRef] [PubMed]

110. Deleault, N.R.; Lucassen, R.W.; Supattapone, S. RNA molecules stimulate prion protein conversion. *Nature* **2003**, *425*, 717–720. [CrossRef] [PubMed]

111. Ma, J. The role of cofactors in prion propagation and infectivity. *PLoS Pathog.* **2012**, *8*, e1002589. [CrossRef] [PubMed]

112. Küffer, A.; Lakkraju, A.K.; Mogha, A.; Petersen, S.C.; Airich, K.; Doucerain, C.; Marpakwar, R.; Bakirci, P.; Senatore, A.; Monnard, A.; et al. The prion protein is an agonistic ligand of the G protein-coupled receptor Adgrg6. *Nature* **2016**, *536*, 464–468. [CrossRef] [PubMed]

113. Tobler, I.; Gaus, S.E.; Deboer, T.; Achermann, P.; Fischer, M.; Rulicke, T.; Moser, M.; Oesch, B.; McBride, P.A.; Manson, J.C. Altered circadian activity rythms and sleep in mice devoid of prion protein. *Nature* **1996**, *380*, 639–642. [CrossRef] [PubMed]

114. Collinge, J.; Whittington, M.A.; Sidle, K.C.; Smith, C.J.; Palmer, M.S.; Clarke, A.R.; Jefferys, J.G.R. Prion protein is necessary for normal synaptic function. *Nature* **1994**, *370*, 295–297. [CrossRef] [PubMed]

115. Coitinho, A.S.; Roesler, R.; Martins, V.R.; Brentani, R.R.; Izquierdo, I. Cellular prion protein ablation impairs behaviour as a function of age. *Neuroreport* **2003**, *14*, 1375–1379. [CrossRef] [PubMed]

116. Walz, R.; Amaral, O.B.; Rockenbach, I.C.; Roesler, R.; Izquierdo, I.; Cavalheiro, E.A.; Martins, V.R.; Brentani, R.R. Increased sensitivity to seizures in mice lacking cellular prion protein. *Epilesia* **1999**, *40*, 1679–1682. [CrossRef]

117. Mouillet-Richard, S.; Ermonval, M.; Chebassier, C.; Laplanche, J.L.; Lehmann, S.; Launay, J.M.; Kellermann, O. Signal transduction through prion protein. *Science* **2000**, *289*, 1925–1928. [CrossRef] [PubMed]

118. Spielhaupter, C.; Schatzl, H.M. PrP^C directly interacts with proteins involved in signalling pathways. *J. Biol. Chem.* **2001**, *276*, 44604–44612. [CrossRef] [PubMed]

119. Bounar, Y.; Zhang, Y.; Goodyer, C.G.; LeBlanc, A. Prion protein protects human neurons against Bax-mediated apoptosis. *J. Biol. Chem.* **2001**, *276*, 39145–39149. [CrossRef] [PubMed]

120. Mitteregger, G.; Vosko, M.; Krebs, B.; Xiang, W.; Kohlmannsperger, V.; Nolting, S.; Hamann, G.F.; Kretzschmar, H.A. The role of the octarepeat region in neuroprotective function of the cellular prion protein. *Brain Pathol.* **2007**, *17*, 174–183. [CrossRef] [PubMed]

121. Mabbott, N.A.; Young, J.; McConnell, I.; Bruce, M.E. Follicular dendritic cell dedifferentiation by treatment with an inhibitor of the lymphotoxin pathway dramatically reduces scrapie susceptibility. *J. Virol.* **2003**, *77*, 6845–6854. [CrossRef] [PubMed]

122. Prinz, M.; Huber, G.; Macpherson, A.J.S.; Heppner, F.L.; Glatzel, M.; Eugster, H.-P.; Wagner, N.; Aguzzi, A. Oral prion infection requires normal numbers of Peyer's patches but not of enteric lymphocytes. *Am. J. Pathol.* **2003**, *162*, 1103–1111. [CrossRef]

123. Glaysher, B.R.; Mabbott, N.A. Role of the GALT in scrapie agent neuroinvasion from the intestine. *J. Immunol.* **2007**, *178*, 3757–3766. [CrossRef] [PubMed]

124. Sigurdson, C.J.; Williams, E.S.; Miller, M.W.; Spraker, T.R.; O'Rourke, K.I.; Hoover, E.A. Oral transmission and early lymphoid tropism of chronic wasting disease PrP^{res} in mule deer fawns (*Odocoileus hemionus*). *J. Gen. Virol.* **1999**, *80*, 2757–2764. [CrossRef] [PubMed]

125. Andreoletti, O.; Berthon, P.; Marc, D.; Sarradin, P.; Grosclaude, J.; van Keulen, L.; Schelcher, F.; Elsen, J.-M.; Lantier, F. Early accumulation of PrP^{Sc} in gut-associated lymphoid and nervous tissues of susceptible sheep from a Romanov flock with natural scrapie. *J. Gen. Virol.* **2000**, *81*, 3115–3126. [CrossRef] [PubMed]

126. Heggebø, R.; Press, C.M.; Gunnes, G.; Lie, K.I.; Tranulis, M.A.; Ulvund, M.; Groschup, M.H.; Landsverk, T. Distribution of prion protein in the ileal Peyer's patch of scrapie-free lambs and lambs naturally and experimentally exposed to the scrapie agent. *J. Gen. Virol.* **2000**, *81*, 2327–2337. [CrossRef] [PubMed]

127. van Keulen, L.J.M.; Schreuder, B.E.G.; Vromans, M.E.W.; Langeveld, J.P.M.; Smits, M.A. Scrapie-associated prion protein in the gastro-intestinal tract of sheep with scrapie. *J. Comp. Pathol.* **1999**, *121*, 55–63. [CrossRef] [PubMed]

128. McCulloch, L.; Brown, K.L.; Mabbott, N.A. Ablation of the cellular prion protein, PrP^C, specifcally on follicular dendritic cells has no effect on their maturation or function. *Immunology* **2013**, *138*, 246–257. [CrossRef] [PubMed]

129. Cashman, N.R.; Loertscher, R.; Nalbantoglu, J.; Shaw, I.; Kascsak, R.J.; Bolton, D.C.; Bendheim, P.E. Cellular isoform of the scrapie agent protein participates in lymphocyte activation. *Cell* **1990**, *61*, 185–192. [CrossRef]

130. Mabbott, N.A.; Brown, K.L.; Manson, J.; Bruce, M.E. T lymphocyte activation and the cellular form of the prion protein. *Immunology* **1997**, *92*, 161–165. [CrossRef] [PubMed]

131. Martínez del Hoyo, G.; López-Bravo, M.; Metharom, P.; Ardavín, C.; Aucouturier, P. Prion protein expression by mouse dendritic cells is restricted to the nonplasmacytoid subsets and correlates with the maturation state. *J. Immunol.* **2006**, *177*, 6137–6142. [CrossRef] [PubMed]

132. Jouvin-Marche, E.; Attuli-Audenis, V.; Aude-Garcia, C.; Rachidi, W.; Zabel, M.; Podevin-Dimster, V.; Siret, C.; Huber, C.; Martinic, M.; Riondel, J.; et al. Overexpression of cellular prion protein induces and antioxidant environment altering T cell development in the thymus. *J. Immunol.* **2006**, *176*, 3490–3497. [CrossRef] [PubMed]

133. Ballerini, C.; Gourdain, P.; Bachy, V.; Blanchard, N.; Levavasseur, E.; Gregoire, S.; Fontes, P.; Aucouturier, P.; Hivroz, C.; Carnaud, C. Funcitonal implication of cellular prion protein in antigen-driven interactions between T cells and dendritic cells. *J. Immunol.* **2006**, *176*, 7254–7262. [CrossRef] [PubMed]

134. Nakato, G.; Hase, K.; Suzuki, M.; Kimura, M.; Ato, M.; Hanazato, M.; Tobiume, M.; Horiuchi, M.; Atarashi, R.; Nishida, N.; et al. Cutting edge: *Brucella abortus* exploits a cellular prion protein on intestinal M cells as an invasive receptor. *J. Immunol.* **2012**, *189*, 1540–1544. [CrossRef] [PubMed]

135. De Almeida, C.J.G.; Chiarini, L.B.; da Silva, J.P.; e Silva, P.M.R.; Martins, M.A.; Linden, R. The cellular prion protein modulates phagocytosis and inflammatory response. *J. Leukoc. Biol.* **2005**, *77*, 238–246. [CrossRef] [PubMed]

136. Tsukamoto, T.; Diringer, H.; Ludwig, H. Absence of autoantibodies against neurofilament proteins in the sera of scarpie infected mice. *Tohoku J. Exp. Med.* **1985**, *4*, 483–484. [CrossRef]

137. Clarke, M.C.; Haig, D.A. Attempts to demonstrate neutralising antibodies in the sera of scrapie-infected animals. *Vet. Rec.* **1966**, *19*, 647–649. [CrossRef]

138. Sassa, Y.; Kataoka, N.; Inoshima, Y.; Ishiguro, N. Anti-PrP antibodies detected at terminal stage of prion-affected mouse. *Cell. Immunol.* **2010**, *263*, 212–218. [CrossRef] [PubMed]

139. Harris, D.A. Trafficking, turnover and membrane topology of PrP. *Br. Med. Bull.* **2003**, *66*, 71–85. [CrossRef] [PubMed]

140. Campana, V.; Sarnataro, D.; Zurzolo, C. The highways and byways of prion protein trafficking. *Trends Cell Biol.* **2005**, *15*, 102–111. [CrossRef] [PubMed]

141. Borchelt, D.R.; Taraboulos, A.; Prusiner, S.B. Evidence for synthesis of scrapie prion proteins in the endocytic pathway. *J. Biol. Chem.* **1992**, *267*, 188–199.

142. Arnold, J.E.; Tipler, C.; Laszlo, L.; Hope, J.; Landon, M.; Mayer, R.J. The abnormal isoform of the prion protein accumulates in late-endosome-like organelles in scrapie-infected mouse brain. *J. Pathol.* **1995**, *176*, 403–411. [CrossRef] [PubMed]

143. Marijanovic, Z.; Caputo, A.; Campana, V.; Zurzolo, C. Identification of an intracellular site of prion conversion. *PLoS Pathog.* **2009**, *5*, e1000426. [CrossRef] [PubMed]

144. Godsave, S.F.; Wille, H.; Kujala, P.; Latawiec, D.; DeArmond, S.J.; Serban, A.; Prusiner, S.B.; Peters, P.J. Cryo-immunogold EM for prions: Towards identification of a conversion site. *J. Neurosci.* **2008**, *28*, 12489–12499. [CrossRef] [PubMed]

145. Beranger, F.; Mange, A.; Goud, B.; Lehmann, S. Stimulation of PrP^C retrograde transport toward the endoplasmic reticulum increases accumulation of PrP^{Sc} in prion -infected cells. *J. Biol. Chem.* **2002**, *277*, 38972–38977. [CrossRef] [PubMed]

146. Kujala, P.; Raymond, C.R.; Romeijn, M.; Godsave, S.F.; van Kasteren, S.I.; Wille, H.; Prusiner, S.B.; Mabbott, N.A.; Peters, P.J. Prion uptake in the gut: Identification of the first uptake and replication sites. *PLoS Pathog.* **2011**, *7*, e1002449. [CrossRef] [PubMed]

147. McGovern, G.; Mabbott, N.A.; Jeffrey, M. Scrapie affects the maturation cycle and immune complex trapping by follicular dendritic cells in mice. *PLoS ONE* **2009**, *4*, e8186. [CrossRef] [PubMed]

148. Goold, R.; Rabbanian, S.; Sutton, L.; Andre, R.; Arora, P.; Moonga, J.; Clarke, A.R.; Schiavo, G.; Jat, P.; Collinge, J.; et al. Rapid cell-surface prion protein conversion revealed using a novel cell system. *Nat. Commun* **2011**, *2*, 281. [CrossRef] [PubMed]

149. Goold, R.; McKinnon, C.; Rabbanian, S.; Collinge, J.; Schiavo, G.; Tabrizi, S.J. Alternative sites of newly formed PrP^{Sc} upon prion conversion on the plasma membrane. *J. Cell Sci.* **2013**, *126*, 3552–3562. [CrossRef] [PubMed]

150. Fraser, H.; Dickinson, A.G. Pathogenesis of scrapie in the mouse: The role of the spleen. *Nature* **1970**, *226*, 462–463. [CrossRef] [PubMed]

151. Fraser, H.; Dickinson, A.G. Studies on the lymphoreticular system in the pathogenesis of scrapie: The role of spleen and thymus. *J. Comp. Pathol.* **1978**, *88*, 563–573. [CrossRef]

152. Horiuchi, M.; Furuoka, H.; Kitamura, N.; Shinagawa, M. Alymphoplasia mice are resistant to prion infection via oral route. *Jpn. J. Vet. Res.* **2006**, *53*, 149–157. [PubMed]

153. Donaldson, D.S.; Else, K.J.; Mabbott, N.A. The gut-associated lymphoid tissues in the small intestine, not the large intestine, play a major role in oral prion disease pathogenesis. *J. Virol.* **2015**, *15*, 9532–9547. [CrossRef] [PubMed]

154. Glaysher, B.R.; Mabbott, N.A. Role of the draining lymph node in scrapie agent transmission from the skin. *Immunol. Lett.* **2007**, *109*, 64–71. [CrossRef] [PubMed]

155. González, L.; Dagleish, M.P.; Bellworthy, S.J.; Sisó, S.; Stack, M.J.; Chaplin, M.J.; Davis, L.A.; Hawkins, S.A.C.; Hughes, J.; Jeffrey, M. Postmortem diagnosis of preclinical and clinical scrapie in sheep by the detection of disease-associated PrP in their rectal mucosa. *Vet. Rec.* **2006**, *158*, 325–331. [CrossRef] [PubMed]

156. Espenes, A.; Press, C.M.; Landsverk, T.; Tranulis, M.A.; Aleksandersen, M.; Gunnes, G.; Benestad, S.L.; Fuglestveit, R.; Ulvund, M.J. Detection of PrP^{Sc} in rectal biopsy and necroscopy samples from sheep with experimental scrapie. *J. Comp. Pathol.* **2006**, *134*, 115–125. [CrossRef] [PubMed]

157. Spraker, T.R.; Gidlewski, T.L.; Balachandran, A.; VerCauteren, K.C.; Creekmore, L.; Munger, R.D. Detection of PrP^{CWD} in postmortem rectal lymphoid tissues in Rocky Mountain elk (*Cervus elaphus nelsoni*) infected with chronic wasting disease. *J. Vet. Diagn. Investig.* **2006**, *18*, 553–557. [CrossRef] [PubMed]

158. Wolfe, L.L.; Spraker, T.R.; González, L.; Dagleish, M.P.; Sirochman, T.M.; Brown, J.C.; Jeffrey, M.; Miller, M.W. PrP^{CWD} in rectal lymphoid tissue of deer (*Odocoileus* spp.). *J. Gen. Virol.* **2007**, *88*, 2078–2082. [CrossRef] [PubMed]

159. Dennis, M.M.; Thomsen, B.V.; Marshall, K.L.; Hall, S.M.; Wagner, B.A.; Salman, M.D.; Norden, D.K.; Gaiser, C.; Sutton, D.L. Evaluation of immunohistochemical detection of prion protein in rectoanal mucosa-associated lymphoid tissue for diagnosis of scrapie in sheep. *Am. J. Vet. Res.* **2009**, *70*, 63–72. [CrossRef] [PubMed]

160. Hilton, D.; Fathers, E.; Edwards, P.; Ironside, J.; Zajicek, J. Prion immunoreactivity in appendix before clinical onset of variant Creutzfeldt-Jakob disease. *Lancet* **1998**, *352*, 703–704. [CrossRef]

161. Hilton, D.A.; Ghani, A.C.; Conyers, L.; Edwards, P.; McCardle, L.; Ritchie, D.; Penney, M.; Hegazy, D.; Ironside, J.W. Prevalence of lymphoreticular prion protein accumulation in UK tissue samples. *J. Pathol.* **2004**, *203*, 733–739. [CrossRef] [PubMed]

162. Hilton, D.A.; Ghani, A.C.; Conyers, L.; Edwards, P.; McCardle, L.; Penney, M.; Ritchie, D.; Ironside, J.W. Accumulation of prion protein in tonsil and appendix: Review of tissue samples. *Br. Med. J.* **2002**, *325*, 633–634. [CrossRef]

163. Gill, O.N.; Spencer, Y.; Richard-Loendt, A.; Kelly, C.; Dabaghian, R.; Boyes, L.; Lineham, J.; Simmons, M.; Webb, P.; Bellerby, P.; et al. Prevelent abnormal prion protein in human appendixes after bovine spongiform encephalopathy epizootic: Large scale survey. *Br. Med. J.* **2013**, *347*, f5675. [CrossRef] [PubMed]

164. Mabbott, N.A.; Donaldson, D.S.; Ohno, H.; Williams, I.R.; Mahajan, A. Microfold (M) cells: Important immunosurveillance posts in the intestinal epithelium. *Mucosal Immunol.* **2013**, *6*, 666–677. [CrossRef] [PubMed]

165. Hase, K.; Kawano, K.; Nochi, T.; Pontes, G.S.; Fukuda, S.; Ebisawa, M.; Kadokura, K.; Tobe, T.; Fujimura, Y.; Kawano, S.; et al. Uptake through glycoprotein 2 of FimH$^+$ bacteria by M cells initiates mucosal immune responses. *Nature* **2009**, *462*, 226–231. [CrossRef] [PubMed]

166. Rios, D.; Wood, M.B.; Li, J.; Chassaing, B.; Gewirtz, A.T.; Williams, I.R. Antigen sampling by intestinal M cells is the principal pathway initiating mucosal IgA production to commensal enteric bacteria. *Mucosal Immunol.* **2016**, *9*, 907–916. [CrossRef] [PubMed]

167. Tahoun, A.; Mahajan, S.; Paxton, E.; Malterer, G.; Donaldson, D.S.; Wang, D.; Tan, A.; Gillespie, T.L.; O'Shea, M.; Rose, A.; et al. Salmonella transforms follicle-associated epithelial cells into M cells to promote intestinal invasion. *Cell Host Microbe* **2012**, *12*, 645–666. [CrossRef] [PubMed]

168. Westphal, S.; Lugering, A.; von Wedel, J.; von Eiff, C.; Maaser, C.; Spahn, T.; Heusipp, G.; Schmidt, M.A.; Herbst, H.; Williams, I.R.; et al. Resistance of chemokine receptor 6-deficient mice to *Yersinia enterocolitica* infection: Evidence on defective M-cell formation in vivo. *Am. J. Pathol.* **2008**, *172*, 671–680. [CrossRef] [PubMed]

169. Kolawole, A.O.; Gonzalez-Hernandez, M.B.; Turula, H.; Yu, C.; Elftman, M.D.; Wobus, C.E. Oral norovirus infection is blocked in mice lacking Peyer's patches and mature M cells. *J. Virol.* **2015**, *90*, 1499–1506. [CrossRef] [PubMed]

170. Gonzalez-Hernandez, M.B.; Liu, T.; Payne, H.C.; Stencel-Baerenwald, J.E.; Ikizler, M.; Yagita, H.; Dermody, T.S.; Williams, I.R.; Wobus, C.E. Efficient norovirus and reovirus replication in the mouse intestine requires microfold (M) cells. *J. Virol.* **2014**, *88*, 6934–6943. [CrossRef] [PubMed]

171. Heppner, F.L.; Christ, A.D.; Klein, M.A.; Prinz, M.; Fried, M.; Kraehenbuhl, J.-P.; Aguzzi, A. Transepithelial prion transport by M cells. *Nat. Med.* **2001**, *7*, 976–977. [CrossRef] [PubMed]

172. Miyazawa, K.; Kanaya, T.; Takakura, I.; Tanaka, S.; Hondo, T.; Watanabe, H.; Rose, M.T.; Kitazawa, H.; Yamaguchi, T.; Katamine, S.; et al. Transcytosis of murine-adapted bovine spongiform encephalopathy agents in an in vitro bovine M cell model. *J. Virol.* **2010**, *84*, 12285–12291. [CrossRef] [PubMed]

173. Takakura, I.; Miyazawa, K.; Kanaya, T.; Itani, W.; Watanabe, K.; Ohwada, S.; Watanabe, H.; Hondo, T.; Rose, M.T.; Mori, T.; et al. Orally administered prion protein is incorporated by M cells and spreads to lymphoid tissues with macrophages in prion protein knockout mice. *Am. J. Pathol.* **2011**, *179*, 1301–1309. [CrossRef] [PubMed]

174. Donaldson, D.S.; Kobayashi, A.; Ohno, H.; Yagita, H.; Williams, I.R.; Mabbott, N.A. M cell depletion blocks oral prion disease pathogenesis. *Mucosal Immunol.* **2012**, *5*, 216–225. [CrossRef] [PubMed]

175. Donaldson, D.S.; Sehgal, A.; Rios, D.; Williams, I.R.; Mabbott, N.A. Increased abundance of M cells in the gut epithelium dramatically enhances oral prion disease susceptibility. *PLoS Pathog.* **2016**, *12*, e1006075. [CrossRef] [PubMed]

176. Beekes, M.; McBride, P.A. Early accumulation of pathological PrP in the enteric nervous system and gut-associated lymphoid tissue of hamsters orally infected with scrapie. *Neurosci. Lett.* **2000**, *278*, 181–184. [CrossRef]

177. Bennet, K.M.; Parnell, E.A.; Sanscartier, C.; Parks, S.; Chen, G.; Nair, M.G.; Lo, D.D. Induction of colonic M cells during intestinal inflammation. *Am. J. Pathol.* **2016**, *186*, 166–179. [CrossRef] [PubMed]

178. Terahara, K.; Yoshida, M.; Igarashi, O.; Nochi, T.; Soares Pontes, G.; Hase, K.; Ohno, H.; Kurokawa, S.; Mejima, M.; Takayama, N.; et al. Comprehensive gene expression profiling of Peyer's patch M cells, villous M-like cells, and intestinal epithelial cells. *J. Immunol.* **2008**, *180*, 7840–7846. [CrossRef] [PubMed]

179. Knoop, K.A.; Kumar, N.; Butler, B.R.; Sakthivel, S.K.; Taylor, R.T.; Nochi, T.; Akiba, H.; Yagita, H.; Kiyono, H.; Williams, I.R. RANKL is necessary and sufficient to initiate development of antigen-sampling M cells in the intestinal epithelium. *J. Immunol.* **2009**, *183*, 5738–5747. [CrossRef] [PubMed]

180. Nagashima, K.; Sawa, S.; Nitta, T.; Tsutsumi, M.; Okamura, T.; Penninger, J.M.; Nakashima, T.; Takayanagi, H. Identification of subepithelial mesenchymal cells that induce IgA and diversify gut microbiota. *Nat. Immunol.* **2017**, *18*, 675–682. [CrossRef] [PubMed]

181. Foster, N.; Macpherson, G.G. Murine cecal patch M cells transport infectious prions in vivo. *J. Infect. Dis.* **2010**, *202*, 1916–1919. [CrossRef] [PubMed]

182. Jeffrey, M.; González, L.; Espenes, A.; Press, C.M.; Martin, S.; Chaplin, M.; Davis, L.; Landsverk, T.; MacAldowie, C.; Eaton, S.; et al. Transportation of prion protein across the intestinal mucosa of scrapie-susceptible and scrapie-resistant sheep. *J. Pathol.* **2006**, *209*, 4–14. [CrossRef] [PubMed]

183. Mishra, R.S.; Basu, S.; Gu, Y.; Luo, X.; Zou, W.-Q.; Mishra, R.; Li, R.; Chen, S.G.; Gambetti, P.; Fujioka, H.; et al. Protease-resistant human prion protein and ferritin are cotransported across Caco-2 epithelial cells: Implications for species barrier in prion uptake from the intestine. *J. Neurosci.* **2004**, *24*, 11280–11290. [CrossRef] [PubMed]

184. Mutoh, M.; Kimura, S.; Takashi-Iwanaga, H.; Hisamoto, M.; Iwanaga, T.; Iida, J. RANKL regulates differentiation of microfold cells in mouse nasopharynx-associated lymphoid tissue (NALT). *Cell Tissue Res.* **2016**, *364*, 175–184. [CrossRef] [PubMed]

185. Nair, V.R.; Franco, L.H.; Zacharia, V.M.; Khan, H.S.; Stamm, C.E.; You, W.; Marciano, D.K.; Yagita, H.; Levine, B.; Shiloh, M.U. Microfold cells actively translocate *Mycobacterium tuberculosis* to initiate infection. *Cell Rep.* **2016**, *16*, 1253–1258. [CrossRef] [PubMed]

186. Kincaid, A.E.; Hudson, K.F.; Richey, M.W.; Bartz, J.C. Rapid transepithelial transport of prions following inhalation. *J. Virol.* **2012**, *86*, 12731–12740. [CrossRef] [PubMed]

187. Elder, A.M.; Henderson, D.M.; Nalls, A.V.; Hoover, E.A.; Kincaid, A.E.; Bartz, J.C.; Mathiason, C.K. Immediate and ongoing detection of prions in the blood of hamsters and deer following oral, nasal and blood inoculations. *J. Virol.* **2015**, *89*, 7421–7424. [CrossRef] [PubMed]

188. Urayama, A.; Concha-Marambio, L.; Khan, U.; Bravo-Alegria, J.; Kharat, V.; Soto, C. Prions efficiently cross the intestinal barrier after oral administration: Study of the bioavailability, and cellular tissue distribution in vivo. *Sci. Rep.* **2016**, *6*, 32338. [CrossRef] [PubMed]

189. Raymond, C.R.; Aucouturier, P.; Mabbott, N.A. In vivo depletion of CD11c+ cells impairs scrapie agent neuroinvasion from the intestine. *J. Immunol.* **2007**, *179*, 7758–7766. [CrossRef] [PubMed]

190. Nakato, G.; Fukuda, S.; Hase, K.; Goitsuka, R.; Cooper, M.D.; Ohno, H. New approach for M-cell-specific molecules by screening comprehensive transcriptome analysis. *DNA Res.* **2009**, *16*, 227–235. [CrossRef] [PubMed]

191. Sakhon, O.S.; Ross, B.; Gusti, V.; Pham, A.J.; Vu, K.; Lo, D.D. M cell-derived vesicles suggest a unique pathway for trans-epithelial antigen delivery. *Tissue Barriers* **2015**, *3*, e1004975. [CrossRef] [PubMed]

192. Delamarre, L.; Pack, M.; Chang, H.; Mellman, I.; Trombetta, E.S. Differential lysosomal proteolysis in antigen-presenting cells determines antigen fate. *Science* **2005**, *307*, 1630–1634. [CrossRef] [PubMed]

193. Bergtold, A.; Desai, D.D.; Gavhane, A.; Clynes, R. Cell surface recycling of internalized antigen permits dendritic cell priming to B cells. *Immunity* **2005**, *23*, 503–514. [CrossRef] [PubMed]

194. Banchereau, J.; Briere, F.; Caux, C.; Davoust, J.; Lebecque, S.; Liu, Y.J.; Pulendran, B.; Palucka, K. Immunobiology of dendritic cells. *Annu. Rev. Immunol.* **2000**, *18*, 767–811. [CrossRef] [PubMed]

195. Liu, M.; MacPherson, G.G. Antigen acquisition by dendritic cells: Intestinal dendritic cells acquire antigen administered orally and can prime naive T cells in vivo. *J. Exp. Med.* **1993**, *177*, 1299–1307. [CrossRef] [PubMed]

196. Cerovic, V.; Houston, S.A.; Westlund, J.; Utriainen, L.; Davison, E.S.; Scott, C.L.; Bain, C.; Joeris, T.; Agace, W.W.; Kroczek, R.A.; et al. Lymph-borne CD8a+ dendritic cells are uniquely able to cross-prime CD8+ T cells with antigen acquired from intestinal epithelial cells. *Mucosal Immunol.* **2014**, *8*, 38–48. [CrossRef] [PubMed]

197. Wu, S.-J.L.; Grouard-Vogel, G.; Sun, W.; Mascola, J.R.; Brachtel, E.; Putvatana, R.; Louder, M.K.; Filgueira, L.; Marovich, M.A.; Wong, H.K.; et al. Human skin Langerhans cells are targets of dengue virus infection. *Nat. Med.* **2000**, *6*, 816–820. [PubMed]

198. Ho, L.-J.; Wang, J.-J.; Shaio, M.-F.; Kao, C.-L.; Chang, D.-M.; Han, S.-W.; Lai, J.-H. Infection of human dendritic cells by dengue virus causes cell maturation and cytokine production. *J. Immunol.* **2001**, *166*, 1499–1506. [CrossRef] [PubMed]

199. Steinman, R.M.; Granelli-Piperno, A.; Pope, M.; Trumpfheller, C.; Ignatius, R.; Arrode, G.; Racz, P.; Tenner-Racz, K. The interaction of immunodeficiency viruses with dendritic cells. *Curr. Top. Microbiol. Immunol.* **2003**, *276*, 1–30. [PubMed]

200. Ho, A.W.; Prabhu, N.; Betts, R.J.; Ge, M.Q.; Dai, X.; Hutchinson, P.E.; Lew, F.C.; Wong, K.L.; Hanson, B.J.; Macary, P.A.; et al. Lung CD103+ dendritic cells efficiently transport infuenza virus to the lymph node and load viral antigen onto MHC class I for presentation to CD8 T cells. *J. Immunol.* **2011**, *187*, 6011–6021. [CrossRef] [PubMed]

201. Wykes, M.; Pombo, A.; Jenkins, C.; MacPherson, G.G. Dendritic cells interact directly with Naive B lymphocytes to transfer antigen and initiate class switching in a primary T-dependent response. *J. Immunol.* **1998**, *161*, 1313–1319. [PubMed]

202. Macpherson, A.J.; Uhr, T. Induction of protective IgA by intestinal dendritic cells carrying commensal bacteria. *Science* **2004**, *303*, 1662–1665. [CrossRef] [PubMed]

203. Saeki, H.; Wu, M.; Olasz, E.; Hwang, S.T. A migratory population of skin-derived dendritic cells expresses CXCR5, responds to B lymphocyte chemoattractant *in vitro*, and co-localizes to B cell zones in lymph nodes in vivo. *Eur. J. Immunol.* **2000**, *30*, 2808–2814. [CrossRef]

204. Berney, C.; Herren, S.; Power, C.A.; Gordon, S.; Martinez-Pomares, L.; Kosco-Vilbois, M.H. A member of the dendritic cell family that enters B cell follicles and stimulates primary antibody responses identified by a mannose receptor fusion protein. *J. Exp. Med.* **1999**, *190*, 851–860. [CrossRef] [PubMed]

205. Leon, B.; Ballesteros-Tato, A.; Browning, J.L.; Dunn, R.; Randall, T.D.; Lund, F.E. Regulation of T(H)2 development by CXCR5+ dendritic cells and lymphotoxin-expressing B cells. *Nat. Immunol.* **2012**, *13*, 681–690. [CrossRef] [PubMed]

206. Huang, F.-P.; Farquhar, C.F.; Mabbott, N.A.; Bruce, M.E.; MacPherson, G.G. Migrating intestinal dendritic cells transport PrP^{Sc} from the gut. *J. Gen. Virol.* **2002**, *83*, 267–271. [CrossRef] [PubMed]

207. Cordier-Dirikoc, S.; Chabry, J. Temporary depletion of CD11c+ dendritic cells delays lymphoinvasion after intraperitoneal scrapie infection. *J. Virol.* **2008**, *82*, 8933–8936. [CrossRef] [PubMed]

208. Sethi, S.; Kerksiek, K.M.; Brocker, T.; Kretzschmar, H. Role of the CD8+ dendritic cell subset in transmission of prions. *J. Virol.* **2007**, *81*, 4877–4880. [CrossRef] [PubMed]

209. Wathne, G.J.; Kissenpfennig, A.; Malissen, B.; Zurzolo, C.; Mabbott, N.A. Determining the role of mononuclear phagocytes in prion neuroinvasion from the skin. *J. Leukoc. Biol.* **2012**, *91*, 817–828. [CrossRef] [PubMed]

210. Iwasaki, A.; Kelsalla, B.A. Localization of distinct Peyer's patch dendritic cell subsets and their recruitment by chemokines macrophage inflammatory protein (MIP)-3, MIP-3ß, and secondary lymphoid organ chemokine. *J. Exp. Med.* **2000**, *191*, 1381–1394. [CrossRef] [PubMed]

211. Castro-Seoane, R.; Hummerich, H.; Sweeting, T.; Tattum, M.H.; Lineham, J.M.; Fernandez de Marco, M.; Brandner, S.; Collinge, J.; Klöhn, P.C. Plasmacytoid dendritic cells sequester high prion titres at early stages of prion infection. *PLoS Pathog.* **2012**, *8*, e1002538. [CrossRef] [PubMed]

212. Yrlid, U.; Cerovic, V.; Milling, S.; Jenkins, C.D.; Zhang, J.; Crocker, P.R.; Klavinskis, L.S.; MacPherson, G.G. Plasmacytoid dendritic cells do not migrate in intestinal or hepatic lymph. *J. Immunol.* **2006**, *177*, 6115–6121. [CrossRef] [PubMed]

213. Burthem, J.; Urban, B.; Pain, A.; Roberts, D.J. The normal cellular prion protein is strongly expressed by myeloid dendritic cells. *Blood* **2001**, *98*, 3733–3738. [CrossRef] [PubMed]

214. Cordier-Dirikoc, S.; Zsürger, N.; Cazareth, J.; Ménard, B.; Chabry, J. Expression profiles of prion and doppel proteins and of their receptors in mo'use splenocytes. *Eur. J. Immunol.* **2008**, *38*, 1–11. [CrossRef] [PubMed]

215. Miyazawa, K.; Kanaya, T.; Tanaka, S.; Takakura, I.; Watanabe, K.; Ohwada, S.; Kitazawa, H.; Rose, M.T.; Sakaguchi, S.; Katamine, S.; et al. Immunohistochemical characterization of cell types expressing the cellular prion protein in the small intestine of cattle and mice. *Histochem. Cell Biol.* **2007**, *127*, 291–301. [CrossRef] [PubMed]

216. Brown, K.L.; Stewart, K.; Ritchie, D.; Mabbott, N.A.; Williams, A.; Fraser, H.; Morrison, W.I.; Bruce, M.E. Scrapie replication in lymphoid tissues depends on PrP-expressing follicular dendritic cells. *Nat. Med.* **1999**, *5*, 1308–1312. [CrossRef] [PubMed]

217. Klein, M.A.; Frigg, R.; Raeber, A.J.; Flechsig, E.; Hegyi, I.; Zinkernagel, R.M.; Weissmann, C.; Aguzzi, A. PrP expression in B lymphocytes is not required for prion neuroinvasion. *Nat. Med.* **1998**, *4*, 1429–1433. [CrossRef] [PubMed]

218. Mohan, J.; Brown, K.L.; Farquhar, C.F.; Bruce, M.E.; Mabbott, N.A. Scrapie transmission following exposure through the skin is dependent on follicular dendritic cells in lymphoid tissues. *J. Dermatol. Sci.* **2004**, *35*, 101–111. [CrossRef] [PubMed]

219. Loeuillet, C.; Lemaire-Vielle, C.; Naquet, P.; Cesbron-Delauw, M.-F.; Gagnon, J.; Cesbron, J.-Y. Prion replication in the hematopoietic compartment is not required for neuroinvasion in scrapie mouse model. *PLoS ONE* **2010**, *5*, e13166. [CrossRef] [PubMed]

220. McCulloch, L.; Brown, K.L.; Bradford, B.M.; Hopkins, J.; Bailey, M.; Rajewsky, K.; Manson, J.C.; Mabbott, N.A. Follicular dendritic cell-specific prion protein (PrP^C) expression alone is sufficient to sustain prion infection in the spleen. *PLoS Pathog.* **2011**, *7*, e1002402. [CrossRef] [PubMed]

221. Michel, B.; Meyerett-Reid, C.; Johnson, T.; Ferguson, A.; Wycoff, C.; Pulford, B.; Bender, H.; Avery, A.; Telling, G.; Dow, S.; et al. Incunabular immunological events in prion trafficking. *Sci. Rep.* **2012**, *2*, 440. [CrossRef] [PubMed]

222. Flores-Lagnarica, A.; Sebti, Y.; Mitchell, D.A.; Sim, R.B.; MacPherson, G.G. Scrapie pathogenesis: The role of complement C1q in scrapie agent uptake by conventional dendritic cells. *J. Immunol.* **2009**, *182*, 1305–1313. [CrossRef]

223. Bradford, B.M.; Brown, K.L.; Mabbott, N.A. Prion pathogenesis is unaltered following down-regulation of SIGN-R1. *Virology* **2016**, *497*, 337–345. [CrossRef] [PubMed]

224. Sowinski, S.; Jolly, C.; Berninghausen, O.; Purbhoo, M.A.; Chauveau, A.; Kohler, K.; Oddos, S.; Eissmann, P.; Brodsky, F.M.; Hopkins, C.; et al. Membrane nanotubes physically connect T cells over long distances presenting a novel route for HIV-1 transmission. *Nat. Cell Biol.* **2008**, *10*, 211–219. [CrossRef] [PubMed]

225. Xu, W.; Santini, P.A.; Sullivan, J.S.; He, B.; Shan, M.; Ball, S.C.; Dyer, W.B.; Ketas, T.J.; Chadburn, A.; Cohen-Gould, L.; et al. HIV-1 evades virus-specific IgG2 and IgA responses by targeting systemic conduits and intestinal B cells via long-range intercellular conduits. *Nat. Immunol.* **2009**, *10*, 1008. [CrossRef] [PubMed]

226. Hase, K.; Kimura, S.; Takatsu, H.; Ohmae, M.; Kawano, S.; Kitamura, H.; Ito, M.; Watarai, H.; Hazelett, C.C.; Yeaman, C.; et al. M-Sec promotes membrane nanotube formation by interacting with Ral and the exocyst complex. *Nat. Cell Biol.* **2009**, *11*, 1427–1432. [CrossRef] [PubMed]

227. Zhu, S.; Victoria, G.S.; Marzo, L.; Ghosh, R.; Zurzolo, C. Prion aggregates transfer through tunneling nanotubes in endocytic vesicles. *Prion* **2015**, *9*, 125–135. [CrossRef] [PubMed]

228. Gousset, K.; Schiff, E.; Langevin, C.; Marijanovic, Z.; Caputo, A.; Browman, D.T.; Chanouard, N.; de Chaumont, F.; Martino, A.; Enninga, J.; et al. Prions hijack tunnelling nanotubes for intercellular spread. *Nat. Cell Biol.* **2009**, *11*, 328–336. [CrossRef] [PubMed]

229. Langevin, C.; Gousset, K.; Costanzo, M.; Richard-Le Goff, O.; Zurzolo, C. Characterization of the role of dendritic cells in prion transfer to primary neurons. *Biochem. J.* **2010**, *431*, 189–198. [CrossRef] [PubMed]

230. Tanaka, Y.; Sadaike, T.; Inoshima, Y.; Ishiguro, N. Characterisation of PrP^{Sc} transmission from immune cells to neuronal cells. *Cell. Immunol.* **2012**, *279*, 145–150. [CrossRef] [PubMed]

231. Fevrier, B.; Vilette, D.; Archer, F.; Loew, D.; Faigle, W.; Vidal, M.; Laude, H.; Raposo, G. Cells release prions in association with exosomes. *Proc. Natl. Acad. Sci. USA* **2004**, *101*, 9683–9688. [CrossRef] [PubMed]

232. Arellano-Anaya, Z.E.; Huor, A.; Leblanc, P.; Lehmann, S.; Provansal, M.; Raposo, G.; Andréoletti, O.; Viette, D. Prion strains are differentially released through the exosomal pathway. *Cell. Mol. Life Sci.* **2015**, *72*, 1185–1196. [CrossRef] [PubMed]

233. Carp, R.I.; Callahan, S.M. In vitro interaction of scrapie agent and mouse peritoneal macrophages. *Intervirology* **1981**, *16*, 8–13. [CrossRef] [PubMed]

234. Carp, R.I.; Callahan, S.M. Effect of mouse peritoneal macrophages on scrapie infectivity during extended in vitro incubation. *Intervirology* **1982**, *17*, 201–207. [CrossRef] [PubMed]

235. Maignien, T.; Shakweh, M.; Calvo, P.; Marce, D.; Sales, N.; Fattal, E.; Deslys, J.-P.; Couvreur, P.; Lasmezas, C.I. Role of gut macrophages in mice orally contaminated with scrapie or BSE. *Int. J. Pharm.* **2005**, *298*, 293–304. [CrossRef] [PubMed]

236. Beringue, V.; Demoy, M.; Lasmezas, C.I.; Gouritin, B.; Weingarten, C.; Deslys, J.-P.; Andreux, J.P.; Couvreur, P.; Dormont, D. Role of spleen macrophages in the clearance of scrapie agent early in pathogenesis. *J. Pathol.* **2000**, *190*, 495–502. [CrossRef]

237. Mabbott, N.A.; Williams, A.; Farquhar, C.F.; Pasparakis, M.; Kollias, G.; Bruce, M.E. Tumor necrosis factor-alpha-deficient, but not interleukin-6-deficient, mice resist peripheral infection with scrapie. *J. Virol.* **2000**, *74*, 3338–3344. [CrossRef] [PubMed]

238. Kitamoto, T.; Muramoto, T.; Mohri, S.; Doh-Ura, K.; Tateishi, J. Abnormal Isoform of Prion Protein Accumulates in Follicular Dendritic Cells in Mice with Creutzfeldt-Jakob Disease. *J. Virol.* **1991**, *65*, 6292–6295. [PubMed]

239. Fraser, H.; Brown, K. Peripheral Pathogenesis of Scrapie in Normal and Immunocompromised Mice. *Anim. Technol.* **1994**, *45*, 21–22.

240. Klein, M.A.; Frigg, R.; Flechsig, E.; Raeber, A.J.; Kalinke, U.; Bluethman, H.; Bootz, F.; Suter, M.; Zinkernagel, R.M.; Aguzzi, A. A crucial role for B cells in neuroinvasive scrapie. *Nature* **1997**, *390*, 687–691. [CrossRef] [PubMed]

241. Fraser, H.; Brown, K.L.; Stewart, K.; McConnell, I.; McBride, P.; Williams, A. Replication of scrapie in spleens of SCID mice follows reconstitution with wild-type mouse bone marrow. *J. Gen. Virol.* **1996**, *77*, 1935–1940. [CrossRef] [PubMed]

242. McFarlin, D.E.; Raff, M.C.; Simpson, E.; Nehlsen, S.H. Scrapie in immunologically deficient mice. *Nature* **1971**, *233*, 336. [CrossRef] [PubMed]

243. Raeber, A.J.; Sailer, A.; Hegyi, I.; Klein, M.A.; Rulicke, T.; Fischer, M.; Brandner, S.; Aguzzi, A.; Weissmann, C. Ectopic expression of prion protein (PrP) in T lymphocytes or hepatocytes of PrP knockout mice is insufficient to sustain prion replication. *Proc. Natl. Acad. Sci. USA* **1999**, *96*, 3987–3992. [CrossRef] [PubMed]

244. McBride, P.; Eikelenboom, P.; Kraal, G.; Fraser, H.; Bruce, M.E. PrP protein is associated with follicular dendritic cells of spleens and lymph nodes in uninfected and scrapie-infected mice. *J. Pathol.* **1992**, *168*, 413–418. [CrossRef] [PubMed]

245. Montrasio, F.; Cozzio, A.; Flechsig, E.; Rossi, D.; Klein, M.A.; Rulicke, T.; Raeber, A.J.; Vosshenrich, C.A.J.; Proft, J.; Aguzzi, A.; Weissmann, C. B-lymphocyte-restricted expression of the prion protein does not enable prion replication in PrP knockout mice. *Proc. Natl. Acad. Sci. USA* **2001**, *98*, 4034–4037. [CrossRef] [PubMed]

246. Kapasi, Z.F.; Burton, G.F.; Schultz, L.D.; Tew, J.G.; Szakal, A.K. Induction of functional follicular dendritic cell development in severe combined immunodeficiency mice. *J. Immunol.* **1993**, *150*, 2648–2658. [PubMed]

247. Chaplin, D.D.; Fu, Y.-X. Cytokine regulation of secondary lymphoid organ development. *Curr. Opin. Immunol.* **1998**, *10*, 289–297. [CrossRef]

248. Mackay, F.; Browning, J.L. Turning off follicular dendritic cells. *Nature* **1998**, *395*, 26–27. [CrossRef] [PubMed]

249. Tumanov, A.V.; Kuprash, D.V.; Lagarkova, M.A.; Grivennikov, S.I.; Abe, K.; Shakhov, A.; Drutskaya, L.N.; Stewart, C.L.; Chervonsky, A.V.; Nedospasov, S.A. Distinct role of surface lymphotoxin epxressed by B cells in the organization of secondary lymphoid tissues. *Immunity* **2002**, *239*, 239–250. [CrossRef]

250. Krautler, N.J.; Kana, V.; Kranich, J.; Tian, Y.; Perera, D.; Lemm, D.; Schwarz, P.; Armulik, A.; Browning, J.L.; Tallquist, M.; et al. Follicular dendritic cells emerge from ubiquitous perivascular precursors. *Cell* **2012**, *150*, 194–206. [CrossRef] [PubMed]

251. Shortman, K.; Liu, Y.-J. Mouse and human dendritic cell subtypes. *Nat. Rev. Immunol.* **2002**, *2*, 151–161. [CrossRef] [PubMed]

252. Mabbott, N.A.; Bailie, J.K.; Kobayashi, A.; Donaldson, D.S.; Ohmori, H.; Yoon, S.-O.; Freedman, A.S.; Freeman, T.C.; Summers, K.M. Expression of mesenchyme-specific gene signatures by follicular dendritic cells: Insights from the meta-analysis of microarray data from multiple mouse cell populations. *Immunology* **2011**, *133*, 482–498. [CrossRef] [PubMed]

253. Fütterer, A.; Mink, K.; Luz, A.; Kosco-Vilbois, M.H.; Pfeffer, K. The lymphotoxin b receptor controls organogenesis and affinity maturation in peripheral lymphoid tissues. *Immunity* **1998**, *9*, 59–70. [CrossRef]

254. Matsumoto, M.; Lo, S.F.; Carruthers, C.J.L.; Min, J.; Mariathasan, S.; Huang, G.; Plas, D.R.; Martin, S.M.; Geha, R.S.; Nahm, M.H.; et al. Affinity maturation without germinal centres in lymphotoxin-a-deficient mice. *Nature* **1996**, *382*, 462–466. [CrossRef] [PubMed]

255. Koni, P.A.; Sacca, R.; Lawton, P.; Browning, J.L.; Ruddle, N.H.; Flavell, R.A. Distinct roles in lymphoid organogenesis for lymphotoxins a and b revealed in lymphotoxin b-deficient mice. *Immunity* **1997**, *6*, 491–500. [CrossRef]

256. Pasparakis, M.; Alexopoulo, L.; Episkopou, V.; Kollias, G. Immune and inflammatory responses in TNFa-deficient mice: A critical requirement for TNFa in the formation of primary B cell follicles, follicular dendritic cell networks and germinal centres, and in the maturation of the humoral immune response. *J. Exp. Med.* **1996**, *184*, 1397–1411. [CrossRef] [PubMed]

257. Tkachuk, M.; Bolliger, S.; Ryffel, B.; Pluschke, G.; Banks, T.A.; Herren, S.; Gisler, R.H.; Kosco-Vilbois, M.H. Crucial role of tumour necrosis factor receptor 1 expression on nonhematopoietic cells for B cell localization within the splenic white pulp. *J. Exp. Med.* **1998**, *187*, 469–477. [CrossRef] [PubMed]

258. Prinz, M.; Montrasio, F.; Klein, M.A.; Schwarz, P.; Priller, J.; Odermatt, B.; Pfeffer, K.; Aguzzi, A. Lymph nodal prion replication and neuroinvasion in mice devoid of follicular dendritic cells. *Proc. Natl. Acad. Sci. USA* **2002**, *99*, 919–924. [CrossRef] [PubMed]

259. Mabbott, N.A.; Mackay, F.; Minns, F.; Bruce, M.E. Temporary inactivation of follicular dendritic cells delays neuroinvasion of scrapie. *Nat. Med.* **2000**, *6*, 719–720. [CrossRef] [PubMed]

260. Montrasio, F.; Frigg, R.; Glatzel, M.; Klein, M.A.; Mackay, F.; Aguzzi, A.; Weissmann, C. Impaired prion replication in spleens of mice lacking functional follicular dendritic cells. *Science* **2000**, *288*, 1257–1259. [CrossRef] [PubMed]

261. Mohan, J.; Bruce, M.E.; Mabbott, N.A. Follicular dendritic cell dedifferentiation reduces scrapie susceptibility following inoculation via the skin. *Immunology* **2005**, *114*, 225–234. [CrossRef] [PubMed]

262. Prinz, M.; Heikenwalder, M.; Junt, T.; Schwarz, P.; Glatzel, M.; Heppner, F.L.; Fu, Y.-X.; Lipp, M.; Aguzzi, A. Positioning of follicular dendritic cells within the spleen controls prion neuroinvasion. *Nature* **2003**, *425*, 957–962. [CrossRef] [PubMed]

263. Mabbott, N.A.; McGovern, G.; Jeffrey, M.; Bruce, M.E. Temporary blockade of the tumour necrosis factor signaling pathway impedes the spread of scrapie to the brain. *J. Virol.* **2002**, *76*, 5131–5139. [CrossRef] [PubMed]

264. Heikenwalder, M.; Federau, C.; von Boehmer, L.; Schwarz, P.; Wagner, M.; Zeller, N.; Haybaeck, J.; Prinz, M.; Becher, B.; Aguzzi, A. Germinal centre B cells are dispensable for prion transport and neuroinvasion. *J. Neuroimmunol.* **2007**, in press. [CrossRef] [PubMed]

265. Helm, S.L.T.; Burton, G.F.; Szakal, A.K.; Tew, J.G. Follicular Dendritic Cells and the Maintenance of IgE Responses. *Eur. J. Immunol.* **1995**, *25*, 2362–2369. [CrossRef] [PubMed]

266. Fu, Y.-X.; Molina, H.; Matsumoto, M.; Huang, G.; Min, J.; Chaplin, D.D. Lymphotoxin-a (LTa) supports development of splenic follicular structure that is required for IgG response. *J. Exp. Med.* **1997**, *185*, 2111–2120. [CrossRef] [PubMed]

267. Fu, Y.-X.; Huang, G.; Wang, Y.; Chaplin, D.D. B lymphocytes induce the formation of follicular dendritic cell clusters in a lymphotoxin a-dependent fashion. *J. Exp. Med.* **1998**, *187*, 1009–1018. [CrossRef] [PubMed]

268. Endres, R.; Alimzhanov, M.B.; Plitz, T.; Futterer, A.; Kosco-Vilbois, M.H.; Nedospasov, S.A.; Rajewsky, K.; Pfeffer, K. Mature follicular dendritic cell networks depend on expression of lymphotoxin b receptor by radioresistant stromal cells and of lymphotoxin b and tumour necrosis factor by B cells. *J. Exp. Med.* **1999**, *189*, 159–168. [CrossRef] [PubMed]

269. Fu, Y.-X.; Huang, G.; Wang, Y.; Chaplin, D.D. Lymphotoxin-a-dependent spleen microenvironment supports the generation of memory B cells and is required for their subsequent antigen-induced activation. *J. Immunol.* **2000**, *164*, 2508–2514. [CrossRef] [PubMed]

270. Aydar, Y.; Sukumar, A.; Szakal, A.K.; Tew, J.G. The influence of immune complex-bearing follicular dendritic cells on the IgM response, Ig class switching, and production of high affinity IgG. *J. Immunol.* **2005**, *174*, 5358–5366. [CrossRef] [PubMed]

271. Heesters, B.A.; Myers, R.C.; Carroll, M.C. Follicular dendritic cells: Dynamic antigen libraries. *Nat. Rev. Immunol.* **2014**, *14*, 495–504. [CrossRef] [PubMed]

272. Kranich, J.; Krautler, N.J.; Heinen, E.; Polymenidou, M.; Bridel, C.; Schildknecht, A.; Huber, C.; Kosco-Vilbois, M.H.; Zinkernagel, R.; Miele, G.; et al. Follicular dendritic cells control engulfment of apoptotic bodies by secreting Mfge8. *J. Exp. Med.* **2008**, *205*, 1293–1302. [CrossRef] [PubMed]

273. Victoratos, P.; Lagnel, J.; Tzima, S.; Alimzhanov, M.B.; Rajewsky, K.; Pasparakis, M.; Kollias, G. FDC-specific functions of p55TNFR and IKK2 in the development of FDC networks and of antibody responses. *Immunity* **2006**, *24*, 65–77. [CrossRef] [PubMed]

274. Zabel, M.D.; Heikenwalder, M.; Prinz, M.; Arright, I.; Schwarz, P.; Kranich, J.; Von Teichman, A.; Haas, K.M.; Zeller, N.; Tedder, T.F.; et al. Stromal complement receptor CD21/35 facilitates lymphoid prion colonization and pathogenesis. *J. Immunol.* **2007**, *179*, 6144–6152. [CrossRef] [PubMed]

275. McBride, P.A.; Schulz-Shaeffer, W.J.; Donaldson, M.; Bruce, M.; Diringer, H.; Kretzschmar, H.A.; Beekes, M. Early spread of scrapie from the gastrointestinal tract to the central nervous system involves autonomic fibers of the splanchnic and vagus nerves. *J. Virol.* **2001**, *75*, 9320–9327. [CrossRef] [PubMed]

276. Glatzel, M.; Heppner, F.L.; Albers, K.M.; Aguzzi, A. Sympathetic innervation of lymphoreticular organs is rate limiting for prion neuroinvasion. *Neuron* **2001**, *31*, 25–34. [CrossRef]

277. Brown, K.L.; Gossner, A.; Mok, S.; Mabbott, N.A. The effects of host age on the transport of complement-bound complexes to the spleen and the pathogenesis of intravenous scrapie infection. *J. Virol.* **2012**, *86*, 1228–1237. [CrossRef] [PubMed]

278. Sim, R.B.; Kishore, U.; Villiers, C.L.; Marche, P.N.; Mitchell, D.A. C1q binding and complement activation by prions and amyloid. *Immunobiology* **2007**, *212*, 355–362. [CrossRef] [PubMed]

279. Mitchell, D.A.; Kirby, L.; Paulin, S.M.; Villiers, C.L.; Sim, R.B. Prion protein activates and fixes complement directly via the classical pathway: Implications for the mechanism of scrapie agent propagation in lymphoid tissue. *Mol. Immunol.* **2007**, *44*, 2997–3004. [CrossRef] [PubMed]

280. Klein, M.A.; Kaeser, P.S.; Schwarz, P.; Weyd, H.; Xenarios, I.; Zinkernagel, R.M.; Carroll, M.C.; Verbeek, J.S.; Botto, M.; Walport, M.J.; et al. Complement facilitates early prion pathogenesis. *Nat. Med.* **2001**, *7*, 488–492. [CrossRef] [PubMed]

281. Mabbott, N.A.; Bruce, M.E.; Botto, M.; Walport, M.J.; Pepys, M.B. Temporary depletion of complement component C3 or genetic deficiency of C1q significantly delays onset of scrapie. *Nat. Med.* **2001**, *7*, 485–487. [CrossRef] [PubMed]

282. Michel, B.; Ferguson, A.; Johnson, T.; Bender, H.; Meyerett-Reid, C.; Pulford, B.; von Teichman, A.; Seelig, D.; Weiss, J.H.; Telling, G.C.; et al. Genetic depletion of complement receptors CD21/35 prevents terminal prion disease in a mouse model of chronic wasting disease. *J. Immunol.* **2012**, *189*, 4520–4527. [CrossRef] [PubMed]

283. Michel, B.; Ferguson, A.; Johnson, T.; Bender, H.; Meyerett-Reid, C.; Wycoff, A.C.; Pulford, B.; Telling, G.C.; Zabel, M.D. Complement protein C3 exacerbates prion disease in a mouse model of chronic wasting disease. *Int. Immunol.* **2013**, *25*, 697–702. [CrossRef] [PubMed]

284. Kane, S.J.; Farley, T.K.; Gordon, E.O.; Estep, J.; Bender, H.R.; Moreno, J.A.; Bartz, J.; Telling, G.C.; Pickering, M.C.; Zabel, M.D. Complement regulatory protein factor H is a soluble prion receptor that potentiates peripheral prion pathogenesis. *J. Immunol.* **2017**, in press. [CrossRef] [PubMed]

285. Kane, S.J.; Swanson, E.; Gordon, E.O.; Rocha, S.; Bender, H.R.; Donius, L.R.; Hannan, J.P.; Zabel, M.D. Relative impact of complement receptors CD21/35 (Cr2/1) on scrapie pathogenesis in mice. *mSphere* **2017**, in press. [CrossRef]

286. Mabbott, N.A.; Bruce, M.E. Complement component C5 is not involved in scrapie pathogenesis. *Immunobiology* **2004**, *209*, 545–549. [CrossRef] [PubMed]

287. Heesters, B.A.; Chatterjee, P.; Kim, Y.A.; Kuligowski, M.P.; Kirchhausen, T.; Carroll, M.C. Endocytosis and recycling of immune complexes by follicular dendritic cells enhances B cell antigen binding and activation. *Immunity* **2013**, *38*, 1164–1175. [CrossRef] [PubMed]

288. Jeffrey, M.; McGovern, G.; Goodsir, C.M.; Brown, K.L.; Bruce, M.E. Sites of prion protein accumulation in scrapie-infected mouse spleen revealed by immuno-electron microscopy. *J. Pathol.* **2000**, *191*, 323–332. [CrossRef]

289. McGovern, G.; Brown, K.L.; Bruce, M.E.; Jeffrey, M. Murine scrapie infection causes an abnormal germinal centre reaction in the spleen. *J. Comp. Pathol.* **2004**, *130*, 181–194. [CrossRef] [PubMed]

290. Sigurdson, C.J.; Barillas-Mury, C.; Miller, M.W.; Oesch, B.; van Keulen, L.J.M.; Langeveld, J.P.M.; Hoover, E.A. PrP^{CWD} lymphoid cell targets in early and advanced chronic wasting disease of mule deer. *J. Gen. Virol.* **2002**, *83*, 2617–2628. [CrossRef] [PubMed]

291. Gunn, M.D.; Ngo, V.N.; Ansel, K.M.; Ekland, E.H.; Cyster, J.G.; Williams, L.T. A B-cell-homing chemokine made in lymphoid follicles activates Burkitt's lymphoma receptor-1. *Nature* **1998**, *391*, 799–803. [CrossRef] [PubMed]

292. Ansel, K.M.; Ngo, V.N.; Hyman, P.L.; Luther, S.A.; Forster, R.; Sedgwick, J.D.; Browning, J.L.; Lipp, M.; Cyster, J. A chemokine-driven feedback loop organizes lymphoid follicles. *Nature* **2000**, *406*, 309–314. [CrossRef] [PubMed]

293. Yu, P.; Wang, Y.; Chin, R.K.; Martinez-Pomares, L.; Gordon, S.; Kosco-Vilbois, M.H.; Cyster, J.; Fu, Y.-X. B cells control the migration of a subset of dendritic cells into B cell follicles via CXC chemokine ligand 13 in a lymphotoxin-dependent fashion. *J. Immunol.* **2002**, *168*, 5117–5123. [CrossRef] [PubMed]

294. Bradford, B.M.; Reizis, B.; Mabbott, N.A. Oral prion disease pathogenesis is impeded in the specific absence of CXCR5-expressing dendritic cells. *J. Virol.* **2017**, *91*, e00124-17. [CrossRef] [PubMed]

295. Jang, M.H.; Sougawa, N.; Tanaka, T.; Hirata, T.; Hiroi, T.; Tohya, K.; Guo, Z.; Umemoto, E.; Ebisuno, Y.; Yang, B.-G.; et al. CCR7 is critically important for migration of dendritic cells in intestinal lamina propria to mesenteric lymph nodes. *J. Immunol.* **2006**, *176*, 803–810. [CrossRef] [PubMed]

296. Levavasseur, E.; Matharom, P.; Dorban, G.; Nakano, H.; Kakiuchi, T.; Carnaud, C.; Sarradin, P.; Aucouturier, P. Experimental scrapie in '*plt*' mice: An assessment of the role of dendritic-cell migration in the pathogenesis of prion diseases. *J. Gen. Virol.* **2007**, *88*, 2353–2360. [CrossRef] [PubMed]

297. Phan, T.G.; Grigorova, I.; Okada, T.; Cyster, J.G. Subcapsular encounter and complement-dependent transport of immune complexes by lymph node B cells. *Nat. Immunol.* **2007**, *8*, 992–1000. [CrossRef] [PubMed]

298. Carrasco, Y.R.; Batista, F.D. B cells acquire particulate antigen in a macrophage-rich area at the boundary between the follicle and the subcapsular sinus of the lymph node. *Immunity* **2007**, *27*, 1–12. [CrossRef] [PubMed]

299. Phan, T.G.; Green, J.A.; Gray, E.E.; Xu, Y.; Cyster, J.G. Immune complex relay by subcapsular sinus macrophages and noncognate B cells drives antibody affinity maturation. *Nat. Immunol.* **2009**, *10*, 786–793. [CrossRef] [PubMed]

300. Cinamon, G.; Zachariah, M.A.; Lam, O.M.; Foss Jr, F.W.; Cyster, J.G. Follicular shuttling of marginal zone B cells facilitates antigen transport. *Nat. Immunol.* **2008**, *9*, 54–62. [CrossRef] [PubMed]

301. Heggebø, R.; Press, C.M.; Gunnes, G.; González, L.; Jeffrey, M. Distribution and accumulation of PrP in gut-associated and peripheral lymphoid tissue of scrapie-affected Suffolk sheep. *J. Gen. Virol.* **2002**, *83*, 479–489.

302. Glatzel, M.; Abela, E.; Maissen, M.; Aguzzi, A. Extraneural pathological prion protein in sporadic Creutzfeldt-Jakob disease. *N. Engl. J. Med.* **2003**, *349*, 1812–1820. [CrossRef] [PubMed]

303. Schlomchik, M.J.; Radebold, K.; Duclos, N.; Manuelidis, L. Neuroinvasion by a Creutzfeldt-Jakob disease agent in the absence of B cells and follicular dendritic cells. *Proc. Natl. Acad. Sci. USA* **2001**, *98*, 9289–9294. [CrossRef] [PubMed]

304. Somerville, R.A.; Birkett, C.R.; Farquhar, C.F.; Hunter, N.; Goldmann, W.; Dornan, J.; Grover, D.; Hennion, R.M.; Percy, C.; Foster, J.; et al. Immunodetection of PrP^{Sc} in spleens of some scrapie-infected sheep but not BSE-infected cows. *J. Gen. Virol.* **1997**, *78*, 2389–2396. [CrossRef] [PubMed]

305. Terry, L.A.; Marsh, S.; Ryder, S.J.; Hawkins, S.A.C.; Wells, G.A.H.; Spencer, Y.I. Detection of disease-specific PrP in the distal ileum of cattle exposed orally to the agent of bovine spongiform encephalopathy. *Vet. Rec.* **2003**, *152*, 387–392. [CrossRef] [PubMed]

306. Wells, G.A.H.; Dawson, M.; Hawkins, S.A.C.; Green, R.B.; Dexter, I.; Francis, M.E.; Simmons, M.M.; Austin, A.R.; Horigan, M.W. Infectivity in the ileum of cattle challenged orally with bovine spongiform encephalopathy. *Vet. Rec.* **1994**, *135*, 40–41. [CrossRef] [PubMed]

307. Hoffmann, C.; Eiden, M.; Kaatz, M.; Keller, M.; Ziegler, U.; Rogers, R.; Hills, B.; Balkema-Buschmann, A.; Van Keulen, L.; Jacobs, J.G.; et al. BSE infectivity in jejunum, ileum and ileocaecal junction of incubating cattle. *Vet. Res.* **2011**, *42*, 21. [CrossRef] [PubMed]

308. Foster, J.D.; Parnham, D.W.; Hunter, N.; Bruce, M. Distribution of the prion protein in sheep terminally affected with BSE following experimental oral transmission. *J. Gen. Virol.* **2001**, *82*, 2319–2326. [CrossRef] [PubMed]

309. Brown, K.L.; Mabbott, N.A. Evidence of subclinical prion disease in aged mice following exposure to bovine spongiform encephalopathy. *J. Gen. Virol.* **2014**, *95*, 231–243. [CrossRef] [PubMed]

310. O'Connor, T.; Frei, N.; Sponarova, J.; Schwarz, P.; Heikenwalder, M.; Agguzi, A. Lymphotxin, but not TNF, is required for prion invasion of lymph nodes. *PLoS Pathog.* **2012**, *8*, e1002867.

311. Heikenwalder, M.; Kurrer, M.O.; Margalith, I.; Kranich, J.; Zeller, N.; Haybaeck, J.; Polymenidou, M.; Matter, M.; Bremer, J.; Jackson, W.S.; et al. Lymphotoxin-dependent prion replication in inflammatory stromal cells of granulomas. *Immunity* **2008**, *29*, 998–1008. [CrossRef] [PubMed]

312. Clarke, M.C. The antibody response of scrapie-affected mice to immunisation with sheep red blood cells. *Res.Vet. Sci.* **1968**, *9*, 595–597. [PubMed]

313. Garfin, D.E.; Stites, D.P.; Perlman, J.D.; Cochran, S.P.; Prusiner, S.B. Mitogen stimulation of splenocytes from mice infected with scrapie agent. *J. Infect. Dis.* **1978**, *138*, 396–400. [CrossRef] [PubMed]

314. Kingsbury, D.T.; Smeltzer, D.A.; Gibbs, C.J.; Gadjusek, D.C. Evidence for Normal Cell-Mediated Immunity in Scrapie-Infected Mice. *Infect. Immun.* **1981**, *32*, 1176–1180.

315. Elleman, C.J. ConA induced suppressor cells in scrapie-infected mice. *Vet. Immunol. Immunopathol.* **1985**, *8*, 79–82. [CrossRef]

316. Gonzalez, L.; Martin, S.; Siso, S.; Konold, T.; Ortiz-Pelaez, A.; Phelan, L.; Goldmann, W.; Stewart, P.; Saunders, G.; Windl, O.; et al. High prevalence of scrapie in a dairy goat herd: Tissue distribution of disease-associated PrP and effect of *PRNP* genotype and age. *Vet. Res.* **2009**, *40*, 65. [CrossRef] [PubMed]

317. Thomsen, B.V.; Schneider, D.A.; O'Rourke, K.I.; Gidlewski, T.; McLane, J.; Allen, R.W.; mcIsaac, A.A.; Mitchell, G.B.; Keane, D.P.; Spraker, T.R.; et al. Diagnostic accuracy of rectal mucosa biopsy testing for chronic wasting disease within white-tailed deer (*Odocoileus virginianus*) herds in North America: Effects of age, sex, polymorphism at *PRNP* codon 96, and disease progression. *J. Vet. Diagn. Intestig.* **2012**, *24*, 878–887. [CrossRef] [PubMed]

318. Monello, R.J.; Powers, J.G.; Hobbs, N.T.; Spraker, T.R.; O'Rourke, K.I.; Wild, M.A. Efficacy of antemortem rectal biopsies to diagnose and estimate prevalence of chronic wasting disease in free-ranging cow elk (*Cervus elaphus nelsoni*). *J. Wildl. Dis.* **2013**, *49*, 270–278. [CrossRef] [PubMed]

319. Van Keulen, L.J.; Schreuder, B.E.; Vromans, M.E.; Langeveld, J.P.; Smits, M.A. Pathogenesis of natural scrapie in sheep. *Arch. Virol. Suppl.* **2000**, *16*, 57–71.

320. Van Keulen, L.J.M.; Vromans, M.E.W.; van Zijderveld, F.G. Ealry and late pathogenesis of natural scrapie infection in sheep. *APMIS* **2002**, *110*, 23–32. [CrossRef] [PubMed]

321. Van Keulen, L.J.M.; Bossers, A.; Van Zijderveld, F.G. TSE pathogenesis in cattle and sheep. *Vet. Res.* **2008**, *39*, 24. [CrossRef] [PubMed]

322. Tabouret, G.; Lacroux, C.; Lugan, S.; Costes, P.; Corbiere, F.; Weisbecker, J.L.; Schelcher, F.; Andréoletti, O. Relevance of oral experimental challenge with classical scrapie in sheep. *J. Gen. Virol.* **2010**, *91*, 2139–2144. [CrossRef] [PubMed]

323. Van Keulen, L.J.M.; Vromans, M.E.W.; Dolstra, C.H.; Bossers, A.; van Zijderveld, F.G. Pathogenesis of bovine spongiform encephalopathy in sheep. *Arch. Virol.* **2008**, *153*, 445–453. [CrossRef] [PubMed]

324. Keane, D.; Barr, D.; Osborn, R.; Langenberg, J.; O'Rourke, K.; Schneider, D.; Bochsler, P. Validation of use of rectoanal mucosa-associated lymphoid tissue for immunohistochemical diagnosis of chronic wasting disease in white-tailed deer (*Odocoileus virginianus*). *J. Clin. Microbiol.* **2009**, *47*, 1412–1417. [CrossRef] [PubMed]

325. Spraker, T.R.; VerCauteren, K.C.; Gidlewski, T.; Schneider, D.A.; Munger, R.; Balachandran, A.; O'Rourke, K.I. Antermortem detection of PrP^{CWD} in preclinical, ranch-raised Rocky Mountain elk (*Cervus elaphus nelsoni*) by biopsy of the rectal mucosa. *J. Vet. Diagn. Intestig.* **2009**, *21*, 15–24. [CrossRef] [PubMed]

326. Hoover, C.E.; Davenport, K.A.; Henderson, D.M.; Denkers, N.D.; Mathiason, C.K.; Soto, C.; Zabel, M.D.; Hoover, E.A. Pathways of prion spread during early chronic wasting disease in deer. *J. Virol.* **2017**, in press. [CrossRef] [PubMed]

327. Huang, F.-P.; Platt, N.; Wykes, M.; Major, J.R.; Powell, T.J.; Jenkins, C.D.; MacPherson, G.G. A discrete subpopulation of dendritic cells transports apoptotic intestinal epithelial cells to T cell areas of mesenteric lymph nodes. *J. Exp. Med.* **2000**, *191*, 435–443. [CrossRef] [PubMed]

328. Houston, S.A.; Cerovic, V.; Thomson, C.; Brewer, J.; Mowat, A.M.; Milling, S. The lymph nodes draining the small intestine and colon are anatomically separate and immunologically distinct. *Mucosal Immunol.* **2016**, *9*, 468–478. [CrossRef] [PubMed]

329. Kimberlin, R.H.; Walker, C.A. Pathogenesis of scrapie in mice after intragastric infection. *Virus Res.* **1989**, *12*, 213–220. [CrossRef]

330. Schmidt, T.H.; Bannard, O.; Gray, E.E.; Cyster, J.G. CXCR4 promotes B cell egress from Peyer's patches. *J. Exp. Med.* **2013**, *210*, 1099–1107. [CrossRef] [PubMed]

331. Turner, M.; Gulbranson-Judge, A.; Quinn, M.E.; Walters, A.E.; MacLennan, I.C.; Tybulewicz, V.L.J. Syk tyrosine kinase is required for the positive selection of immature B cells into the recirculating B cell pool. *J. Exp. Med.* **1997**, *186*, 2013–2021. [CrossRef] [PubMed]

332. Suzuki, K.; Grigorova, I.; Phan, T.G.; Kelly, L.M.; Cyster, J.G. Visualizing B cell capture of cognate antigen from follicular dendritic cells. *J. Exp. Med.* **2009**, *206*, 1485–1493. [CrossRef] [PubMed]

333. Mok, S.W.; Proia, R.L.; Brinkmann, V.; Mabbott, N.A. B cell-specific S1PR1 deficiency blocks prion dissemination between secondary lymphoid organs. *J. Immunol.* **2012**, *188*, 5032–5040. [CrossRef] [PubMed]

334. Andreoletti, O.; Litaise, C.; Simmons, H.; Corbiere, F.; Lugan, S.; Costes, P.; Schelcher, F.; Viette, D.; Grassi, J.; Lacroux, C. Highly efficient prion transmission by blood transfusion. *PLoS Pathog.* **2012**, *8*, e1002782. [CrossRef] [PubMed]

335. Douet, J.Y.; Lacroux, C.; Litaise, C.; Lugan, S.; Corbiere, F.; Arnold, M.; Simmons, H.; Aron, N.; Costes, P.; Tillier, C.; et al. Mononucleated blood cell populations display different abilities to transmit prion disease by the transfusion route. *J. Virol.* **2016**, *90*, 3439–3445. [CrossRef] [PubMed]

336. Edwards, J.C.; Moore, S.J.; Hawthorne, J.A.; Neale, M.H.; Terry, L.A. PrP(Sc) is associated with B cells in the blood of scrapie-infected sheep. *Virology* **2010**, *405*, 110–119. [CrossRef] [PubMed]

337. Mathiason, C.K.; Hayes-Klug, J.; Hays, S.A.; Powers, J.; Osborn, D.A.; Dahmes, S.J.; Miller, K.V.; Warren, R.J.; Mason, G.L.; Telling, G.C.; et al. B cells and platelets harbour prion infectivity in the blood of deer infected with chronic wasting disease. *J. Virol.* **2010**, *84*, 5097–5107. [CrossRef] [PubMed]

338. Sisó, S.; González, L.; Jeffrey, M.; Martin, S.; Chianini, F.; Steele, P. Prion protein in kidneys of scrapie-infected sheep. *Vet. Rec.* **2006**, *159*, 327–328. [CrossRef] [PubMed]

339. Gomez-Nicola, D.; Schetters, S.T.T.; Perry, V.H. Differential role of CCR2 in the dynamics of microglia and perivascular macrophages during prion disease. *Glia* **2014**, *62*, 1041–1052. [CrossRef] [PubMed]

340. Armstrong, R.A.; Cairns, N.J.; Ironside, J.W.; Lantos, P.L. Does the neuropathology of human patients with variant Creutzfeldt-Jakob disease reflect haematogenous spread of the disease. *Neurosci. Lett.* **2003**, *348*, 37–40. [CrossRef]

341. Felten, S.Y.; Felten, D.L. Innervation of Lymphoid Tissue. In *Psychoneuroimmunology*, 2nd ed.; Academic Press Inc.: Cambridge, MA, USA, 1991; pp. 27–69.

342. Beekes, M.; Baldauf, E.; Diringer, H. Sequential appearance and accumulation of pathognomonic markers in the central nervous system of hamsters orally infected with scrapie. *J. Gen. Virol.* **1996**, *77*, 1925–1934. [CrossRef] [PubMed]

343. Baldauf, E.; Beekes, M.; Diringer, H. Evidence for an alternative direct route of access for the scrapie agent to the brain bypassing the spinal cord. *J. Gen. Virol.* **1997**, *78*, 1187–1197. [CrossRef] [PubMed]

344. Beekes, M.; McBride, P.A.; Baldauf, E. Cerebral targeting indicates vagal spread of infection in hamsters fed with scrapie. *J. Gen. Virol.* **1998**, *79*, 601–607. [CrossRef] [PubMed]

345. McBride, P.A.; Beekes, M. Pathological PrP is abundant in sympathetic and sensory ganglia of hamsters fed with scrapie. *Neurosci. Lett.* **1999**, *265*, 135–138. [CrossRef]

346. Flores-Lagnarica, A.; Meza-Perez, S.; Calderon-Amador, J.; Estrada-Garcia, T.; Macpherson, G.; Saeland, S.; Steinman, R.M.; Flores-Romo, L. Network of dendritic cells within the muscular layer of the mouse. *Proc. Natl. Acad. Sci. USA* **2005**, *102*, 19039–19044. [CrossRef] [PubMed]

347. Muller, P.A.; Koscso, B.; Rajani, G.M.; Stevanovic, K.; Berres, M.-L.; Hashimoto, D.; Mortha, A.; Leboeuf, M.; Li, X.-M.; Mucida, D.; et al. Crosstalk between muscularis macrophages and enteric neurones regulates gastrointestinal motility. *Cell* **2014**, *158*, 300–313. [CrossRef] [PubMed]

348. Aucouturier, P.; Geissmann, F.; Damotte, D.; Saborio, G.P.; Meeker, H.C.; Kascsak, R.; Kascsak, R.; Carp, R.I.; Wisniewski, T. Infected splenic dendritic cells are sufficient for prion transmission to the CNS in mouse scrapie. *J. Clin. Investig.* **2001**, *108*, 703–708. [CrossRef] [PubMed]

349. Sigurdson, C.J.; Heikenwalder, M.; Manco, G.; Barthel, M.; Schwarz, P.; Stecher, B.; Krautler, N.J.; Hardt, W.-D.; Seifert, B.; MacPherson, A.J.S.; et al. Bacterial colitis increases susceptibility to oral prion pathogenesis. *J. Infect. Dis.* **2009**, *199*, 243–252. [CrossRef] [PubMed]

350. Dickinson, A.G.; Fraser, H.; McConnell, I.; Outram, G.W. Mitogenic Stimulation of the Host Enhances Susceptibility to Scrapie. *Nature* **1978**, *272*, 54–55. [CrossRef] [PubMed]

351. Bremer, J.; Heikenwalder, M.; Haybaeck, J.; Tiberi, C.; Krautler, N.J.; Kurrer, M.O.; Aguzzi, A. Repetitive immunization enhances the susceptibility of mice to peripherally administered prions. *PLoS ONE* **2009**, *4*, e7160. [CrossRef] [PubMed]

352. Heikenwalder, M.; Zeller, N.; Seeger, H.; Prinz, M.; Klöhn, P.-C.; Schwarz, P.; Ruddle, N.H.; Weissmann, C.; Aguzzi, A. Chronic lymphocytic inflammation specifies the organ tropism of prions. *Science* **2005**, *307*, 1107–1110. [CrossRef] [PubMed]

353. Ligios, C.; Sigurdson, C.; Santucciu, C.; Carcassola, G.; Manco, G.; Basagni, M.; Maestrale, C.; Cancedda, M.G.; Madau, L.; Aguzzi, A. PrP^{Sc} in mammary glands of sheep affected by scrapie and mastitis. *Nat. Med.* **2005**, *11*, 1137–1138. [CrossRef] [PubMed]

354. Valleron, A.-J.; Boelle, P.-Y.; Will, R.; Cesbron, J.-Y. Estimation of epidemic size and incubation time based on age characteristics of vCJD in the United Kingdom. *Science* **2001**, *294*, 1726–1728. [CrossRef] [PubMed]

355. Diack, A.B.; Head, M.W.; McCutcheon, S.; Boyle, A.; Knight, R.; Ironside, J.W.; Manson, J.C.; Will, R.G. Variant CJD. 18 years of research and surveillance. *Prion* **2014**, *2014*, 286–295. [CrossRef] [PubMed]

356. Bishop, M.T.; Hart, P.; Aitchison, L.; Baybutt, H.N.; Plinston, C.; Thomson, V.; Tuzi, N.L.; Head, M.W.; Ironside, J.W.; Will, R.G.; et al. Predicting susceptibility and incubation time of human-to-human transmission of vCJD. *Lancet Neurol.* **2006**, *5*, 393–398. [CrossRef]

357. Brown, K.L.; Stewart, K.; Bruce, M.E.; Fraser, H. Severly combined immunodeficient (SCID) mice resist infection with bovine spongiform encephalopathy. *J. Gen. Virol.* **1997**, *78*, 2707–2710. [CrossRef] [PubMed]

358. Srivastava, S.; Makarava, N.; Katorcha, E.; Savtchenko, R.; Brossmer, R.; Baskakov, I.V. Post-conversion sialylation of prions in lymphoid tissues. *Proc. Natl. Acad. Sci. USA* **2015**, *112*, E6654–E6662. [CrossRef] [PubMed]

359. Boelle, P.-Y.; Cesbron, J.-Y.; Valleron, A.-J. Epidemiological evidence of higher susceptibility to vCJD in the young. *BMC Infect. Dis.* **2004**, *4*, 1–7. [CrossRef] [PubMed]

360. Gibson, K.L.; Wu, Y.-C.; Barnett, Y.; Duggan, O.; Vaughan, R.; Kondeatis, E.; Nilsson, B.-O.; Wikby, A.; Kipling, D.; Dunn-Walters, D.K. B-cell diversity decreases in old age and is correlated with poor health status. *Aging Cell* **2009**, *8*, 18–25. [CrossRef] [PubMed]

361. Henson, S.M.; Akbar, A.N. Memory T-cell homeostasis and senescence during aging. *Adv. Exp. Med. Biol.* **2010**, *684*, 189–197. [PubMed]

362. Bradford, B.M.; Crocker, P.R.; Mabbott, N.A. Peripheral prion disease pathogenesis is unaltered in the absence of sialoadhesin (Siglec-1/CD169). *Immunology* **2014**, *143*, 120–129. [CrossRef] [PubMed]

363. St. Rose, S.; Hunter, N.; Matthews, D.; Foster, J.; Chase-Topping, M.E.; Kruuk, L.E.B.; Shaw, D.J.; Rhind, S.M.; Will, R.G.; Woolhouse, M.E.J. Comparative evidence for a link between Peyer's patch development and susceptibility to transmissible spongiform encephalopathies. *BMC Infect. Dis.* **2006**, *6*, 5. [CrossRef] [PubMed]

364. St. Rose, S.G.; Hunter, N.; Foster, J.D.; Drummond, D.; McKenzie, C.; Parnham, D.; Will, R.G.; Woolhouse, M.E.J.; Rhind, S.M. Quantification of Peyer's patches in Cheviot sheep for future scrapie pathogenesis studies. *Vet. Immunol. Immunopathol.* **2007**, *116*, 163–171. [CrossRef] [PubMed]

365. Brown, K.L.; Wathne, G.J.; Sales, J.; Bruce, M.E.; Mabbott, N.A. The effects of host age on follicular dendritic cell status dramatically impair scrapie agent neuroinvasion in aged mice. *J. Immunol.* **2009**, *183*, 5199–5207. [CrossRef] [PubMed]

366. Kobayashi, A.; Donaldson, D.S.; Erridge, C.; Kanaya, T.; Williams, I.R.; Ohno, H.; Mahajan, A.; Mabbott, N.A. The functional maturation of M cells is dramatically reduced in the Peyer's patches of aged mice. *Mucosal Immunol.* **2013**, *6*, 1027–1037. [CrossRef] [PubMed]

367. Mebius, R.E.; Kraal, G. Structure and function of the spleen. *Nat. Rev. Immunol.* **2005**, *5*, 606–616. [CrossRef] [PubMed]

368. Turner, V.M.; Mabbott, N.A. Ageing adversely affects the migration and function of marginal zone B cells. *Immunology* **2017**, *151*, 349–362. [CrossRef] [PubMed]

369. Spraker, T.R.; VerCauteren, K.C.; Gidlewski, T.L.; Munger, R.D.; Walter, W.D.; Balachandran, A. Impact of age and sex of Rocky Mountain elk (*Cervus elaphus nelsoni*) on follicle counts from rectal mucosal biopsies for preclinical detection of chronic wasting disease. *J. Vet. Diagn. Intestig.* **2009**, *21*, 868–870. [CrossRef] [PubMed]

370. Geremia, C.; Hoeting, J.A.; Wolfe, L.L.; Galloway, N.L.; Antolin, M.F.; Spraker, T.R.; Miller, M.W.; Hobbs, N.T. Age and repeated biopsy influence antemortem PrP^{CWD} testing in mule deer (*Odocoileus hemionus*) in Colorado, USA. *J. Wildl. Dis.* **2015**, *51*, 801–810. [CrossRef] [PubMed]

371. Outram, G.W.; Dickinson, A.G.; Fraser, H. Developmental maturation of susceptibility to scrapie in mice. *Nature* **1973**, *241*, 536–537. [CrossRef] [PubMed]

372. Ierna, M.I.; Farquhar, C.F.; Outram, G.W.; Bruce, M.E. Resistance of neonatal mice to scrapie is associated with inefficient infection of the immature spleen. *J. Virol.* **2006**, *80*, 474–482. [CrossRef] [PubMed]

373. Hunter, N.; Houston, F.; Foster, J.; Goldmann, W.; Drummond, D.; Parnham, D.; Kennedy, I.; Green, A.; Stewart, P.; Chong, A. Susceptibility of young sheep to oral infection with bovine spongiform encephalopathy decreases significantly after weaning. *J. Virol.* **2012**, *86*, 11856–11862. [CrossRef] [PubMed]

374. Ano, Y.; Sakudo, A.; Uraki, R.; Sato, Y.; Kono, J.; Sugiura, K.; Yokoyama, T.; Itohara, S.; Nakayama, H.; Yukawa, M.; et al. Enhanced enteric invasion of scrapie agents into the villous columnar epithelium via maternal immunoglobulin. *Int. J. Mol. Med.* **2010**, *26*, 845–851. [PubMed]

375. Korth, C.; May, B.C.; Cohen, F.E.; Prusiner, S.B. Acridine and phenothiazine derivatives as pharmocotherapeutics for prion disease. *Proc. Natl. Acad. Sci. USA* **2001**, *98*, 9836–9841. [CrossRef] [PubMed]

376. Collins, S.J.; Lewis, V.; Brazier, M.; Hill, A.F.; Fletcher, A.; Masters, C.L. Quinacrine does not prolong survival in a murine Creutzfeldt-Jakob disease model. *Ann. Neurol.* **2002**, *52*, 503–506. [CrossRef] [PubMed]

377. Collinge, J.; Gorham, M.; Hudson, F.; Kennedy, A.; Keogh, G.; Pal, S.; Rossor, M.; Rudge, P.; Siddique, D.; Spyer, M.; et al. Safety and efficacy of quinacrine in human prion disease (PRION-1 study): A patient-preference trial. *Lancet Neurol.* **2009**, *8*, 334–344. [CrossRef]

378. Geschwind, M.D.; Kuo, A.L.; Wong, K.S.; Haman, A.; Devereux, G.; Raudabaugh, B.J.; Johnson, D.Y.; Torres-Chae, C.C.; Finley, R.; Garcia, P.; et al. Quinacrine treatment for sporadic Creutzfeldt-Jakob disease. *Neurology* **2013**, *81*, 2015–2023. [CrossRef] [PubMed]

379. Farquhar, C.; Dickinson, A.; Bruce, M. Prophylactic potential of pentosan polysulphate in transmissible spongiform encephalopathies. *Lancet* **1999**, *353*, 117. [CrossRef]

380. Doh-ura, K.; Ishiwaki, K.; Murakami-Kubo, I.; Sasaki, K.; Mohri, S.; Race, R.; Iwaki, T. Treatment of transmissible spongiform encephalopathy by intraventricular drug infusion in animal models. *J. Virol.* **2004**, *78*, 4999–5006. [CrossRef] [PubMed]

381. Tsuboi, Y.; Doh-Ura, K.; Yamada, T. Continuous intraventricular infusion of pentosan polysulfate: Clinical trial against prion diseases. *Neuropathology* **2009**, *29*, 632–636. [CrossRef] [PubMed]

382. Tagliavini, F.; Forloni, G.; Colombo, L.; Rossi, G.; Girola, L.; Canciani, B.; Angretti, N.; Giampaolo, L.; Pressini, E.; Awan, T.; et al. Tetracycline affects abnormal properties of synthetic PrP peptides and PrP^{Sc} in vitro. *J. Mol. Biol.* **2000**, *300*, 1309–1322. [CrossRef] [PubMed]

383. Forloni, G.; Iussich, S.; Awan, T.; Colombo, L.; Angeretti, N.; Girola, L.; Bertani, I.; Poli, G.; Caramelli, M.; Grazia Bruzzone, M.; et al. Tetracyclines affect prion infectivity. *Proc. Natl. Acad. Sci. USA* **2002**, *99*, 10849–10854. [CrossRef] [PubMed]

384. Haik, S.; Marcon, G.; Tettamanti, M.; Welaratne, A.; Giaccone, G.; Azimi, S.; Pietrini, V.; Fabrequettes, J.R.; Imperiale, D.; Cesaro, P.; et al. Doxycycline in Creutzfeldt-Jakob disease: A phase 2, randomised, double-blind, placebo-controlled trial. *Lancet Neurol.* **2014**, *13*, 150–158. [CrossRef]

385. Dickinson, A.G.; Fraser, H.; McConnell, I.; Outram, G.W.; Sales, D.I.; Taylor, D.M. Extraneural competition between different scrapie agents leading to loss of infectivity. *Nature* **1975**, *253*, 556. [CrossRef] [PubMed]

386. Dickinson, A.G.; Fraser, H.; Meikle, V.M.H.; Outram, G.W. Competition between different scrapie agents in mice. *Nat. New Biol.* **1972**, *237*, 244–245. [CrossRef] [PubMed]

387. Manuelidis, L. Vaccination with an attenuated Creutzfeldt-Jakob disease strain prevents expression of a virulent agent. *Proc. Natl. Acad. Sci. USA* **1998**, *95*, 2520–2525. [CrossRef] [PubMed]

388. Diaz-Espinoza, R.; Morales, R.; Concha-Marambio, L.; Moreno-Gonzalez, I.; Moda, F.; Soto, C. Treatment with a non-toxic, self-replicating anti-prion delays or prevents prion disease in vivo. *Mol. Psychiatry* **2017**. [CrossRef] [PubMed]

389. Moreno, J.A.; Radford, H.; Peretti, D.; Steinert, J.R.; Verity, N.; Martin, M.G.; Halliday, M.; Morgan, J.; Dinsdale, D.; Ortori, C.A.; et al. Sustained translational repression by eIF2aP mediates prion neurodegeneration. *Nature* **2012**, *485*, 507–511. [PubMed]

390. Moreno, J.A.; Halliday, M.; Molloy, C.; Radford, H.; Verity, N.; Axten, J.M.; Ortori, C.A. Oral treatment targeting the unfolded protein response prevents neurodegeneration and clinical disease in prion-infected mice. *Sci. Transl. Med.* **2013**, *5*, 206ra138. [CrossRef] [PubMed]

391. Mabbott, N.A. Prospects for safe and effective vaccines against prion diseases. *Exp. Rev. Vaccines* **2015**, *14*, 1–4. [CrossRef] [PubMed]

392. Roettger, Y.; Du, Y.; Bacher, M.; Zerr, I.; Dodel, R.; Back, J.-P. Immunotherapy in prion disease. *Nat. Rev. Neurol.* **2013**, *9*, 98–105. [CrossRef] [PubMed]

393. Heppner, F.L.; Musahl, C.; Arrighi, I.; Klein, M.A.; Rulicke, T.; Oesch, B.; Zinkernagel, R.M.; Kalinke, U.; Aguzzi, A. Prevention of scrapie pathogenesis by transgenic expression of anti-prion protein antibodies. *Science* **2001**, *294*, 178–182. [CrossRef] [PubMed]

394. White, A.R.; Enever, P.; Tayebi, M.; Mushens, R.; Lineham, J.; Brandner, S.; Anstee, D.; Collinge, J.; Hawke, S. Monoclonal antibodies inhibit prion replication and delay the development of prion disease. *Nature* **2003**, *422*, 80–83. [CrossRef] [PubMed]

395. Goñi, F.; Knudsen, E.; Schreiber, F.; Scholtzova, H.; Pankiewicz, J.; Carp, R.; Meeker, H.C.; Rubenstein, R.; Brown, D.R.; Sy, M.S.; et al. Mucosal vaccination delays or prevents prion infection via the oral route. *Neuroscience* **2005**, *133*, 413–421. [CrossRef] [PubMed]

396. Goñi, F.; Chablagoity, J.A.; Prelli, F.; Schreiber, F.; Scholtzova, H.; Chung, E.; Kascsak, R.; Brown, D.R.; Sigurdsson, E.M.; Wisniewski, T. High titres of mucosal and systemic anti-PrP antibodies abrogates oral prion infection in mucosal vaccinated mice. *Neuroscience* **2008**, *153*, 679–686. [CrossRef] [PubMed]

397. Sonati, T.; Reimann, R.R.; Falsig, J.; Baral, P.K.; O'Connor, T.; Hornemann, S.; Yaganoglu, S.; Li, B.; Herrmann, U.S.; Wieland, B.; et al. The toxicity of antiprion antibodies is mediated by the flexible tail of the prion protein. *Nature* **2013**, *501*, 102–106. [CrossRef] [PubMed]

398. Asante, E.A.; Smidak, M.; Grimshaw, A.; Houghton, R.; Tomlinson, R.; Jeelani, A.; Jakubcova, T.; Hamdan, S.; Richard-Londt, A.; Lineham, J.M.; et al. A naturally occuring variant of the human prion protein completely prevents prion disease. *Nature* **2015**, *522*, 478–481. [CrossRef] [PubMed]

pathogens

MDPI

Review

The Role of the Mammalian Prion Protein in the Control of Sleep

Amber Roguski [1] and Andrew C. Gill [1,2,*]

[1] The Roslin Institute and Royal (Dick) School of Veterinary Sciences, University of Edinburgh, Easter Bush Veterinary Centre, Edinburgh EH25 9RG, UK; agproguski@gmail.com

[2] School of Chemistry, Joseph Banks Laboratories, University of Lincoln, Green Lane, Lincoln, Lincolnshire LN6 7DL, UK

* Correspondence: angill@lincoln.ac.uk; Tel.: +44-(0)-1522-835-258; Fax: +44-(0)-131-651-9105

Received: 28 September 2017; Accepted: 13 November 2017; Published: 17 November 2017

Abstract: Sleep disruption is a prevalent clinical feature in many neurodegenerative disorders, including human prion diseases where it can be the defining dysfunction, as in the case of the "eponymous" fatal familial insomnia, or an early-stage symptom as in certain types of Creutzfeldt-Jakob disease. It is important to establish the role of the cellular prion protein (PrP^C), the key molecule involved in prion pathogenesis, within the sleep-wake system in order to understand fully the mechanisms underlying its contribution to both healthy circadian rhythmicity and sleep dysfunction during disease. Although severe disruption to the circadian rhythm and melatonin release is evident during the pathogenic phases of some prion diseases, untangling whether PrP^C plays a role in circadian rhythmicity, as suggested in mice deficient for PrP^C expression, is challenging given the lack of basic experimental research. We provide a short review of the small amount of direct literature focused on the role of PrP^C in melatonin and circadian rhythm regulation, as well as suggesting mechanisms by which PrP^C might exert influence upon noradrenergic and dopaminergic signaling and melatonin synthesis. Future research in this area should focus upon isolating the points of dysfunction within the retino-pineal pathway and further investigate PrP^C mediation of pinealocyte GPCR activity.

Keywords: prion; sleep; circadian rhythm; melatonin; serotonin

---

1. Introduction

The highly conserved prion protein (PrP) is encoded by the *PRNP* gene (human cytogenic location 20p12) [1] and exists predominately in two conformationally different isoforms [2]. The cellular prion protein (PrP^C) has an α-helical structure and is expressed highly within the central and peripheral nervous systems [3] and localized to neuronal and glial cell membranes [4]. PrP^C expression has also been identified in tissues beyond the nervous systems, including the intestine, heart and lymph nodes [5]. This widespread distribution of apparently functional PrP^C, coupled with its highly conserved nature, suggests that it has either one important role or multiple, context-dependent roles throughout the body. Total knockout of PrP^C is not associated with major deleterious phenotypes [6], suggesting that the latter hypothesis is more likely to be true. Indeed, PrP^C has been implicated in, among other things, immune response modulation [7], myelin maintenance [8], mitochondrial homeostasis [9] and signal transduction as reviewed recently by Castle and Gill [4]. A final possibility is that PrP^C does indeed have a single role in a key biochemical pathway that impacts other physiological processes that PrP^C is proposed to control; in this possibility, PrP^C may be involved only in modulating activity of the top-level system rather than turning it on or off, thereby explaining its apparent redundancy and context-dependent functions. Of relevance to this review is the finding that, in models of PrP^C dysfunction or knockout, marked alterations to circadian rhythms occur [10] suggesting a role for PrP^C in one of the most essential processes for life: the sleep/wake (or active/rest) cycle.

The prion protein is one of a handful of naturally expressed proteins that can misfold into specific pathogenic isoforms and, hence, become integral to neurodegenerative phenotypes. The prion protein's conformationally-altered isoform is known as PrP^{Sc}, which is enriched in β-sheet structure, is insoluble and which assembles into amyloid plaques and fibrils [11]. The conversion from PrP^C to PrP^{Sc} and the subsequent aggregation and oligomerization of PrP^{Sc} results in severe, and ultimately fatal, neurodegenerative prion diseases, also known as transmissible spongiform encephalopathies or TSEs [11,12]. Prion diseases can occur sporadically, due to infection, or as the result of genetic influence [13]. The exact mechanism underlying sporadic conversion from PrP^C to PrP^{Sc} is unknown, but it is believed that this mechanism is expedited when PrP^C contains a disease-initiating mutation. However, following the initiation of an infection (e.g., by ingesting contaminated meat) exogenous PrP^{Sc} is believed to be used as a template, upon which the functional endogenous PrP^C misfolds in a process of autocatalytic conversion [11,12]. The incubation period between inoculation and disease onset is prolonged in prion diseases [11], providing a potential window for neuroprotective interventions [14], but this relies on knowledge of the molecular mechanisms responsible for neuronal loss so that intervention can be targeted effectively and specifically. Identifying such mechanisms is complicated by the fact that different prion disease types (or strains) target pathology to rather different areas of the brain and the information that causes neuropathological targeting is believed to be encoded in the structure of PrP^{Sc}. Although multimeric forms of PrP^{Sc} may be toxic to neurons directly, by activating signaling pathways that lead to apoptosis or necrosis, it is also possible that the neuropathology that arises with PrP^{Sc} propagation occurs in combination with loss of function of PrP^C [15]. Indeed, during pathogenesis of a prion disease, PrP^C expression has been suggested to be downregulated [15,16]. Herein we review whether there may be role for PrP^C loss of function in the sleep-wake cycle, a key biological system that is compromised to different extents during various prion diseases.

2. Sleep Dysfunction during Prion Disease Pathogenesis

Observations of human prion disease patients provided the first links between PrP^C/PrP^{Sc} and sleep. In the genetic prion disease fatal familial insomnia (FFI), the "eponymous" clinical feature is severely disturbed sleep, characterized by anxiolytic-resistant insomnia, circadian rhythm dysfunction, sleep fragmentation and altered arousal. Polysomnographic studies show a decrease in total sleep time, decreased REM sleep and loss of REM atonia [17]. Central sleep apneas and decreased slow wave sleep are also common features of FFI clinicopathology [18]. These extensive symptoms are likely to be the result of dysfunction across the various systems regulating sleep.

To variable degrees, sleep disturbances are also evident in other prion diseases, including various human prion diseases, although they are not currently considered part of the clinical diagnostic criteria. Nevertheless, in a recent study almost 90% of sporadic Creutzfeldt-Jakob disease (spCJD) patients reported sleep dysfunction during clinical evaluation, making sleep disturbance more prevalent than any other diagnostic criteria for CJD [19]. Sleep disturbances were also prevalent clinical complaints in all familial CJD patients examined during a separate study [20]. Certain animal models of prion disease also result in disrupted sleep patterns; rats inoculated with various prion strains have pronounced slow wave sleep decreases [21], rhesus monkeys infected with the human prion disease kuru show complete loss of REM sleep and disrupted sleep stage cycling [22], whilst mice inoculated with the murine prion disease RML show alterations in rest period activity from extremely early in the incubation period [23]. Interestingly, it has been reported that patients with Gerstmann-Sträussler-Sheinker disease (a genetic prion disease characterized predominantly by ataxia and pyramidal dysfunction) do not exhibit sleep alterations [24,25], suggesting that sleep dysfunction is specific to particular prion strains. Given that neuropathology in different prion strains is targeted to different regions of the brain, it follows that molecular or cellular alterations in specific brain regions may underlie some of the prion-induced sleep abnormalities. It is pertinent, therefore, to consider the brain areas that control the different aspects of normal sleep and how these are affected during prion pathogenesis.

The slow wave oscillations of deep, non-REM (NREM sleep) are generated, synchronized and stabilized by the thalamocortical network [26,27], whilst fluctuations in thalamocortical excitability produces the hallmark deflections visualized on EEG traces during NREM sleep: the K-complex and sleep spindle [28]. Cortically-generated K-complexes occur both spontaneously and in response to sensory stimulation, acting to monitor environmental stimuli during reduced states of consciousness and to enhance sleep stability [28]. The K-complex waveform is reflected in thalamic activity, with thalamic neurons reinforcing the K-complex slow oscillation [26]. In response to the K-complex, thalamic reticular neurons generate their own oscillation—known as the sleep spindle—which is thought to be instrumental in thalamocortical plasticity, cognition and memory function [29]. One of the most striking polysomnographic observations in FFI and fCJD patients is a reduction or absence of the thalamocortical K-complex and sleep spindle oscillations [19,30–32]. Prion-induced dysregulation of slow wave sleep, sleep instability and the loss of K-complex and sleep spindle oscillations (as well as the memory and cognitive impairments seen in prion diseases) are therefore likely due to the gross neurodegeneration of the thalamocortical network as the disease process reaches the clinical phase [33]. Support for this rationale comes from mouse models of FFI: mice expressing PrP^C carrying the D178N mutation (that causes FFI in humans in combination with methionine expressed at codon 129) develop thalamic pathology and exhibit disrupted circadian rhythmicity of sleep and motor activity. Both over-expressing [34] and knock-in [35] FFI transgenic mice are phenotypically similar to human FFI patients, with the overexpressing mice exhibiting abnormal REM-sleep transitioning, loss of sleep spindles, reduced slow wave activity and decreased sleep continuity. Anatomically, these mice also exhibited thalamic degeneration, which, as in human patients, is responsible for the observed breakdown of sleep architecture. However, changes in circadian-regulated motor activity of these mice, including decreased dark-phase activity compared to controls [34], are not readily rationalized by thalamic degeneration and knock-in FFI mice also show significant decreases in dark-phase motor activity compared to controls [35].

Symptoms of insomnia reported by prion disease patients can be explained by the reduction in both sleep stability and sleep maintenance associated with thalamocortical degeneration. It is also likely that the dysregulation of circadian rhythmicity seen in prion disease contributes to symptoms of insomnia. A key organ responsible for the modulation of sleep patterns through synthesis of the circadian hormone melatonin is the pineal gland, and it has been demonstrated that non-prion patients with insomnia exhibit reduced pineal gland volume [36] as well as significantly decreased nocturnal plasma melatonin levels [37]. The pineal gland is a site of high level expression of PrP^C [4]. Whilst this may render the pineal gland highly susceptible to infection by prions, it also raises the possibility that reduced levels of functional PrP^C during a prion infection [15] may be involved in sleep dysfunction. Thus, the mechanisms of circadian dysfunction evident in prion diseases require further basic science investigation beyond clinical reporting, as currently there is little known of how a prion infection can dysregulate such an essential system.

3. Circadian Rhythm, Homeostatic Sleep Pressure and Melatonin

The sleep-wake cycle is driven by two factors: circadian rhythm and homeostatic sleep pressure [38,39]. Homeostatic sleep pressure is best described as the feeling of sleepiness, such that the longer one goes without sleep, the more tired one becomes. The driving mechanism of homeostatic sleep pressure is thought to be the result of chemical build-up, such as increased levels of adenosine [40], within the brain. Slow wave oscillations during wakefulness and sleep are regulated homeostatically, with slow wave activity decreasing over the sleep period and increasing as time awake increases [41,42].

Similar to homeostatic sleep pressure, circadian rhythmicity is the result of endogenous processes, with every cell exhibiting its own intrinsic, oscillatory, circadian rhythm [43]. The oscillations in cellular gene expression, caused by a negative feedback transcription/translation loop, have a natural period of around 24 h, but are then synchronized to the exogenous light/dark cycle [43]. The circadian rhythm is also modulated, to a degree, by other external factors such as exercise and food consumption [44] as well as internal factors including core body temperature and the menstrual cycle [45].

The main regulatory hormone of the circadian rhythm is melatonin [39]. Melatonin is part of the tryptophan metabolic pathway [42] and it is synthesized and secreted by both central nervous system (CNS) tissues and organs throughout the body [46]. Within the CNS, melatonin production is entrained to the light/dark cycle, with the hormone acting as a mediator between hypothalamic nuclei and target tissues, relaying diurnal and seasonal timings to the body [47]. As with PrP^C, the variety of organs capable of melatonin synthesis is suggestive of important, contextual roles for this hormone, including regulation of seasonal and circadian rhythms, reproductive function, modulation of neurotransmission and antioxidant effects [47].

The complex, multi-nuclei pathway for melatonin synthesis, depicted schematically in Figure 1, begins with blue light excitation of photosensitive, melanopsin-containing retinal ganglion cells (RGCs). The RGC axons form the retino-hypothalamic tract which projects to the hypothalamic suprachiasmatic nucleus (SCN), providing glutamatergic excitation of SCN neurons. The SCN is the central pacemaker that fine-tunes the body's circadian rhythm, conveying the light/dark fluctuations of the external environment to various tissues, including the brain, via oscillatory activity in order to coordinate metabolic function and homeostasis accordingly [48]. These SCN oscillations are generated in response to RCG neurotransmission by a transcriptional-translational feedback loop between clock genes [49].

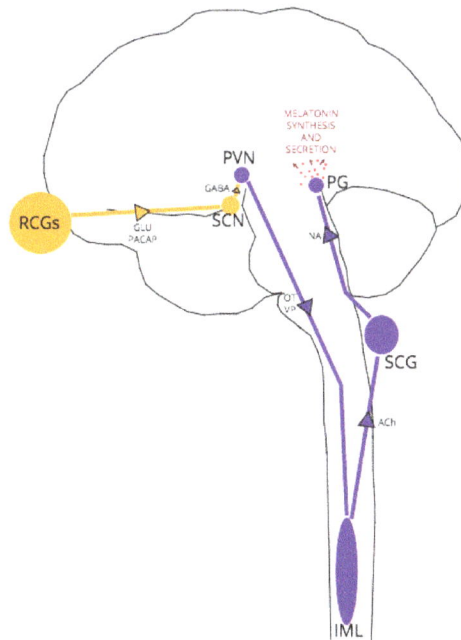

Figure 1. The Retino-Pineal Pathway. Structures and pathways in yellow represent light-active regions, those in dark blue represent dark-active regions. Arrows represent direction of signaling; associated text represents main neurotransmitters in the pathway. Region abbreviations: RCGs *retinal ganglion cells*; SCN *suprachiasmatic nucleus*; PVN *paraventricular nucleus*; IML *intermediolateral nucleus*; SCG *superior cervical ganglion*; PG *pineal gland*. Neurotransmitter abbreviations: GLU *glutamate*; PACAP *pituitary adenylate-cyclase activating peptide*; GABA *gamma-aminobutyric acid*; OT *oxytocin*; VP *vasopressin*; ACh *acetylcholine*; NA *noradrenaline*.

In all melatonin-synthesizing organisms, light acts as a zeitgeber (environmental cue) for melatonin secretion, with peak melatonin levels arising in the middle of the night's sleep at approximately 100 times the daytime melatonin level [39]. In daylight, GABAergic SCN projections synapse to paraventricular (PVN) neurons, inhibiting further excitation of the pathway. At night, without RCG/SCN excitation, PVN neurons are disinhibited, leading in turn to excitation of thoracic intermediolateral (IML) neurons and superior cervical ganglia (SCG) neurons. Noradrenergic SCG neurons project to the pineal gland, which expresses both α1 and β1-adrenoceptors [50]. Noradrenaline-adrenoceptor binding then activates adenylate cyclase and the melatonin synthesis cascade within pinealocytes.

Pinealocytes provide the majority of circulating melatonin, with immediate melatonin release into the blood (bypassing the blood brain barrier) and into CSF circulation at the point of the pineal recess and third ventricle [51]. Melatonin travels via CSF and blood to target tissues expressing G-protein coupled melatonin receptors MT1 and MT2 [52]. MT1 binding has been associated with metabolic regulation, whilst MT2 binding is associated with circadian rhythmicity [46]. Expression of both melatonin receptors is widespread throughout the brain, with localization to neurons of the SCN, thalamus and hippocampus, amongst other structures [41].

The pineal gland is not the only source of melatonin; extrapineal melatonin synthesis is thought to occur in a number of organs that express melatonin-synthesizing enzymes, including the heart, liver and placenta [46]. Indeed, the gut expresses melatonin levels 400 times greater than that of the pineal gland, and extrapineal melatonin synthesis has been shown to occur independently of the photoperiod, with an absence of diurnal fluctuations [53]. In contrast to the pineal-synthesized melatonin, which is immediately released into CSF and blood circulation, melatonin synthesized in tissues beyond the CNS is retained within cells [46]. Intracellular melatonin is correlated with its local anti-inflammatory/oxidative effects; in neonates with respiratory distress syndrome, melatonin treatment significantly reduces pro-inflammatory cytokine and nitrate levels [54], whilst melatonin levels in the elderly are inversely correlated with Charlson Comorbidity Index score, indicating a protective role against chronic disease [55].

Abolition of endogenous CNS melatonin has far-reaching consequences for the brain and body; reports of impaired wound healing [56], disrupted circadian organization of the SCN [57], abolition of clock gene expression in adipose tissue [58] as well as altered bone metabolism leading to bone loss [59] provide evidence for the protective function of melatonin in health. The effects of pinealectomy upon sleep are equally as wide-ranging. Long-term changes post-pinealectomy in humans include reduced sleep efficiency and increased switching between sleep stages, indicating a role for melatonin in sleep stability [60]. Pinealectomized rats show a significant decrease in REM-sleep theta power compared to controls [61]. Intriguingly, there are also reports of the loss of hippocampal CA1 and CA3 neurons post pinealectomy in rats [62], which can be reversed by administering exogenous melatonin; since MT1/MT2-expressing hippocampal CA1 and CA3 neurons [63] generate cortical theta oscillations, the theta reduction in pinealectomized rats may be, in part, the result of the hippocampal pyramidal neuron loss.

Extrapolating the influence of melatonin upon the CNS and body is challenging, given how intertwined the circadian rhythm is with whole-system function. It is important to identify whether prion disease-related sleep/wake dysfunctions are the result of circadian rhythm anomalies caused by PrP^C loss of function from the pineal gland, whether anomalies can be explained by neurodegeneration caused by PrP^{Sc}-mediated neurotoxicity or whether both affects play a role. This requires a consideration of the link between PrP^C, circadian rhythms and melatonin signaling.

4. A Role for PrP^C in the Regulation of Melatonin Synthesis

Early research revealed that PrP mRNA expression peaks during the circadian dark phase at approximately 14 h (zeitgeber time) [64]. This significant increase in PrP mRNA expression precedes melatonin synthesis, with pineal melatonin levels increasing to a peak from 18 to 20 h (zeitgeber time) [65]. Whilst the apparent correlation between PrP^C and melatonin levels may be just that—correlative rather than causative—there is some evidence that PrP^C levels impact directly or indirectly on melatonin levels and that the high level of PrP^C expression in the pineal glands may play a role in regulating melatonin synthesis.

Early studies of PrP knockout (PrP$^{-/-}$) mice highlighted similar sleep disruptions to those reported in human prion disease cases, such as increased sleep fragmentation and altered slow wave activity [10]. It was noted that PrP$^{-/-}$ mice exhibited a significantly longer circadian phase than controls (23.9 h vs. 23.3 h, respectively), as well as demonstrating inverse dark phase activity compared to controls; control mice carried out the majority of their wheel-running during the first half of the dark phase, whilst PrP$^{-/-}$ mice had increased activity towards the end of the dark phase [10]. These findings suggest PrP$^{-/-}$ mice experience a phase shift in their circadian timings, akin to that seen in delayed phase sleep disorders, with circadian period elongation and delay in activity timings [66]. It was also noted that PrP$^{-/-}$ mice exhibited alterations to intrinsic circadian rhythms. Normally, in the complete absence of light, the circadian rhythm deregulates and becomes "free-running", which manifests in incremental shifts in peak melatonin release from night to day [67], since intrinsic circadian rhythms in mice are slightly greater than 24 h. When placed in complete darkness, control mice exhibited this free-running circadian rhythm, using motor activity as an indicator, whilst, confusingly, the PrP$^{-/-}$ mice maintained 24 h rhythmicity despite no light entrainment [10]. Further to this, PrP$^{-/-}$ mice demonstrate non-cyclical, phase-shifted melatonin release in comparison to controls, with no increase in nighttime plasma melatonin level but increased levels of melatonin during the day relative to controls [68].

The above observations beg the question of whether PrP^C functions as part of the pathway that turns light cues into melatonin and suggest that disruption of this pathway might be responsible, in part, for circadian abnormalities during prion disease pathogenesis. Progressive circadian disruption is observed in FFI patients, evidenced by decreasing plasma melatonin levels with disease advancement and complete loss of circadian rhythm by end-stage FFI [18,69]. A case study of one FFI patient reported normal function of the SCN, as evidenced by normal core body temperature rhythms and appropriate sleep stage timing, but dissociation of the circadian rhythms for melatonin and cortisol [69]. This suggests normal light/dark entrainment and SCN oscillatory activity, but a "functional interference" somewhere in the retino-pineal tract between the SCN and pineal gland function [69].

Reports of circadian phase-shifts and dysregulated melatonin expression in PrP^C dysfunction, supported by the above experimental and clinical findings [10,68], suggest a delay in the synthesis or release of melatonin. Establishing the point at which PrP^C interacts with the melatonin-synthesis pathway, whether at the pinealocyte or at structure upstream in the circuit, is therefore essential for revealing the mechanism underlying the abnormal circadian cycling.

As indicated earlier and depicted in Figure 1, functionally, between the SCN and pineal gland lie three structures: the hypothalamic paraventricular nucleus (PVN), the intermediolateral nucleus (IML) of the thoracic spinal column and the superior cervical ganglion (SCG) at the top of the sympathetic chain. The main neurotransmitters released by each structure are vasopressin/oxytocin, acetylcholine and noradrenaline respectively [47]. The fact that melatonin is still synthesized at healthy levels in models of PrP$^{-/-}$ and early-stage TSEs, albeit with an altered expression pattern, suggests that the structures are capable of neurotransmission of reasonable fidelity, but the phasing may be compromised in some way. It has been demonstrated that noradrenaline dysregulation is implicated in TSE pathology; for example, after oral and intraperitoneal prion inoculation, PrP^{Sc} neuroinvasion follows a selective route from the periphery via sympathetic nerves to the brain [70,71]. Noradrenergic cell death [72] and dramatically altered noradrenaline levels in specific brain regions (cerebral cortex, cerebellum, pons) and plasma following intraperitoneal or intracerebral inoculation [71–74] have also been demonstrated in experimental models of prion disease. However, these effects on noradrenaline expression and neurotransmission have not been demonstrated in PrP$^{-/-}$ models, which suggests the specific targeting of the noradrenergic system is the result of PrP^{Sc} pathology, not a loss of PrP^{C} function in disease.

If neurotransmission throughout the retino-pineal tract is unchanged in PrP$^{-/-}$, then the point of melatonin dysregulation or synthesis delay must be at the SCG-pinealocyte synapse, or due to a dysfunction at some point downstream and within the melatonin-synthesis signaling pathway (see Figure 2). Despite experimental evidence for PrP mRNA upregulation within the pinealocyte during the subjective night [75], no studies have specifically explored PrP^{C} interactions or co-localization at the pinealocyte. One scenario that might explain the melatonin phase shift exhibited by clinical and experimental models of prion dysfunction is if compromised PrP^{C} function causes dysregulation at the point of serotonin acetylation within the melatonin synthesis pathway. Serotonin is acetylated by the enzyme aralkylamine N-acetyltransferase (AA-NAT), whose activity acts as the rate-limiting step of melatonin synthesis [76]. The AA-NAT gene contains a cAMP-responsive element (cre) within its promoter region which, when stimulated by increased levels of cAMP, drives enzymatic expression and activity [77]. Levels of pinealocyte cAMP increase in response to noradrenergic, and to a lesser degree, dopaminergic, binding respectively with adrenergic and D$_1$ receptors [77,78].

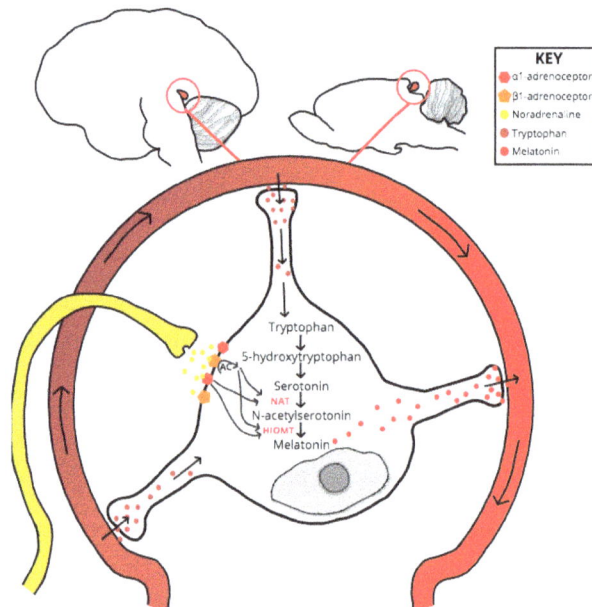

Figure 2. Location of the Pineal Gland in the Human (left) and Mouse (right) Brain and Pinealocyte Melatonin Synthesis. The essential amino acid tryptophan is uptaken by pinealocytes from the surrounding vasculature and is converted to 5-hydroxytryptophan via a process of hydroxylation. Decarboxylation of 5-hydroxytryptophan gives rise to serotonin. SCG afferents, which project to the pineal adjacent to capillaries, release noradrenaline into the pineal perivascular space, where the neurotransmitter binds to α1B and β1-adrenoceptors on the pinealocyte membrane. Coincident activation of adrenoceptors increases cAMP levels, in turn inducing enzyme expression increases of *N-acetyltransferase (NAT)*—the rate-limiting step of the synthesis pathway- and *acetylserotonin O-methyltransferase (HIOMT)*. Melatonin synthesis is dependent upon serotonin acetylation by *NAT* and *HIOMT* methylation activity of *N*-acetylserotonin. Once synthesized the indolamine is secreted into the circulation, where it has a biological half-life of ~45 min in humans and ~20 min in rats, before hepatic metabolism and urinal metabolite excretion.

AA-NAT activity determines melatonin synthesis lag [79], and a significant association between AA-NAT polymorphisms and delayed phase sleep disorder has been shown [76]. Intriguingly, PrP^C has continually been proposed to act as a cell-surface scaffolding protein and in this role it may influence AA-NAT activity indirectly through modulating the activity of pinealocyte G protein-coupled receptor (GPCR) complexes. The GPCRs can be divided according to whether they couple with the $G_{\alpha i}$ protein subunit, which inhibits cAMP production and adenylyl cyclase activity, or with the $G_{\alpha s}$ subunit, which inversely stimulates cAMP and adenylyl cyclase activity. Of the main GPCRs expressed by the pinealocyte, only the β-adrenergic and D_1 GPCRs are $G_{\alpha s}$-coupled [80,81]; it is, therefore, plausible that PrP^C associates with one or other of these two GPCRs (or both) as a scaffold protein and influences their signaling cascade in pinealocytes. Supporting this, it has been demonstrated recently that PrP^C co-localizes with D_1 receptors, and there is selective impairment of cAMP signaling in response to D_1 stimulation in PrP$^{-/-}$ mice [78]. A lack of PrP^C may act to reduce adenylyl cyclase activity, resulting in an increased time for cAMP concentrations to reach a supra-threshold level. This would result in dysregulation of AA-NAT, since the length of time taken to achieve optimal acetylation of serotonin increases due to decreased AA-NAT enzymatic activity. Future investigations for a role of PrP^C within the circadian rhythm and regulation of melatonin synthesis should therefore investigate protein-protein interactions of PrP^C with noradrenergic and dopaminergic receptors, as well as analysis of analytes and catalysts within the melatonin synthesis pathway, with a particular focus on levels of serotonin, *N*-acetylserotonin and AA-NAT.

5. Conclusions

Clinical observation and experimental models of prion disease demonstrate clear circadian dysfunction, suggesting a role for PrP^C within the synthesis or regulation of melatonin release in health. Drawing upon the limited literature, we propose that PrP^C acts as a scaffold protein at the pinealocyte, associating with the $G_{\alpha s}$-coupled β-adrenergic and D_1 GPCRs. In this hypothesis, PrP^C regulates the cAMP signaling cascade downstream of β-adrenergic/D_1 activation, and in the absence of PrP^C (as evidenced by PrP$^{-/-}$ models), a phase shift in melatonin synthesis results due to delayed AA-NAT enzymatic activity.

Acknowledgments: We acknowledge scientists at Roslin Institute for useful discussion and Sandra McCutcheon for proof reading the text. No specific funding was used for the review.

Author Contributions: A.R. performed the literature review, formulated hypotheses and wrote the first draft of the manuscript. A.C.G. supervised the work, edited the text and generated a final manuscript version. Both authors agreed the final manuscript content.

Conflicts of Interest: The authors declare no conflicts of interest, financial or otherwise.

References

1. Sparkes, R.S.; Simon, M.; Cohn, V.H.; Fournier, R.E.; Lem, J.; Klisak, I.; Heinzmann, C.; Blatt, C.; Lucero, M.; Mohandas, T.; et al. Assignment of the human and mouse prion protein genes to homologous chromosomes. *Proc. Natl. Acad. Sci. USA* **1986**, *83*, 7358–7362. [CrossRef] [PubMed]
2. Colby, D.W.; Prusiner, S.B. Prions. *Cold Spring Harb. Perspect. Biol.* **2011**, *3*, a006833. [CrossRef] [PubMed]
3. Wulf, M.A.; Senatore, A.; Aguzzi, A. The biological function of the cellular prion protein: An update. *BMC Biol.* **2017**, *15*, 34. [CrossRef] [PubMed]
4. Castle, A.R.; Gill, A.C. Physiological functions of the cellular prion protein. *Front. Mol. Biosci.* **2017**, *4*, 19. [CrossRef] [PubMed]
5. Peralta, O.A.; Eyestone, W.H. Quantitative and qualitative analysis of cellular prion protein (prp(c)) expression in bovine somatic tissues. *Prion* **2009**, *3*, 161–170. [CrossRef] [PubMed]
6. Bueler, H.; Fischer, M.; Lang, Y.; Bluethmann, H.; Lipp, H.P.; DeArmond, S.J.; Prusiner, S.B.; Aguet, M.; Weissmann, C. Normal development and behaviour of mice lacking the neuronal cell-surface prp protein. *Nature* **1992**, *356*, 577–582. [CrossRef] [PubMed]

7. Bakkebo, M.K.; Mouillet-Richard, S.; Espenes, A.; Goldmann, W.; Tatzelt, J.; Tranulis, M.A. The cellular prion protein: A player in immunological quiescence. *Front. Immunol.* **2015**, *6*, 450. [CrossRef] [PubMed]

8. Bremer, J.; Baumann, F.; Tiberi, C.; Wessig, C.; Fischer, H.; Schwarz, P.; Steele, A.D.; Toyka, K.V.; Nave, K.A.; Weis, J.; et al. Axonal prion protein is required for peripheral myelin maintenance. *Nat. Neurosci.* **2010**, *13*, U310–U319. [CrossRef] [PubMed]

9. Miele, G.; Jeffrey, M.; Turnbull, D.; Manson, J.; Clinton, M. Ablation of cellular prion protein expression affects mitochondrial numbers and morphology. *Biochem. Biophys. Res. Commun.* **2002**, *291*, 372–377. [CrossRef] [PubMed]

10. Tobler, I.; Gaus, S.E.; Deboer, T.; Achermann, P.; Fischer, M.; Rulicke, T.; Moser, M.; Oesch, B.; McBride, P.A.; Manson, J.C. Altered circadian activity rhythms and sleep in mice devoid of prion protein. *Nature* **1996**, *380*, 639–642. [CrossRef] [PubMed]

11. Prusiner, S.B. Prions. *Proc. Natl. Acad. Sci. USA* **1998**, *95*, 13363–13383. [CrossRef] [PubMed]

12. Annus, A.; Csati, A.; Vecsei, L. Prion diseases: New considerations. *Clin. Neurol. Neurosurg.* **2016**, *150*, 125–132. [CrossRef] [PubMed]

13. Brandner, S.; Jaunmuktane, Z. Prion disease: Experimental models and reality. *Acta Neuropathol.* **2017**, *133*, 197–222. [CrossRef] [PubMed]

14. Fraser, J.R. What is the basis of transmissible spongiform encephalopathy induced neurodegeneration and can it be repaired? *Neuropathol. Appl. Neurobiol.* **2002**, *28*, 1–11. [CrossRef] [PubMed]

15. Mays, C.E.; Kim, C.; Haldiman, T.; van der Merwe, J.; Lau, A.; Yang, J.; Grams, J.; Di Bari, M.A.; Nonno, R.; Telling, G.C.; et al. Prion disease tempo determined by host-dependent substrate reduction. *J. Clin. Investig.* **2014**, *124*, 847–858. [CrossRef] [PubMed]

16. Llorens, F.; Ansoleaga, B.; Garcia-Esparcia, P.; Zafar, S.; Grau-Rivera, O.; Lopez-Gonzalez, I.; Blanco, R.; Carmona, M.; Yague, J.; Nos, C.; et al. Prp mrna and protein expression in brain and prp(c) in csf in creutzfeldt-jakob disease mm1 and vv2. *Prion* **2013**, *7*, 383–393. [CrossRef] [PubMed]

17. Llorens, F.; Zarranz, J.J.; Fischer, A.; Zerr, I.; Ferrer, I. Fatal familial insomnia: Clinical aspects and molecular alterations. *Curr. Neurol. Neurosci. Rep.* **2017**, *17*, 30. [CrossRef] [PubMed]

18. Portaluppi, F.; Cortelli, P.; Avoni, P.; Vergnani, L.; Maltoni, P.; Pavani, A.; Sforza, E.; Degli Uberti, E.C.; Gambetti, P.; Lugaresi, E. Progressive disruption of the circadian rhythm of melatonin in fatal familial insomnia. *J. Clin. Endocrinol. Metab.* **1994**, *78*, 1075–1078. [PubMed]

19. Kang, P.; de Bruin, G.S.; Wang, L.H.; Ward, B.A.; Ances, B.M.; Lim, M.M.; Bucelli, R.C. Sleep pathology in creutzfeldt-jakob disease. *J. Clin. Sleep Med.* **2016**, *12*, 1033–1039. [CrossRef] [PubMed]

20. Givaty, G.; Maggio, N.; Cohen, O.S.; Blatt, I.; Chapman, J. Early pathology in sleep studies of patients with familial creutzfeldt-jakob disease. *J. Sleep Res.* **2016**, *25*, 571–575. [CrossRef] [PubMed]

21. Bassant, M.H.; Cathala, F.; Court, L.; Gourmelon, P.; Hauw, J.J. Experimental scrapie in rats: First electrophysiological observations. *Electroencephalogr. Clin. Neurophysiol.* **1984**, *57*, 541–547. [CrossRef]

22. Bert, J.; Vuillon-Cacciuttolo, G.; Balzamo, E.; De Micco, P.; Gambarelli, D.; Tamalet, J.; Gastaut, H. Experimental kuru in the rhesus monkey: A study of eeg modifications in the waking state and during sleep. *Electroencephalogr. Clin. Neurophysiol.* **1978**, *45*, 611–620. [CrossRef]

23. Steele, A.D.; Jackson, W.S.; King, O.D.; Lindquist, S. The power of automated high-resolution behavior analysis revealed by its application to mouse models of huntington's and prion diseases. *Proc. Natl. Acad. Sci. USA* **2007**, *104*, 1983–1988. [CrossRef] [PubMed]

24. Pierangeli, G.; Bono, F.; Aguglia, U.; Maltoni, P.; Montagna, P.; Lugaresi, E.; Quattrone, A.; Cortelli, P. Normal sleep-wake and circadian rhythms in a case of gerstmann-straussler-sheinker (gss) disease. *Clin. Auton. Res.* **2004**, *14*, 39–41. [CrossRef] [PubMed]

25. Provini, F.; Vetrugno, R.; Pierangeli, G.; Cortelli, P.; Rizzo, G.; Filla, A.; Strisciuglio, C.; Gallassi, R.; Montagna, P. Sleep and temperature rhythms in two sisters with p102l gerstmann-straussler-scheinker (gss) disease. *Sleep Med.* **2009**, *10*, 374–377. [CrossRef] [PubMed]

26. Amzica, F.; Steriade, M. Cellular substrates and laminar profile of sleep k-complex. *Neuroscience* **1998**, *82*, 671–686. [CrossRef]

27. McCormick, D.A.; Bal, T. Sleep and arousal: Thalamocortical mechanisms. *Annu. Rev. Neurosci.* **1997**, *20*, 185–215. [CrossRef] [PubMed]

28. Jahnke, K.; von Wegner, F.; Morzelewski, A.; Borisov, S.; Maischein, M.; Steinmetz, H.; Laufs, H. To wake or not to wake? The two-sided nature of the human k-complex. *Neuroimage* **2012**, *59*, 1631–1638. [CrossRef] [PubMed]

29. Clawson, B.C.; Durkin, J.; Aton, S.J. Form and function of sleep spindles across the lifespan. *Neural Plast.* **2016**, *2016*, 6936381. [CrossRef] [PubMed]

30. Cohen, O.S.; Chapman, J.; Korczyn, A.D.; Warman-Alaluf, N.; Orlev, Y.; Givaty, G.; Nitsan, Z.; Appel, S.; Rosenmann, H.; Kahana, E.; et al. Characterization of sleep disorders in patients with e200k familial creutzfeldt-jakob disease. *J. Neurol.* **2015**, *262*, 443–450. [CrossRef] [PubMed]

31. Ferrillo, F.; Plazzi, G.; Nobili, L.; Beelke, M.; De Carli, F.; Cortelli, P.; Tinuper, P.; Avoni, P.; Vandi, S.; Gambetti, P.; et al. Absence of sleep eeg markers in fatal familial insomnia healthy carriers: A spectral analysis study. *Clin. Neurophysiol.* **2001**, *112*, 1888–1892. [CrossRef]

32. Gemignani, A.; Laurino, M.; Provini, F.; Piarulli, A.; Barletta, G.; d'Ascanio, P.; Bedini, R.; Lodi, R.; Manners, D.N.; Allegrini, P.; et al. Thalamic contribution to sleep slow oscillation features in humans: A single case cross sectional EEG study in fatal familial insomnia. *Sleep Med.* **2012**, *13*, 946–952. [CrossRef] [PubMed]

33. Piao, Y.S.; Kakita, A.; Watanabe, H.; Kitamoto, T.; Takahashi, H. Sporadic fatal insomnia with spongiform degeneration in the thalamus and widespread prpsc deposits in the brain. *Neuropathology* **2005**, *25*, 144–149. [CrossRef] [PubMed]

34. Bouybayoune, I.; Mantovani, S.; Del Gallo, F.; Bertani, I.; Restelli, E.; Comerio, L.; Tapella, L.; Baracchi, F.; Fernandez-Borges, N.; Mangieri, M.; et al. Transgenic fatal familial insomnia mice indicate prion infectivity-independent mechanisms of pathogenesis and phenotypic expression of disease. *PLoS Pathog.* **2015**, *11*, e1004796. [CrossRef] [PubMed]

35. Jackson, W.S.; Borkowski, A.W.; Faas, H.; Steele, A.D.; King, O.D.; Watson, N.; Jasanoff, A.; Lindquist, S. Spontaneous generation of prion infectivity in fatal familial insomnia knockin mice. *Neuron* **2009**, *63*, 438–450. [CrossRef] [PubMed]

36. Bumb, J.M.; Schilling, C.; Enning, F.; Haddad, L.; Paul, F.; Lederbogen, F.; Deuschle, M.; Schredl, M.; Nolte, I. Pineal gland volume in primary insomnia and healthy controls: A magnetic resonance imaging study. *J. Sleep Res.* **2014**, *23*, 274–280. [CrossRef] [PubMed]

37. Hajak, G.; Rodenbeck, A.; Staedt, J.; Bandelow, B.; Huether, G.; Ruther, E. Nocturnal plasma melatonin levels in patients suffering from chronic primary insomnia. *J. Pineal Res.* **1995**, *19*, 116–122. [CrossRef] [PubMed]

38. Borbely, A.A. A two process model of sleep regulation. *Hum. Neurobiol.* **1982**, *1*, 195–204. [PubMed]

39. Carter, M.D.; Juurlink, D.N. Melatonin. *Can. Med. Assoc. J.* **2012**, *184*, 1923. [CrossRef] [PubMed]

40. Bjorness, T.E.; Dale, N.; Mettlach, G.; Sonneborn, A.; Sahin, B.; Fienberg, A.A.; Yanagisawa, M.; Bibb, J.A.; Greene, R.W. An adenosine-mediated glial-neuronal circuit for homeostatic sleep. *J. Neurosci.* **2016**, *36*, 3709–3721. [CrossRef] [PubMed]

41. Achermann, P.; Borbely, A.A. Sleep homeostasis and models of sleep regulation. In *Principles and Practice of Sleep Medicine*, 5th ed.; Kryger, M.H., Roth, T., Dement, W.C., Eds.; Elsevier: Saint Louis, MO, USA, 2011; pp. 431–444.

42. Bersagliere, A.; Achermann, P. Slow oscillations in human non-rapid eye movement sleep electroencephalogram: Effects of increased sleep pressure. *J. Sleep Res.* **2010**, *19*, 228–237. [CrossRef] [PubMed]

43. Patel, V.R.; Ceglia, N.; Zeller, M.; Eckel-Mahan, K.; Sassone-Corsi, P.; Baldi, P. The pervasiveness and plasticity of circadian oscillations: The coupled circadian-oscillators framework. *Bioinformatics* **2015**, *31*, 3181–3188. [CrossRef] [PubMed]

44. Shibata, S.; Sasaki, H.; Ikeda, Y. Chrono-nutrition and chrono-exercise. *Nihon Rinsho* **2013**, *71*, 2194–2199. [PubMed]

45. Baker, F.C.; Driver, H.S. Circadian rhythms, sleep, and the menstrual cycle. *Sleep Med.* **2007**, *8*, 613–622. [CrossRef] [PubMed]

46. Acuna-Castroviejo, D.; Escames, G.; Venegas, C.; Diaz-Casado, M.E.; Lima-Cabello, E.; Lopez, L.C.; Rosales-Corral, S.; Tan, D.X.; Reiter, R.J. Extrapineal melatonin: Sources, regulation, and potential functions. *Cell. Mol. Life Sci.* **2014**, *71*, 2997–3025. [CrossRef] [PubMed]

47. Simonneaux, V.; Ribelayga, C. Generation of the melatonin endocrine message in mammals: A review of the complex regulation of melatonin synthesis by norepinephrine, peptides, and other pineal transmitters. *Pharmacol. Rev.* **2003**, *55*, 325–395. [CrossRef] [PubMed]

48. Hastings, M.H.; Brancaccio, M.; Maywood, E.S. Circadian pacemaking in cells and circuits of the suprachiasmatic nucleus. *J. Neuroendocrinol.* **2014**, *26*, 2–10. [CrossRef] [PubMed]

49. Bonmati-Carrion, M.A.; Arguelles-Prieto, R.; Martinez-Madrid, M.J.; Reiter, R.; Hardeland, R.; Rol, M.A.; Madrid, J.A. Protecting the melatonin rhythm through circadian healthy light exposure. *Int. J. Mol. Sci.* **2014**, *15*, 23448–23500. [CrossRef] [PubMed]

50. Gupta, B.B.; Spessert, R.; Vollrath, L. Molecular components and mechanism of adrenergic signal transduction in mammalian pineal gland: Regulation of melatonin synthesis. *Indian J. Exp. Biol.* **2005**, *43*, 115–149. [PubMed]

51. Leston, J.; Harthe, C.; Mottolese, C.; Mertens, P.; Sindou, M.; Claustrat, B. Is pineal melatonin released in the third ventricle in humans? A study in movement disorders. *Neurochirurgie* **2015**, *61*, 85–89. [CrossRef] [PubMed]

52. Liu, J.; Clough, S.J.; Hutchinson, A.J.; Adamah-Biassi, E.B.; Popovska-Gorevski, M.; Dubocovich, M.L. Mt1 and mt2 melatonin receptors: A therapeutic perspective. *Annu. Rev. Pharmacol. Toxicol.* **2016**, *56*, 361–383. [CrossRef] [PubMed]

53. Chen, C.Q.; Fichna, J.; Bashashati, M.; Li, Y.Y.; Storr, M. Distribution, function and physiological role of melatonin in the lower gut. *World J. Gastroenterol.* **2011**, *17*, 3888–3898. [CrossRef] [PubMed]

54. Gitto, E.; Reiter, R.J.; Cordaro, S.P.; La Rosa, M.; Chiurazzi, P.; Trimarchi, G.; Gitto, P.; Calabro, M.P.; Barberi, I. Oxidative and inflammatory parameters in respiratory distress syndrome of preterm newborns: Beneficial effects of melatonin. *Am. J. Perinatol.* **2004**, *21*, 209–216. [CrossRef] [PubMed]

55. Gallucci, M.; Flores-Obando, R.; Mazzuco, S.; Ongaro, F.; Di Giorgi, E.; Boldrini, P.; Durante, E.; Frigato, A.; Albani, D.; Forloni, G.; et al. Melatonin and the charlson comorbidity index (cci): The treviso longeva (trelong) study. *Int. J. Biol. Markers* **2014**, *29*, e253–e260. [CrossRef] [PubMed]

56. Ozler, M.; Simsek, K.; Ozkan, C.; Akgul, E.O.; Topal, T.; Oter, S.; Korkmaz, A. Comparison of the effect of topical and systemic melatonin administration on delayed wound healing in rats that underwent pinealectomy. *Scand. J. Clin. Lab. Investig.* **2010**, *70*, 447–452. [CrossRef] [PubMed]

57. Yu, G.D.; Rusak, B.; Piggins, H.D. Regulation of melatonin-sensitivity and firing-rate rhythms of hamster suprachiasmatic nucleus neurons: Constant light effects. *Brain Res.* **1993**, *602*, 191–199. [CrossRef]

58. De Farias Tda, S.; de Oliveira, A.C.; Andreotti, S.; do Amaral, F.G.; Chimin, P.; de Proenca, A.R.; Leal, F.L.; Sertie, R.A.; Campana, A.B.; Lopes, A.B.; et al. Pinealectomy interferes with the circadian clock genes expression in white adipose tissue. *J. Pineal Res.* **2015**, *58*, 251–261. [CrossRef] [PubMed]

59. Egermann, M.; Gerhardt, C.; Barth, A.; Maestroni, G.J.; Schneider, E.; Alini, M. Pinealectomy affects bone mineral density and structure—An experimental study in sheep. *BMC Musculoskelet. Disord.* **2011**, *12*, 271. [CrossRef] [PubMed]

60. Slawik, H.; Stoffel, M.; Riedl, L.; Vesely, Z.; Behr, M.; Lehmberg, J.; Pohl, C.; Meyer, B.; Wiegand, M.; Krieg, S.M. Prospective study on salivary evening melatonin and sleep before and after pinealectomy in humans. *J. Biol. Rhythm.* **2016**, *31*, 82–93. [CrossRef] [PubMed]

61. Fisher, S.P.; Sugden, D. Endogenous melatonin is not obligatory for the regulation of the rat sleep-wake cycle. *Sleep* **2010**, *33*, 833–840. [CrossRef] [PubMed]

62. De Butte, M.; Pappas, B.A. Pinealectomy causes hippocampal ca1 and ca3 cell loss: Reversal by melatonin supplementation. *Neurobiol. Aging* **2007**, *28*, 306–313. [CrossRef] [PubMed]

63. Lacoste, B.; Angeloni, D.; Dominguez-Lopez, S.; Calderoni, S.; Mauro, A.; Fraschini, F.; Descarries, L.; Gobbi, G. Anatomical and cellular localization of melatonin mt1 and mt2 receptors in the adult rat brain. *J. Pineal Res.* **2015**, *58*, 397–417. [CrossRef] [PubMed]

64. Cagampang, F.R.; Whatley, S.A.; Mitchell, A.L.; Powell, J.F.; Campbell, I.C.; Coen, C.W. Circadian regulation of prion protein messenger rna in the rat forebrain: A widespread and synchronous rhythm. *Neuroscience* **1999**, *91*, 1201–1204. [CrossRef]

65. Nakahara, D.; Nakamura, M.; Iigo, M.; Okamura, H. Bimodal circadian secretion of melatonin from the pineal gland in a living cba mouse. *Proc. Natl. Acad. Sci. USA* **2003**, *100*, 9584–9589. [CrossRef] [PubMed]

66. Micic, G.; de Bruyn, A.; Lovato, N.; Wright, H.; Gradisar, M.; Ferguson, S.; Burgess, H.J.; Lack, L. The endogenous circadian temperature period length (tau) in delayed sleep phase disorder compared to good sleepers. *J. Sleep Res.* **2013**, *22*, 617–624. [CrossRef] [PubMed]

67. Verwey, M.; Robinson, B.; Amir, S. Recording and analysis of circadian rhythms in running-wheel activity in rodents. *J. Vis. Exp.* **2013**, *113*, 1–8. [CrossRef] [PubMed]

68. Brown, D.R.; Nicholas, R.S.; Canevari, L. Lack of prion protein expression results in a neuronal phenotype sensitive to stress. *J. Neurosci. Res.* **2002**, *67*, 211–224. [CrossRef] [PubMed]

69. Dauvilliers, Y.; Cervena, K.; Carlander, B.; Espa, F.; Bassetti, C.; Claustrat, B.; Laplanche, J.L.; Billiard, M.; Touchon, J. Dissociation in circadian rhythms in a pseudohypersomnia form of fatal familial insomnia. *Neurology* **2004**, *63*, 2416–2418. [CrossRef] [PubMed]

70. Beekes, M.; Baldauf, E.; Diringer, H. Sequential appearance and accumulation of pathognomonic markers in the central nervous system of hamsters orally infected with scrapie. *J. Gen. Virol.* **1996**, *77 Pt 8*, 1925–1934. [CrossRef] [PubMed]

71. Pollera, C.; Bondiolotti, G.; Formentin, E.; Puricelli, M.; Mantegazza, P.; Bareggi, S.; Poli, G.; Ponti, W. Plasma noradrenalin as marker of neuroinvasion in prion diseases. *Vet. Res. Commun.* **2007**, *31* (Suppl. 1), 249–252. [CrossRef] [PubMed]

72. Yun, S.W.; Choi, E.K.; Ju, W.K.; Ahn, M.S.; Carp, R.I.; Wisniewski, H.M.; Kim, Y.S. Extensive degeneration of catecholaminergic neurons to scrapie agent 87v in the brains of im mice. *Mol. Chem. Neuropathol.* **1998**, *34*, 121–132. [CrossRef] [PubMed]

73. Bassant, M.H.; Fage, D.; Dedek, J.; Cathala, F.; Court, L.; Scatton, B. Monoamine abnormalities in the brain of scrapie-infected rats. *Brain Res.* **1984**, *308*, 182–185. [CrossRef]

74. Bondiolotti, G.; Rossoni, G.; Puricelli, M.; Formentin, E.; Lucchini, B.; Poli, G.; Ponti, W.; Bareggi, S.R. Changes in sympathetic activity in prion neuroinvasion. *Neurobiol. Dis.* **2010**, *37*, 114–117. [CrossRef] [PubMed]

75. Su, A.I.; Wiltshire, T.; Batalov, S.; Lapp, H.; Ching, K.A.; Block, D.; Zhang, J.; Soden, R.; Hayakawa, M.; Kreiman, G.; et al. A gene atlas of the mouse and human protein-encoding transcriptomes. *Proc. Natl. Acad. Sci. USA* **2004**, *101*, 6062–6067. [CrossRef] [PubMed]

76. Hohjoh, H.; Takasu, M.; Shishikura, K.; Takahashi, Y.; Honda, Y.; Tokunaga, K. Significant association of the arylalkylamine n-acetyltransferase (aa-nat) gene with delayed sleep phase syndrome. *Neurogenetics* **2003**, *4*, 151–153. [PubMed]

77. Baler, R.; Covington, S.; Klein, D.C. The rat arylalkylamine n-acetyltransferase gene promoter. Camp activation via a camp-responsive element-ccaat complex. *J. Biol. Chem.* **1997**, *272*, 6979–6985. [CrossRef] [PubMed]

78. Beckman, D.; Santos, L.E.; Americo, T.A.; Ledo, J.H.; de Mello, F.G.; Linden, R. Prion protein modulates monoaminergic systems and depressive-like behavior in mice. *J. Biol. Chem.* **2015**, *290*, 20488–20498. [CrossRef] [PubMed]

79. Gauer, F.; Poirel, V.J.; Garidou, M.L.; Simonneaux, V.; Pevet, P. Molecular cloning of the arylalkylamine-n-acetyltransferase and daily variations of its mrna expression in the syrian hamster pineal gland. *Brain Res. Mol. Brain Res.* **1999**, *71*, 87–95. [CrossRef]

80. Green, S.A.; Holt, B.D.; Liggett, S.B. Beta 1- and beta 2-adrenergic receptors display subtype-selective coupling to Gs. *Mol. Pharmacol.* **1992**, *41*, 889–893. [PubMed]

81. Konig, B.; Gratzel, M. Site of dopamine d1 receptor binding to Gs protein mapped with synthetic peptides. *Biochim. Biophys. Acta* **1994**, *1223*, 261–266. [CrossRef]

Review

Prion Strains and Transmission Barrier Phenomena

Angélique Igel-Egalon, Vincent Béringue, Human Rezaei and Pierre Sibille *

Virologie et Immunologie Moléculaires, INRA, Université Paris-Saclay, UR892, 78350 Jouy-en-Josas, France; Angelique.egalon@inra.fr (A.I.-E.); vincent.beringue@inra.fr (V.B.); human.rezaei@inra.fr (H.R.)
* Correspondence: pierre.sibille@inra.fr

Received: 23 October 2017; Accepted: 26 December 2017; Published: 1 January 2018

Abstract: Several experimental evidences show that prions are non-conventional pathogens, which physical support consists only in proteins. This finding raised questions regarding the observed prion strain-to-strain variations and the species barrier that happened to be crossed with dramatic consequences on human health and veterinary policies during the last 3 decades. This review presents a focus on a few advances in the field of prion structure and prion strains characterization: from the historical approaches that allowed the concept of prion strains to emerge, to the last results demonstrating that a prion strain may in fact be a combination of a few quasi species with subtle biophysical specificities. Then, we will focus on the current knowledge on the factors that impact species barrier strength and species barrier crossing. Finally, we present probable scenarios on how the interaction of strain properties with host characteristics may account for differential selection of new conformer variants and eventually species barrier crossing.

Keywords: prion strain; species barrier; strain adaptation; zoonosis; Darwinian evolution; deformed templating; structural elementary brick

1. Introduction

Scientists have been intensively working on prion diseases, nevertheless, several aspects of this transmissible neurodegenerative affection remain obscure. Among these black boxes, the biophysical support for prion strain variation and the species barrier have stand as one of the last accessible achievements. Although being known for a long time, scrapie and related diseases really went in the light in the 90's: society urged scientists for answers to the questions raised by the sudden outbreak of mad cow disease (caused by Bovine Spongiform Encephalopathy (BSE) prions), which could result in a Creutzfeldt-Jakob Disease (CJD)—like disease in humans. At that time, the potential transmission of BSE from cattle to humans had been assessed persistently, leading to perplexing and even opposite results (sometimes coming from the very same lab) [1,2]. Rather than the consequence of an urgent need for results, these inconsistent data underlined the complexity of the addressed question that we are still trying to answer for the last 30 years. This review aims at presenting the main knowledge and the latest milestones in the field of prion strains and species barrier phenomena, both topics being intimately linked.

Prion diseases are fatal neurodegenerative disorders. At the beginning of the 20th century, Drs Creutzfeldt and Jakob first described the pathology in humans as a sporadic disease. Then the disease was identified as the main responsible for an epidemic of neurodegenerative cases among the Fore population of Papua New Guinea, leading the scientists (Gajdusek, Gibbs and Alpers) to propose an infectious spread of the disease (due to endocannibalistic rituals) [3]. Soon after, Griffith proposed a self-replicating model for the related scrapie disease [4] but the formal conceptualization of prion as a protein only infectious agent, responsible for the misfolding of the host cellular prion protein into a pathologic conformer had to wait for Prusiner's seminal work (1982) [5]. The prevalence of the disease in human population is rather low (~1 case per million people per year, mostly among aged population). Some cases (~15%) are genetically linked, due to point mutation in the prion protein gene (*PRNP*). This disease is also long known to affect ruminants, including sheep and goats with scrapie, cattle with bovine spongiform encephalopathy and cervids with chronic wasting diseases (CWD). No crossing of the species barrier between human and ruminant prions have been reported until the implementation of new biophysical parameters in the process for recycling the livestock carcasses into the ruminant alimentary chain: these modifications resulted in incomplete inactivation of the BSE prions and paved the way for this unconventional agent to cross barrier species and to spread in humans in an outbreak known as the mad cow disease and the variant CJD (vCJD), respectively. And prompted scientists to further study the propensity of this agent to adapt from one host to the other.

2. Experimental Prion Transmission

2.1. Early Cases of Prion Interspecies Transmission

The first report on experimental prion transmission in animals focused on reproducing disease-specific clinical signs by inoculating infected brain extract [6]. The studied parameters were the incubation time until disease end stage, the nature of the clinical signs and the anatomic distribution of the lesions that were reported on a score profile. The first experiments have been reported in sheep [7]. Then, Pattison and colleagues reported several successful experimental inoculations of sheep scrapie to goats [8,9]. By that time, it was found that some prions could pass from one species to the other (e.g., mink to small ruminants [10]). As susceptible to infection with most prion strains, the bank vole turned out to act like a "universal acceptor" [11–13] (see also Sections 4.1 and 4.2). Conversely, several studies reported on the difficulty to pass prions from one species to others (e.g., certain scrapie isolates to cattle [14]).

With the development of transgenic mouse engineering, expression of foreign PrP (in the presence or in the absence of endogenous mouse PrP) considerably enlarged the possibilities for studying zoonotic transmission of prions. These approaches proved to be versatile, since it was demonstrated that development of prion pathology relied solely on the presence of convertible PrP [1,15–17]. In many cases, these experimental setups made emerge the idea that almost every prion could adapt to almost every PrP substrate, provided that some critical parameters (presented below) have been set up in order to adapt the strain to its new host PrP.

2.2. Emergence of the Prion Strain Concept

Interestingly they also noticed some reproducibility of the observed clinical signs, depending on the inoculated isolate. The conclusion of their papers proposed that "certain "strains" of the scrapie agent will produce the nervous syndrome, while others will produce the scratching syndrome". Soon in the prion scientific history appeared the fact that prion agents could share common strain features with conventional DNA-encoded pathogens. These prion strain features could be distinguished from each other based on a number of parameters such as incubation time and titration, which were remarkably reproducible among strains in a given host. Addition of anatomical data such as the localization and intensity of vacuolation allowed to isolate and further characterize prion strains, that were secondarily used to study their adaptation when passing from a host to the other [18]. With the refinements in biochemical and biophysical analysis methods, several parameters are now available for the extensive study on prions that will be detailed in the following sections.

3. From Prion Strain Characterization

3.1. First Approaches

3.1.1. Incubation Time, Clinical Signs, Vacuolation Tissue Tropism

Prion diseases are first characterized by clinical signs observed on the affected individuals. In humans, after a long pre-clinical period is ended, affected individuals usually complain with vague sensory feelings, such as depression. Progressive motor paralysis and dementia then rapidly follow. Cerebellar ataxia is often found in the course of the disease [19]. In animals, clinical signs include progressive ataxia as well but some features like itching and scratching seem to be animal-specific. Behaviour modifications, including aggressiveness or enhanced tameness are also recorded [9]. But these general features are markedly influenced by the strain of prion infecting individuals [8,20]. Distribution and abundance of the lesions in specific brain areas, which appeared to be remarkably stable, were used to score and attribute a "lesion" profile to prion strains [16]. Incubation period for a given strain in a given host was highly reproducible and rapidly served as the first criterion for the characterization of prion strains [21]: for instance, 263 K strain kills golden hamster in 65 days, whereas 139H strain kills the same species in 130 days (after intracranial inoculation). It is worth mentioning that inoculum dilution influences the incubation time [22,23]. Thus, a combination of criterions is necessary for efficient discrimination.

Immunoreactivity of antibodies reacting with the diseased form of the PrP (PrP^{Sc}) allowed for significant refinement in the prion strain characterization [20]. Although prions ultimately accumulate in the Central Nervous System (CNS) of their host, peripheral accumulation may also be associated with some prion strains, contrasting with the ones that are only detected in the brain: in this respect, the most relevant tissues turned out to be the secondary lymphoid organs such as lymph nodes and spleen [24]. This lymphotropism may be used for discrimination between two strains that otherwise would look identical in the brain tissue [25].

3.1.2. Biophysical Parameters (Circular Dichroism, Infra Red, …)

Several biophysical approaches have been employed for elucidating the parameters underneath the biological strain phenomenon. Due to its major insolubility and to the highly heterogeneous aspects of the prion material, all conventional approaches (X-ray crystallography, nuclear magnetic resonance (NMR)) mainly revealed unsuccessful in providing good quality crystals or homogenous solutions for the determination of a pathological PrP^{Sc} structure. Though, infra-red approaches first identified strain differences [26]: absorption spectra from Hyper, Drowsy and 263 K hamster strains revealed a difference between the Drowsy strain and the two others. These results were interpreted as differences in β-sheet secondary structures. Using a combination of several investigation techniques, the structural properties of two sorts of fibrils formed under different experimental conditions was further shown to correspond to differential folding patterns of β strands [27]. NMR imaging proved limited in exploring the PrP^{Sc} structure because of insolubility matters. Conversely, Hydrogen/Deuterium exchange proved helpful with solid NMR approaches [28]. Mass spectrometry using acetylation of accessible lysines in PrP^{Sc} assemblies recently added some arguments in favour with a β-solenoid form of the pathologic prion structure [29]. In addition, the approach was able to reveal structural differences between several common prion strains. Notably, Sc237 hamster prion strain is supposed to have an N-terminal fragment reacting less with the core prion protein than the others hamster strains tested. This view is further supported by the fact that Sc237 infectivity is less sensitive to PK-digestion (partial resistance of the PrP^{Sc} protease digestion is used as the gold standard for the detection of infected samples) than the others.

3.1.3. Biochemical Methods (Western Blot, Resistance to Chaotropic Agents, Conformation-Dependent Immunoassays ...)

At first, proteinase-K resistance of the pathological PrP^{Sc} form served as a diagnostic tool, since most of the antibodies raised against the PrP could not discriminate between normal PrP^{C} and pathological PrP^{Sc} forms of the prion protein in western blot. The PrP^{Sc} isoform partially resists to the cleavage by the most common proteases, among which proteinase K (PK): three major conformation have been reported according to the protease-resistant core size observed by western blotting: 21 kDa (type-1), 19 kDa (type-2) or 8 kDa with some specific antibodies specially recognizing some of these types such as 12B2, a type 1-specific antibody [30]. This distribution of prion types is however generally not exclusive: several CJD cases are actually a mix of T1 and T2 [31]. Whether this co-occurrence aroused from biochemical reasons or just by chance remains to be addressed.

The size of the resistant-core depends on the prion strain and its evaluation still remains the gold standard of prion analysis. Antigenic epitope mapping of PrP^{Sc} raised against different prion strains showed specific immunoreactivity [32], indicating conformational differences within PrP^{Sc} assemblies. Even if several monoclonal antibodies (mAbs) have been raised against PrP antigens from various species, in most cases however, mAbs were poor discriminants and cross reactivities were often recorded between mouse, hamster, human and most common ruminants, as it is the case for the conformational antibody 15B3 [33]. What's more, these mAbs would hardly discriminate between normal and pathological PrPs, which would have been of enormous interest for diagnostic or laboratory purposes.

In addition to providing antibodies more or less selective for a given PrP protein, alternative approaches were designed for the study and conformational screening of prion strains [34]. Known as the conformation-dependent immunoassay (CDI), this technique depends on differential recognition of unmodified PrP^{Sc} or altered PrP^{Sc} epitopes to determine a ratio that can be used for direct quantitation of prion in a sample. This technique shows that more than 90% of PrP^{Sc} present in sCJD patients are PK-sensitive [35]; moreover, this ratio may reveal variations from strain to strain, as observed for eight characterized hamster-prion strains [34]. In particular, this approach allowed for a better discrimination between prion strains that otherwise would have been indistinguishable using incubation time and PK-resistance as analysis criterion. However, this approach still failed to discriminate between closely related prions, although showing that PrP^{Sc} assemblies present different degree of stability in presence of chaotropic treatments that reveal epitopes [36].

Post translational protein modifications have been proposed to account for the PrP strain-to-strain variations. For instance, prion protein contains two glycosylation sites located in the structured C-terminal part of the protein. Both N-glycosylation sites are conserved in the *PrnP* gene among species, suggesting that N-glycans play an important role in the protein function. These sites, however, are not systematically glycosylated, as shown by the 3 detected bands in western blots performed with anti-PrP antibodies. Glycosylation pattern changes from one strain to the other, with variations in the relative abundance of the Di-, Mono- and Un-glycosylated forms and even the normal PrP^{C} glycosylation is differentially affected by depending on the prion strain [37]. However, glycosylation deficiency at either one or other glycosylation sites does not alter the susceptibility of the host to scrapie prions [38]. Moreover, strain characters were not modified when these glycosylation mutant mice were used for bio assays. This result was later supported by our work in vitro reporting that several PrP glycosylation mutants are faithful templates for PMCA (protein misfolding cyclic amplification, see Section 3.2.1) [39]. Other studies however reported that sialic acids that are deposited on glycan chains may significantly account in the prion replication: desialylated PrP^{Sc} was mainly found and eliminated in the liver while normal prions were targeted toward secondary lymphoid organs [40]. In the meantime, the desialylation of prions is reported to reduce the species barrier [41]. These data are particularly relevant with respect to cross-species transmission fate.

3.2. New Insights

3.2.1. Templating Activity

In 1996, Prusiner's team published the observation that FFI (Familial Fatal Insomnia) or sporadic CJD (sCJD) inoculation to a mouse expressing a chimeric human-mouse PrP gene reproduced both PK-resistant 19 kD deglycosylated band pattern for FFI and the 21 kD band pattern for the sCJD, respectively [42]. This basic observation paved the way to the concept of templating activity of prions. This concept has been explored in the field of yeast and fungi prions (the yeast prions will voluntarily not be documented here) but had been curiously neglected until the very recent years, when Baskakov's team reported conformational switches within individual amyloids [43]. In this work, two strains of fibrils made from the identical recombinant hamster PrP showed various individual characteristics, derived only from the different conditions of formation: R strain was obtained under rotation of the monomers and displayed straight shape and polymorphous (twisted or not twisted morphology), while S strain was obtained under shaking and displayed a curvy simple line. When incubating these different fibril strains with heterologous mouse recombinant PrP monomers, they observed that whatever the original strain, the fibrils adopted the straight complex forms of the R phenotype. In addition, FTIR (Fourrier–transform Infra-Red) spectroscopy properties of the daughter fibrils were similar to the ones of the R phenotype. By contrast, R and S fibrils incubated with the homologous hamster monomers yielded the expected parental forms (microscopy and FTIR). Thus, monomer origin is able to imprint a new conformation and new properties to a given inoculum, resulting in the change or adaptation of the strain to its new host. However, these data also suggest that the host PrP could restrict rather than enlarge the conformational panel available for PrP^{Sc}.

Owing to their very long incubation periods, prion diseases remain difficult to study in vivo. Although cell-based systems have been developed in several laboratories, they proved to be difficult to set up, particularly because this approach was not possible to implement to every kind of prion strain. Despite these difficulties, Weissman's groups produced seminal data on prion strain adaptation and selection in vitro [44–48]. They first showed that neuron cell lines chronically infected by two prion strains (RML and 22 L) could be derived into several different cell lines with their own response to various other prion strains [44]. Although they could somehow stabilize cell-adapted prion features different from that of brain–adapted prions, they observed the occurrence of a "Darwinian selection" that allowed for the transition from one prion to the other [45]. Further selection could even be achieved using selective prion drugs inhibitors such as swainsonin, demonstrating that the strain features observed finely depend on the prion production conditions [45,48]. In such a context, Weissmann and colleagues further enforced the concept of quasi-species proposed by Collinge [49], consisting of a major component and many variants, which are constantly being generated and selected against in a particular environment: changing of conditions may result in the selection of a new variant with different features.

Several aspects of PMCA and RT-QuIC (real time quaking-induced conversion) amplifications have been studied and used for prion detection in body fluids or for assessing and validating decontamination procedures that will not be emphasized here. A focus on some of these topic has however been recently published [50]. PMCA approach was first described by C. Soto in 2001 [51] and involves the cyclic amplification of minute amounts of infectious material diluted in a brain lysate containing solubilized forms of PrP^C. The repetition of a few seconds ultrasonication bursts followed by an incubation period at 37 °C produces the exponential transconformation of the PrP^C present in the normal brain lysate into a PK-resistant form. PrP^{Sc} is then detected on Dot or Western blot after PK-digestion. Practically, amplification factors and titration capabilities have been reached that extend far beyond that obtained with bioassays: 10^{12} amplifications were routinely obtained with laboratory scrapie strains [52] and the system allows for the amplification of a large number of strains, sometimes to a lesser extent, though. This reduced amplification level could however be largely compensated after several rounds of amplification. The difficulty to amplify sporadic MM1

CJD, appeared as a notable exception until Safar and colleagues reported that a modified PMCA using unglycosylated PrP was able to selectively amplify type 1 sCJD [53].

Castilla and colleagues proved the relatively high strain fidelity of the PMCA-driven prion amplification regarding currently available tools (incubation time, brain lesion scores, western blot profile, PK-resistance) [54,55]. Thus this in vitro amplification tool proved to be a valuable tool for the assessment of cross barrier crossing, a main advantage over bioassays being the extreme shortening of the time required for experimentation. Species barrier crossing could be demonstrated between mouse prion and "cervid" model mice, which would have required much more time with the bio assay [56].

PMCA has been described using brain lysate of transgenic mice as a source of healthy PrP^C. But versatility of the system may be further increased by the use of cultured cell lysates. This was used in our lab to further dissect the requirements for PrP^{Sc} conversion [57,58]. Additional experiments determined that the conversion and amplification of recombinant PrP protein could occur in the presence of only RNA or phospholipids as adjuvant molecules [55,56,59–62].

Simultaneously, RT-QuIC was developed as another technique for the in vitro amplification of prion conversion [63]. In brief, recombinant PrP molecules are driven to fibrillation through alternative shaking and incubating with thioflavin-T (ThT) as a fluorescent marker: an increase in fluorescence emission, that could be observed in real time, is the sign of amyloid formation and PrP conversion. This approach proved to be sensitive enough to detect prion particles within blood cells [64] and diverse body fluids [65]. An important difference between PMCA and RT-QuIC is that the recombinant PrP that is converted in QuIC experiments is poorly infectious, whereas PMCA amplified products are usually as highly infectious as the inoculum. Noteworthy, this technique does not faithfully replicate the species barrier phenomenon that is recorded in target animals. Nonetheless, RT-QuIC has proved useful in discriminating among prion strains: Bank Vole PrP could be converted by almost every strain tested in RT-QuIC experiment. Lag phase and final fluorescence signals could be used for discrimination between different prion strains, though [13]. These observations parallel those made with transgenic mice expressing bank vole PrP [66]. On the contrary, the use of several different substrates could be used as a screen to differentiate between closely similar strains: atypical L-type BSE and classical BSE for instance [67].

A fluorescent approach to discriminate between prion strains was provided by a chemist group from Sweden using oligo-/poly-thiophene derivates [68]. Murine scrapie and CWD have been compared for excitation/emission spectra as well as fluorescence life-time of a few compounds: this parameter is modified in response to conformational restriction of the thiophene backbone following interaction with the different aggregates. All these methods could provide refined tools to differentiate strains that are difficult to by strain typing in animals.

3.2.2. Size Distribution of Aggregates (Quaternary Structure)

Biophysical approaches focused on size-distribution analysis of the prion particles. If not entirely carried on primary or secondary sequence, strain information should be somehow related to the tertiary or quaternary structure of the PrP^{Sc} assemblies. Several studies have already pointed the necessary role played by the PrP structure in the pathological process of prion transconformation [2,26,55,69,70]. Sedimentation velocity centrifugation in density gradients proved to be a valuable tool to separate and analyse prion fragments according to their size and/or shape, while preserving as much as possible the "natural" multimerization state of the prion particles and minimizing artefacts due to improper membrane solubilisation [71,72]: this later point is crucial for the reliability of the technique, since the presence of residual membrane lipids would modify the assemblies' apparent density and lead to improper interpretation of the data. Sedimentation velocity-based fractionation will discriminate dense heavy aggregates that will sediment to the bottom of the gradient from the lighter fractions containing small aggregates or particles of low density. This technique allowed the precise discrimination between several ovine and hamster strains (Figure 1). The disconnection between infectivity level of the fractions (monitored by bioassay) and the PrP^{Sc} abundance (estimated by western blot) specifically for the 'fast' ovine and hamster strains constituted a striking finding of this approach [71,72]. These experiments led to the view that prions are formed of a strain-specified collection of non-uniform PrP^{Sc} assemblies with specific activities.

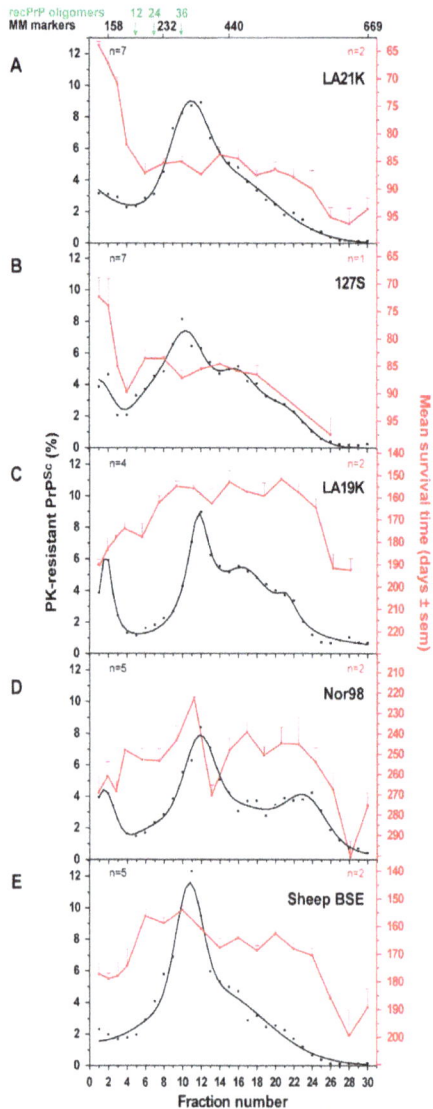

Figure 1. Brain homogenates from tg338 mice infected with LA21K (**A**); 127S (**B**); LA19K (**C**); Nor98 (**D**) and sheep BSE (**E**) were solubilized and fractionated by sedimentation velocity. Fractions collected from the gradient were analysed for PK-resistant PrP^{Sc} content (**black line**) and for infectivity (**red line**). For each fraction, the percentage of total PK-resistant PrP^{Sc} detected on the immunoblot is presented (**left axis**). For each fraction of each strain, infectivity was determined by measuring mean survival times in reporter tg338 mice (mean ± SEM; **right, red axis**). The sedimentation peaks of standard molecular mass markers (MM markers) are indicated on the top of the graph. From [71].

3.3. When Two Strains Look the Same

Box 1. Of prion isolates, strains and types.

Of prion isolates, strains and types

- Isolate: we refer here as to biological material that has been obtained through sampling of infected individuals;
- Strain: the term corresponds to a defined prion population isolated from one specified animal, with regards to the precision of the investigation technique: from basic observations (clinical signs incubation time and so on) to fine biochemical and biophysical parameters that are now becoming precise enough to allow for the discrimination of quasi-species within one strain; for the sake of simplicity, one regularly and erroneously omit the name of the host from which the strain has been originally isolated, even though a totally different prion population may have been selected when passed to the new host.
- Type: refers more particularly to a combination of biochemical parameters (mainly to the size of the unglycosylated PrP^{Sc} fragment after proteinase K partial digestion) that are independent from the host.

The prion phenotype characterization is as precise as the accuracy of observation tools used. In some cases, and despite the combination of several investigation methods, two strains may not be discriminated (Box 1). For instance, it has been observed that some scrapie strains could display the same pattern than that of BSE . This raised the question whether BSE could in fact originate from a scrapie strain [73]. Later on, the identification of an atypical L-type BSE in bovine raised again doubts on the potential transmission of BSE to small ruminants [74]. First described within a flock in Italy, its behaviour in cattle is very different from that of the classical BSE (presence of amyloid plaques, low non-glycosylated PrP^{Sc} fragment). Upon passage on ovinized animals (VRQ allele), however, this strain turned out to be fully similar to C-BSE, which prompted the investigators to speculate that the mad cow outbreak of the late 80's could have arisen from a passage by the small ruminants. In fact, these two strains, although they look virtually the same in the ovinized mice, keep their species characteristics, since they still can be differentiated when further back-passaged on bovinized mice (our unpublished observations). In addition, L-BSE could be differentiated from the classical one through its high propensity to colonize lymphoid compartments (our personal observations). Thus, study of prion replication in the lymphoid tissue and back-passage experiments (or to another intermediate species) can be useful to discriminate between truly identical strains and closely resembling strains.

4. To the Study of Species Barrier in Prion Transmission

What we currently know from the species barrier crossing is that the phenomenon can take a long time to be observed: adapted prions usually kill all the inoculated animals within a few weeks or months, out of a total incubation time that could range between 60 days for the fastest and up to the entire life of the animals (over 700 days for some mice); by contrast, non-adapted prions usually have incomplete attack rates and incubation times greatly increased and much more variable as compared to the original strain. During the adaptation process, however, these hallmarks faint and newly adapted prions recover full attack rate and reduced as well as highly reproducible incubation time, although several other factors may have been changed in the new host (final incubation time, tissue distribution, PK resistance . . .).

Strain-to-strain variation is often associated to the passing of one species to the other. In some instances, this can even result in the adaptation of several different strains: inoculation of transmissible mink encephalopathy (TME) to hamsters resulted in the selection of two different strains, depending on the dilution of the inoculum [75] (see Section 4.3 Coinfections). Recent work report on the influence of PrP^C expression level for the selection of different prion populations (see Section 4.4.1 PrP expression level) [76]. This is highly questioning in regards with the zoonotic and epidemic risks of such diseases (BSE, variant CJD, for instance).

4.1. Some Great Examples

Illustrating the complexity of the species barrier in frame with the prion strain characterization is a difficult task: a few examples are provided below, that have been chosen for the historical role they played in the prion field or because of their highly possible or demonstrated impact on human health (in terms of zoonotic risk).

Historically, one of the first prion species barrier to be studied with laboratory rodents has been the hamster to mouse transmission: early in the 70's Kimberlin noticed that a hamster scrapie strain named 263 K hardly passed on mice [22,77]. However, years later, asymptomatic replication of hamster prions was demonstrated in mice [78,79]. Despite the absence of clinical signs, half of the inoculated mice had detectable levels of PrP^{Sc} in the brain and also showed the presence of specific prion disease lesions in the brain. This work was the first to further include the search for prion specific signature in the target organ with or without clinical signs for the evaluation of the barrier species. In seminal experiments, the use of transgenic mice expressing hamster PrP allowed hamster prion transmission to mice, indicating that PrP amino acid differences contributed to the species barrier [80]. Then, PMCA approaches confirmed that mouse (RML strain) prions could progressively be adapted to hamster species [81]. The in vitro adaptation could be achieved within 4–6 rounds (~2 weeks, while in vivo adaptation would have required more than 3 years). Similarly, hamster prions (263 K strain) could be adapted to mouse. Interestingly, both directions of adaptation yielded prions showing fine variations in incubation time/PK-resistance/brain deposition pattern/glycosylation profile, etc. These results suggest that although very similar to what is observed in vivo, PMCA-driven adaptation process is different from what is observed in living conditions.

Mad cow disease in the 90s triggered very intense research on the mechanisms involved in the crossing of the species barrier for some prion strains. Investigating the capacity of BSE prions to propagate in new host species led to the initial conclusion that the BSE agent was not able to infect mice expressing only human PrP (HuPrP$^{+/+}$ PrnP$^{0/0}$, valine allele at position 129) [1]. Looking closer, however, revealed that the BSE prions finally needed more time than CJD prions to install and replicate in mice expressing human PrP. The same lab detected transmission of BSE to macaque or cats, while their PrnP$^{0/0}$ HuPrP$^{+/+}$ still were alive (after more than 500 days post-inoculation) [2]. Thus, BSE prions were more and more amenable to infect humanized animals. The year after, Bruce and colleagues reported that both BSE and vCJD agents replicated similarly in certain lines of conventional mice and induced the same strain phenotype, thus strengthening the link between the epizootic burst of BSE and the outbreak of British and French vCJD cases [82]. A few years later, Prusiner's team provided another evidence that BSE and vCJD were etiologically linked [83]: they observed that bovinized mice developed similar pathologies when inoculated with either BSE or vCJD agents (incubation time, lesion distribution and score, PK-resistance profile), these profiles being absolutely different from those observed after scrapie infection. These observations confirm that two agents isolated from different species could eventually be the same strain. Other lines of transgenic mice overexpressing the Met allele of human PrP were developed and further provide evidence of the intricate link between BSE and vCJD. In two lines, the disease occurred at incomplete penetrance and was mostly subclinical [84,85]. However, one of the lines showed a higher attack rate extraneurally in the spleen tissue (see below).

The cross-species capacities and zoonotic potential of CWD prions are another emerging public health concern. This disease affecting cervids is known for long but got only recently media coverage when concerns about the passage from wild ruminants to human started to get conceivable. While CWD was reported in the early 90's to mainly circulate between captive wild cervids, [86] a warning was raised against the possible epidemic extension of the disease, which truly occurred in the USA during the last 3 decades (reviewed by Watts [87]) and even recently popped up in

Scandinavia [88]. Considering the data obtained with scrapie and particularly with the fact that some prion strains apparently can circulate without species barrier, it was of importance to determine to what extent this CWD agent is confined to wild cervids. It is already known that CWD can adapt to certain strains of mice, expressing high level of murine PrP [89]. After one passage, strain seems to be stabilized, with mice repeatedly succumbing 220 days after inoculation, all other features of the disease were looking similar to the genuine CWD, including the spleen tropism. Recent data further evidenced a threat of possible transmission to humans: Herbst and colleagues described the different behaviours of two different CWD strains—CDW1 strain could pass to hamsters but not to mice. By contrast H95+ strain could infect efficiently the mice, while the hamster was less susceptible [90]. Humans' natural resistance to CWD infection has been hypothesized to rely on a specific amino acid stretch in the β2 loop of the PrP; by swapping this domain with that of elk PrP, Kurt and colleagues rendered the mice expressing this mutant PrP fully susceptible to CWD [91]. Notably, susceptibility to CJD was inversely reduced. A very recent work presented to the Neuroprion meeting reports that CWD can be passed orally to cynomolgus macaques [92].

Conversely, there are host species that are susceptible to almost every strain: for example, bank vole rodent turned out to stand as a mammal particularly tolerant to many prions [93], including notably sporadic CJD [11]. Transmission of these prions to bank voles results in a disease at full attack rate, with little or no species barrier and in a tempo similar to transgenic mice overexpressing human PrP. Bank vole and human PrP amino acid sequence are differing by 12%. This indicates that the bank vole PrP conformation is *per se* prone to conversion by human CJD prions. Recent transmission of a library of prion strains to transgenic mice expressing bank vole PrP (M109) further support the view that bank vole PrP can be converted by many abnormal PrP^{Sc} conformations. To strengthen the demonstration, so-called 'resistant' animals could be rendered compliant with prion infection after introgression of a genuine or modified prion protein—for instance, the Drowsy strain affecting Syrian hamsters but not Wild-Type (WT) mice could readily infect mice expressing a chimeric hamster-mouse PrP [94].

The species barriers depend in part on host's PrP primary structure and in part on the strain itself. Up to date, every one of these species barriers has shown they could be crossed, provided that enough time, correct animal or organ target and passage numbers have been taken into account: hamster prions can pass onto mice [95]; TME can pass to cattle [14]; BSE and vCJD both can adapt to Guinea Pig [96]; CWD transmission from cervids to human is every day more probable, at least under certain specified conditions [97]: CWD can pass in vitro with PMCA to human brain homogenate but efficiency is greatly enhanced if amplification/adaptation of the CWD prion has been previously performed through PMCA using CWD template [98]; experimental infections to non-primate monkeys were also efficient [99,100]. In conclusion, the strength of the species barrier is mainly dependent on the parameters that are addressed: a closer look to subclinical disease either centrally or peripherally may lead to the assumption that there is finally no absolute species barrier, at least to the experimental level.

What are the factors that render prions adaptable to their new host? What features allow BSE to stand today as the sole prion that adapts easily to other species? Despite significant efforts to improve our knowledge, the answer is still largely unknown. However, several leads have been followed that may help building models for the structure of what we call PrP^{Sc}. Some of these models will be evoked at the end of the paper but the few following sections will first focus on the experimental parameters that demonstrated their influence on the strength of the barrier species.

4.2. Importance of Primary Sequence, Aminoacid Polymorphism

The primary sequence of the PrP stands naturally in front line, since it is accepted by almost all the scientific community that this endogenous protein is the only responsible for the disease. The primary sequence (allelic variations, point mutations) would obviously be the main vector ruling the host's susceptibility to a given strain. With the introduction of transgenic animals, it became easy to test whether one given PrP sequence could account for the susceptibility to prion diseases:

several studies report the acquired susceptibility of mice following expression of recombinant or chimeric hamster protein [101,102]. Homology between the inoculated prion and the host's PrP looks like a prerequisite [80,103,104]. In the latter case, transgenic rabbits expressing ovine PrP were fully susceptible to scrapie, showing that the rabbit environment is not, *per se*, incompatible with prion transconformation, although the animal is known to be naturally resistant to prion infection. However, with the notable example of the bank vole being a universal acceptor, the common view is that structural compatibility between host PrP and the infecting prion strain governs the cross-species transmission of prions.

Polymorphism within one species have a dramatic impact on the susceptibility of the host (see for instance how ovine polymorphism is governing scrapie transmission [105,106]). The extreme cases are illustrated by the spontaneous prion conversion attributable to familial point mutation of the human PrP gene responsible for genetic CJD or FFI (for a review, see [107]), or by the I109 M point mutation affecting the bank vole PrP: mice expressing I109 Bank vole PrP spontaneously develop a prion disease within 4 months of age [108]. This phenomenon does not strictly apply to the species barrier paradigm. However, it had been shown that the mutations reported in humans seem to cluster in several groups depending on their ability to cross species barrier and infect mice [109]. In that particular case, species barrier was regarded as a tool for the characterization of different prions strains. From an epidemiological and clinical point of view, it is relevant to consider these point mutations, since they condition full sensitivity or resistance to prion disease. The homozygous methionine or valine in position 129 of the human PrP is determinant for the sensitivity to CJD [110]. To date, all clinical cases of vCJD have only occurred in patients homozygous for methionine at codon 129 [111]. However the V129 genotype does not protect against vCJD, despite full protection against BSE [112] Polymorphism in position 219 (E/K) is also associated with resistance to prion infection in Asian population [113]. More recently a G/V polymorphism has also been reported to be responsible for full resistance to Kuru infection and was proposed to result from naturally-driven selection process that increased resistance against Kuru in the exposed population [114].

In addition to governing intra species susceptibility to prion diseases PrP polymorphism greatly influences the susceptibility of the host to exogenous prions. Noteworthy however, this polymorphism has an impact on the population genetics, with breeds being historically more susceptible to scrapie than others because of genotype variations in positions 136/154/171. The M/V polymorphism at position 129 for human PrP similarly controls the susceptibility of individuals to vCJD [110]. Green et al. report that the elk prion codon 132 polymorphism controls cervid and scrapie prion propagation [115]. In brief, transgenic mice expressing the elk mutant L132 are resistant to CWD prion inoculation (no clinical sign at up to 600 days post infection). However, the authors report the detection of small amounts of PK-resistant PrP^{Sc} in the brains of the animals inoculated with M/M132 inoculum but not with M/L132. This suggests that L132 genotype is not resistant to infection but rather requires adaptation passages in order to select a fully adapted prion. It is worth noting that in contrast to elk prions, hamster prions adapted easily to the 132 L elk PrP. Using the Bank vole as a model, Agrimi's team observed that I/M109 heterozygote animals incorporated equal quantities of both allelic forms of PrP in the prion particles [116]. Thus, at least in some cases, it seems that polymorphism does not entirely rule the susceptibility to prions.

Beside natural polymorphism, experiments have been performed to test for the mutual influence of different PrP molecules to diverse strains of prions. Several experiments of cross species inoculation have been performed by Kimberlin and colleagues to assess the infectious potential of scrapie prions to infect mice, hamsters, etc. [117–119]. Later, when the first transgenic mice started to be available, the transgenic PrP was co expressed along with the endogenous one. It rapidly appeared that the expression of both the endogenous and the transgene could result in odd responses to prion infection. Chesebro's team published several reports mentioning the interference of heterologous PrP on the accumulation of prions in a cell culture assay [120,121]. A single amino acid substitution could be responsible for drastic inhibition of prion production in cell culture. But other studies also report on the influence of additional factors for the efficient replication of prions [122]. These observations paved the way for a long series of publications mentioning a hypothetical protein X as cofactor in the prion mechanism [123]. But recent in vitro conversion experiments invalidated this hypothesis [124].

4.3. Co-Infections

The influence of co-infection, although not strictly relevant in the species barrier topic, is questioning: since several prion strains can infect a given host, what could a multiple infection look like? One would logically suppose for example that the fastest strain would take over the slowest strain. Surprisingly it happened to be quite the opposite: fast strain 22A inoculated to mice that had first received a slow 22C strain showed delayed incubation periods [125]; when a Hyper/Drowsy combination of TME prions was peripherally inoculated to hamster, the fastest Hyper strain progression was delayed when co-administered with the slowest strain Drowsy [126]. The authors proposed a competition between both strains for a limiting PrP resource [127].This competition did not occur when both prions were simultaneously inoculated at different locations [128].

4.4. Influence of Expression Level and Post Translational Modifications

Considering there is a competition for substrate between two different strains, then the amount of available PrP certainly influences the susceptibility, or to a lesser extent the kinetic of the disease. In addition, the several post translational modifications that undergoes the PrP^C during its journey toward the membrane may also influence the susceptibility/convertibility of PrP to prions.

4.4.1. PrP Expression Level

The influence of PrP^C expression level has been assessed on incubation time, infectivity titre and lesion distribution in 3 strains of mice, whose expression level varied from ½ to 8x [129]. The incubation period was reduced when PrP^C level was increased: during an exponential phase, rapid accumulation of PK-resistant PrP^{Sc} occurs, followed by a plateau phase, whose length is proportional to the PrP^C expression level. The authors suggest the accumulation of toxic forms of the PrP that ultimately induce the clinical signs once a certain threshold has been reached. Recently, our group reported the characterization of three independent prions selected from a single scrapie isolate inoculated to transgenic mice expressing various levels of strictly homologous VRQ allelic forms of ovine PrP. The strains were mainly phenotyped according to their PK-digestion profile and their incubation duration [76]. Upon transmission of a 21 K isolate to overexpressing mice (>3.5×), a new 19 K phenotype is progressively selected, with frequency and PrP levels raising accordingly. A third phenotype (21 K fast) is reported to emerge in a stochastic fashion that outcompetes 19 K prions in high expressor hosts.

Although these studies were done in homotypic transmission context, they highlight the key role of PrP^C levels in the disease tempo and in the prion selection and emergence. It is likely that PrP^C would similarly be at play in heterotypic transmission events.

4.4.2. Secondary Modifications: Glycosylation, Sialylation, Protease Digestion . . .

Glycosylations and other post translational processings have been shown to play a significant role in the transmission and adaptation of prions to a new host. Recent work with PrP mutated to the first or the second glycosylation site dramatically increased or suppressed the species barrier upon infection with 2 human prions (MM2CJD and vCJD) or 263 K hamster scrapie [130]. It is worth noting, however, that the amino acid mutations designed for glycosylation alterations may also account for the observed phenotype [131]. In vitro conversion experiments using hamster prions in presence of various mouse PrP constructs suggest that heterologous conversion favours unglycosylated PrP incorporation, while autologous conversion results in the usual 3-banded profile [132]. Glycosylations may be further modified through sialylation of the sugars. It has recently been shown that sialylation of prion protein could modify the species barrier [41]: while normal mouse brain homogenate needed more than

4 rounds of PMCA to reach a steady state level of amplification using a 263 K hamster strain, the same desialylated brain homogenate reaches the plateau in one single round. Reciprocally, hamster brain homogenate failed to amplify 22L or ME7 mouse strains even after 10 rounds but upon desialylation, the PMCA reaches the plateau in 3–4 rounds. It is furthermore shown that sialylation participates in a host/tissue and cell-specific manner to the regulation of PrP [40,133]. Sialylation process is shown to occur even after the PrP has been converted [40,134].

GlycoPhosphatidyl Inositol (GPI) anchor, which attaches PrP to the membrane through its C-terminus participates also to the conformational landscape of the prions upon infection: upon passage on GPI$^{-/-}$ mice, most of the prions retained their specific characteristics when passaged back to their original host, except CWD which gain in PK resistance and chaotropic [GdnHCl]$_{1/2}$.stability [135].

4.5. Prion Route May Influence Prion Transmission Fate

Lymphoid tropism of some prion strains has been described for a long time [24]. Prions have been shown to replicate and accumulate in follicular dendritic cells (FDC) from the germinal centres of lymph nodes and the spleen. Several other immune cells (like B-cells, macrophages) have been shown to carry prion infectivity, however, it has been demonstrated that FDCs are necessary and sufficient for prion replication in the spleen [136]. Of note, the FDCs are not of lymphoid nor myeloid origin, despite their pivotal role in the initiation and maintenance of immune response: rather, they derive from sub-endothelial cells and differentiate upon trophic interactions with B-cells [137]. Macrophages and B-cells also stain positive for PrP^{Sc} but they are mainly involved in PrP^{Sc} scavenging and transport, respectively [138,139]. Therefore, lymphoid compartment will have a balanced influence on peripherally acquired prion fate: in one hand, macrophage and in particular splenic scavenging functions will actively degrade or neutralize prion infectivity, while, on the other hand, FDCs will actively replicate prions that are then shuttled to the terminal nerves ending near the germinal centres [140]. Thus, the resulting lymphoid tropism could be the net result from this balance, as claimed recently by Bartz and colleagues [141], who were able to detect Drowsy infectivity within hours after inoculation but also reported that these prions disappeared thereafter as a consequence of an increased susceptibility to proteases (in vitro PK assay). Prion replication in periphery could otherwise be dictated by strain features. It is for instance well known that vCJD, as opposed to other strains of CJD remarkably replicate within the human lymphoid system [142]. Thus, oral contamination by the vCJD strain, which targets the Peyer's patches beneath the jejunal epithelium, is much more prone to occur than with other sporadic forms of CJD. However, recent analysis of the last documented Kuru case (in 2003) reveals that despite a certain food borne contamination, the Kuru patient did not exhibit the marked vCJD-typical colonization of the digestive tract. This suggests that peripheral pathogenesis of Kuru is similar to that seen in classical CJD rather than vCJD [143], although a more recent study finally considered that sCJD and vCJD accumulated similarly in the lymphoid system [144]. Thus, strain lymphotropism does not necessarily reflect the preferred inoculation route. This question had been addressed with comparison of oral and intracerebral routes of BSE inoculation in macaque as a model of species barrier transmission [145,146] and the general conclusions that may be derived from these studies is that lymphoid tropism does not facilitate, *per se*, the crossing of barrier species.

4.6. Immune Status, Age of the Host

It is for long known that immune status and host age or developmental stage influence the PrP expression in target organs such as the brain but also and particularly the lymphoid organs [147–149]. As a consequence, susceptibility of young sheep to oral contamination with BSE is drastically decreased after weaning [150]. On the other side of the lifespan, aged mice infected intraperitoneally with RML prions show significant longer incubation time than their younger littermates [151]. This is to be related to the decrease of follicular germinal centres with age [152]. Thus, depending on the age at the time of infection, prion replication may be significantly affected, with obvious effects on the crossing (or not) of the species barrier. For instance, the transmission of BSE was much more efficient on young mice, while older animals remained free of clinical sign all along their lifespan [153]. Noteworthy, some aged mice could still replicate at visible levels the prion in their spleen. Thus, in that particular case, the modification of the species barrier should be regarded as a consequence of a receptor abundance modulation. Still, these observations are of interest when considering the exposed groups within the whole population.

4.7. Cell/organ Selectivity

As evoked in the previous section, prions can replicate and their expression could be drastically modulated in several organs outside the central nervous system. In addition to that, peripheral organs may, *per se*, have a different behaviour with respect to the invading prion and the inoculation route. These observations have been made in our laboratory when we monitored the fate of a scrapie prion strain following inoculation by intracranial or intraperitoneal routes [154]: the disease greatly differed in clinical signs, abnormal prion protein levels and neuropathology. In another study, we monitored brain and spleen of ovinized and humanized mice for the presence of infectivity or PK resistant PrP after inoculation with hamster sc237, CWD or BSE [85]. Overall, the three strains, which are not transmissible to either ovine or human PrP mice, did not indeed replicate efficiently in the brains of the inoculated animals: for instance, only 2 out of 29 ovine mice infected with CWD were positive by western blot at the end of their lifespan. By contrast, spleens of these animals were almost consistently positive (Figure 2) and from the early time of infection onward. Overall, the spleen appeared 9–10 fold more permissive than the brain to foreign prions. The reason for such tissue-dependent strength of the species barrier remains to be determined. Spleen PrP^C might be more prone to heterotypic conversion than brain PrP^C, due to conformational variations. The spleen environment might constitute a better niche, due to prolonged possibility of interactions between brain and spleen (through axonal terminations located close to the germinal centres) or presence of co-factors, such as complement. Ultimately, absence of cell prion toxicity outside the brain could also account for an efficient replication of PrP^{Sc}. Whatever the reasons, these features allow prion extending its host range. It also provides and experimental explanation to the high number of asymptomatic individuals exposed to BSE agent in the UK population showing pronounced accumulation of PrP^{Sc} in their lymphoid tissue [155].

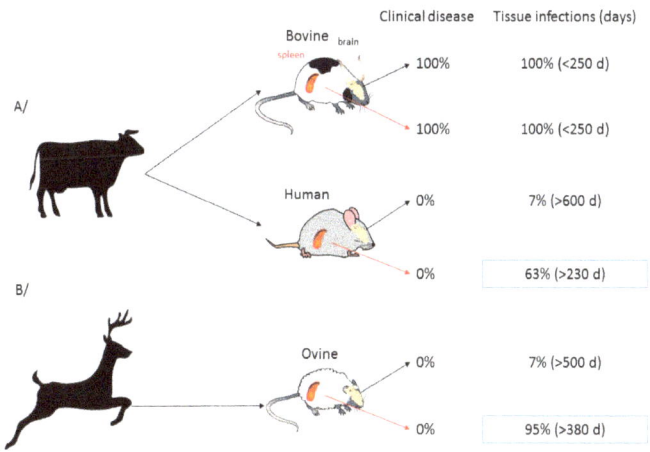

Figure 2. The spleen is much more permissive than the brain to the passage of heterologous prions: percentage of diseased animals and western Blot-positive tissue of BSE (**A**) or CWD (**B**) prion agents inoculated intraperitoneally to bovinized, humanized or ovinized mice. Spleen and brain were collected at the death of the animals. Tissue infection was diagnosed upon the detection of PrP^{Sc} by western blot. From [85].

Other hypotheses involve a tissue specific clearance metabolism as responsible for the different strain tropism [141]. As mentioned previously in a former section, the authors describe the successful PMCA amplification of Drowsy TME in the spleen of peripherally-inoculated hamsters, although this strain was not supposed to replicate in that tissue. They conclude that the strain selectivity against Drowsy prions in the lymphoid organs is a consequence of the strain-specific efficient removal of the infectious material. This hypothesis however does not seem to be valid when brain prion distribution was concerned [156,157]: both studies argue at 20 years interval and with different tools against a brain tissue selectivity. Cell culture experiments have also been set up in order to evaluate strain selectivity but the overall resistance of cell lines or even primary cultures to prion infection remains to be overridden [158].

5. Consequences on Prion Adaptation to New Host

Following a substantial species barrier, different scenarios may be possible: **(i) Silent passing**—The prion may be silently passing as it does when Sc237 is inoculated to ovine PrP mice (see previous section). Several studies have reported that some strains require very long adaptation periods and iterative passaging in order to successfully replicate in one given host [159–161]. Three or more reinoculation steps are often needed to get successful isolation of poor prion transmitters. In such cases, the receptor animals also play a pivotal role, in particular with respect to the amount of target PrP^C that could be expressed. In some instances, the resulting prions retain the features of the parental strain [159]. In other instances, the isolated prions differ significantly from the source inoculum [161]. **(ii) Progressive evolution—Mutational events.** In that case, evolution could concern more than two species: for instance, when questioning the origin of the BSE and vCJD epidemics that hit Western Europe a few decades ago. An atypical form of BSE (BASE) has been shown to evolve into a form that is indistinguishable from BSE in wild-type mice [162]. Others described the progressive evolution from a H-BSE to a classical BSE when the strain was serially inoculated to mice [163]. It has also been reported that passage through an intermediate small ruminant allows BSE to better adapt to humanized mice [164,165]. Overall, BSE prions, whether atypical or classical, seem to display a unique behaviour with regards to species barrier, since it is the only prion strain capable of adaptation to a great variety of hosts [166]. In other situations, evolution may be more abrupt (incubation duration deceased rapidly between 1st and 2nd passage: see for instance, the emergence of T1-Ov and T2-Ov strains following experimental passage of MM2 CJD to ovine PrP mice) [167]. The progressive evolution of prions was further supported by Weissmann's group, who argued for a Darwinian evolution of prions, through silent (or not) "mutations" [45]. These data were supported by in vitro experiments that establish prion evolution induced by chemical prionostatic drugs and selection of prion 'quasi species' from an initially homogenous prion strain [168]. As mentioned earlier, other groups support the concept of a 'portfolio of conformations' for a given strain, that could match (or not) with a portfolio of possible conformations for the receptor PrP [49,169]. The observation that H-BSE phenotype is lost upon passage to hamster mice and restored when passaged back to bovine fits this model [170]. An evolution of the concept emerged from the observation that two strains could too closely look similar to be distinguished, albeit showing different incubation times. In addition, these strains could constantly evolve form one strain to the other, rendering them impossible to clone [171].

5.1. De Novo Synthesis and In Vitro Assessment of Species Barrier

The experiments reported in the previous sections required the massive use of laboratory animals for the evaluation of species barrier. However, even with the use of transgenic mice that considerably shorten the incubation time in comparison to what is recorded with the target animals, the time needed to obtain such data with 4 and even 5 successive passages is extremely long, not to mention the high number of animals to be included . . . For that reason, alternative in vitro methods were eagerly needed. Most of them were conducted using PMCA or QuIC methods. As a recent and comprehensive review has recently been published on the topic [172], we will mainly focus on the latest highlights of in vitro transmission barrier studies.

The ability of PMCA to create de novo prions was certainly one of the most interesting contribution of this technique [62]. Besides the definitive proof of the proteinaceous origin of the prion disease, it allowed to determine a few RNA and phospholipid cofactors that are crucially needed for the conversion to take place. But several other cofactors have been evaluated, whose influence in strain selection appeared to be determinant [173,174]. In that later case, removal of phosphatidylserine provoked the phenotypic convergence of the 3 strains tested, as well as a 5 $\log_{10}$ reduction in infectivity. This phenotypic convergence has been reported elsewhere [175], suggesting that the in vitro environment needs to be better controlled in order to properly mimic what is observed in vivo. In addition to these fidelity problems, several studies mentioned the reduction or absence of infectivity resulting from the QuIC amplification of prions [176,177].

5.2. In Vitro Assessment of High Species Barrier

Barria et al. reported the generation of cervid prions that can replicate on human matrix, provided that they have been previously submitted to 2 rounds of PMCA using cervid brain lysate [98]. Conversely, PMCA amplification of 263K scrapie using partially deglycosylated PrP matrix produced a mixture of classical and atypical PrP^{Sc} profiles that suggested the development of a new strain. If true, however, this new strain was fully restored to the classical 263 K profile after a series of 10 rounds in classical conditions [178]. PMCA was also used to generate in vitro prions using PrP from mammals known to be naturally resistant to the disease [179–181]. In addition to prove that allegedly resistant animals produced transconformation-prone PrP proteins, these studies established that BSE could induce in vitro the conversion of these so-called resistant PrP and that this PrP^{Sc} was fully infectious when administered to the target animal. Similarly, PMCA was used to render mice susceptible to hamster prions and vice versa [95]. While attempting to reduce the number of passages needed for the adaptation of scrapie strains to ovine mice, it was noted that some strains were compliant with a PMCA shortcut, while others produced prions with divergent features [182]. PMCA should therefore not be regarded as a fully effective method to faithfully reproduce the prions obtains with bioassays.

6. Proposed Mechanisms for PrP Conversion, Strain Determination and Species Barrier Crossing

Despite this significant amount of experimental data, we still are waiting for some unifying model for prion conversion. Several hypotheses have been proposed for the transconformation and the elongation of infectious prions following the introduction of a seed. A first generation of models, based on a Prusiner's hypothesis [183] invokes the faithful reproduction of a template. This model however does not account for the emergence of variant prions strains from a given cloned parent. Therefore, a mechanism should be at work to allow this diversity to occur.

6.1. Prion Diversity from a Structural Portfolio and Selection of Mutants upon Species Barrier Crossing

Collinge and Clarke proposed in 2007 a model where prions are described as a panel of thermodynamically favourable conformations, referred to as a portfolio, of which some structures may or may not be selected when passing from one host to another [49]. According to this hypothesis, the strength of the transmission barrier reflects the overlap between the available portfolios of a given primary PrP sequence in two different hosts: the larger they overlap, the lower the species barrier. Some questions still remain to be addressed: in particular, what were the mechanisms at work in the generation of this primary diversity and why the resulting phenotype always displays clonal properties. In addition, observations with promiscuous strains like BSE or universal acceptors like the Bank vole question this view: for instance, one should expect from the bank vole that accepts almost every prion to display a conformational portfolio that includes most of the prion strains portfolios; if true, then no change should be seen when the strains will be passed from an host to the bank vole, which eventually was not the case [12]. When passed on bank vole, most of the strains showed an evolution in their characters but each time the character that emerged was alone, as if every other conformational state was drastically silenced (although it could express in the context of another inoculated prion). Therefore, we still needed a model that fits the puzzling data. The following sections will delineate the factors that have been taken into account in the available studies.

6.2. Deformed Templating

Based on observations published in 2012, Makarava and colleagues proposed a model in which recombinant hamster fibrils induce PrP^C to form what they call an atypical PrP^{res} [184]. In that model, two major successive steps are at work for the induction of PrP^{Sc} and the generation of a new strain of prion: in their experiment, the first step results from the formation of atypical PrP^{res} triggered by 0.5 M GdnHCl. This atypical PrP^{res} was detected in the brain of mice at their end of life but was not associated to any clinical sign. Then, the PrP^{res} could trigger the slow conversion of PrP^C into PrP^{Sc}. The kinetics for the production of both PrP^{res} and PrP^{Sc} are quite different: generation of atypical PrP^{res} is RNA independent and its FTIR structure closely resembles that of parental fibrils; the second step, which was named "deformed templating" is more stochastic and less described. It is postulated that the strain structural diversity is acquired depending on the environmental constraints. The authors added that once formed, PrP^{Sc} does not require atypical PrP^{res} anymore and outcompetes its rival thanks to favourable kinetic constants (Figure 3). Since these two steps are independent, they could occur in separate animals. This hypothesis would thus explain why the crossing of the species barrier may be achieved even after several successive passages in recipient animals in which no classical PrP^{Sc} could be detected. According to Makarava, though, one should be able to detect the atypical PrP^{res}. At the moment, this was confirmed in the case of 263 K scrapie prions [185]. In addition, the authors propose that, in contrast to the first step, the rate of deformed templating is not influenced by PrP^C concentration. This model does not address however the mechanisms that concur to the production of this atypical PrP^{res}.

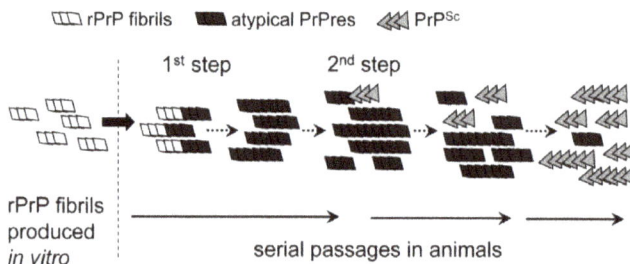

Figure 3. Schematic presentation of the mechanism, illustrating genesis of PrP^{Sc} triggered by rPrP fibrils. In a first step, rPrP fibrils seeded atypical PrP^{res}, a transmissible form of PrP that replicates silently without causing clinical disease. Replication of atypical PrP^{res} occasionally produces PrP^{Sc} in seeding events that appears to be rare and stochastic as described for a deformed templating mechanism. PrP^{Sc} replicates faster than atypical PrP^{res} and eventually replaces it during serial passages. The two forms atypical PrP^{res} and PrP^{Sc} can be distinguished after PK treatment via staining Western blot analyses with discriminating antibodies. Atypical PrP^{res}, alternative self-replicating state of prion protein; PrP^{Sc}, prion protein scrapie isoform; rPrP, recombinant prion protein. From [185].

In an attempt to obtain insight into the quaternary structure of the PrP^{Sc} assemblies, our team recently published data obtained using velocity sedimentation gradients on urea-denatured and refolded purified PrP^{Sc} associated to the assessment of their specific infectivity [186]: we demonstrated the existence of stable packs of oligomeric subunits (suPrP) that encode the main strain structural determinants: when PrP^{Sc} aggregates were denatured under increasing concentrations of urea, the velocity sedimentation gradients evolved from large polydisperse aggregates toward the generation of small elements, presumably trimers that were named suPrP. Upon dialysis refolding, the velocity sedimentation gradients identified condensation of refolded aggregates (rfPrP) but with a different distribution from that before denaturation. SuPrP bricks turned out to be fully PK-sensitive and unable to template infectivity either in vitro or in bio assays. However, upon condensation the suPrP bricks regained full infectivity and PK-resistance properties of the parental strain. One of the most important findings was the fact that suPrP, rfPrP and PrP^{Sc} shared a dynamic equilibrium: upon dilution of 263 K PrP^{Sc} in physiological buffer, a rapid decrease of the light scattered by the oligomer solutions showed a significant reduction in the size of the particles, resulting from the dissociation of the PrP^{Sc} into suPrP. When local suPrP concentration was restored and urea removed, condensation of the suPrP into rfPrP could be observed by western blot and infectivity restored as assessed with PMCA.

Thus, the results presented in this work suggest the existence of two organization levels within prion assemblies (Figure 4), one suPrP oligomeric subunit (that could contain 3–5 monomers) and a meta assembly that gathers the suPrP subunits and supports the strain infectivity level and structural conformation features. Whether the suPrP pre-exist to the PrP^{Sc} before being included in the elongating polymer or the suPrP results from the incorporation of PrP^{C} into the polymer remains to be addressed. The former hypothesis however implies that PrP^{C} and suPrP shall be separated in normal conditions.

This mechanistic proposition for the generation of elementary infectious prion bricks that co-exist as an equilibrium with larger assemblies is compatible with the portfolio model of Collinge. We propose the initial coexistence of several structurally different prions within a single brain homogenate [76]: the emergence of a new strain after prion inoculation to a strictly homologous recipient animal results from a difference in PrP^{C} expression level between inoculum donor and the recipient transgenic mice; the original 21 K strain may be favoured in low PrP expressor animals because the elementary brick could be more efficient at recruiting PrP^{C} at low concentration, while the bricks that lead to 19 K phenotype would benefit the advantage of high PrP expression. Alternatively, stochastic events mays also produce a third strain that could outcompete the two others.

It is probable that in a heterologous transmission, prion inoculum will first depolymerize just after injection and produce the main suPrP that is observed in the gradient experiments. Then, the elementary bricks would have to recruit PrP^{C} that may or may not accommodate the suPrP. This phenomenon could be highly stochastic, the probability that host PrP adopts a conformational state compatible with the foreign suPrP should be related to the proximity of prion strain and host. The generated assemblies could be rapidly stabilized and amplified, thus producing an infection with no apparent species barrier; conversely, when host PrP could not fit the topological constraints imposed by inoculum suPrP, the process would need longer time to produce and test pseudo stabilized oligomers or new suPrP that would ultimately emerge as a new prion strain.

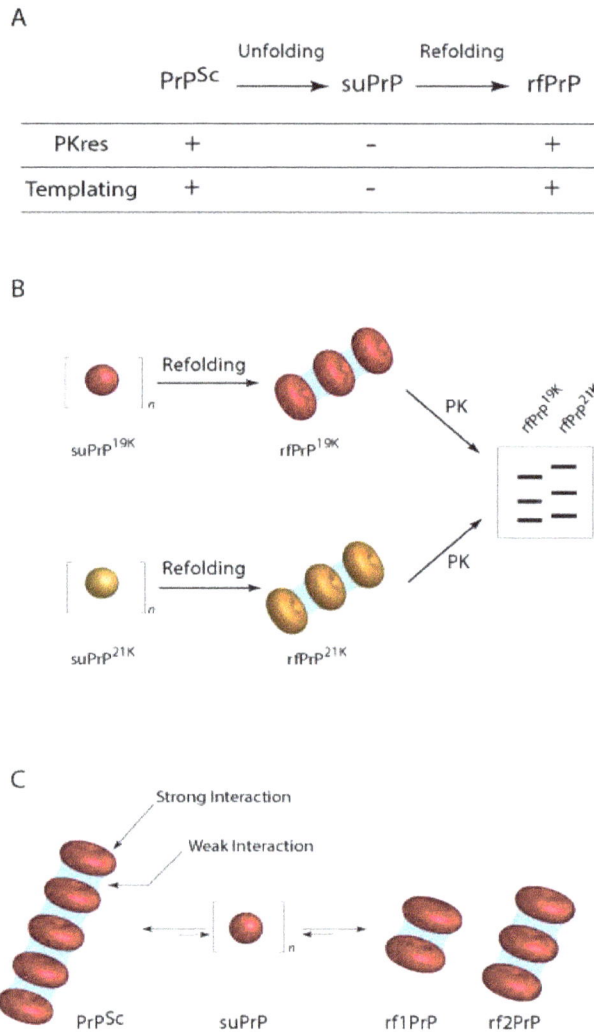

Figure 4. The role of suPrP in the dynamics of PrP^Sc assemblies. (**A**) Evolution of PK resistance and templating propensities of different types of PrP assemblies obtained after sequential unfolding and refolding of the parental prion. PrP^Sc is the native prion; suPrP is the elementary oligomeric PrP subunit; and rfPrP is the refolded conformer formed after the polymerization of suPrP. The process of conversion of suPrP into rfPrP requires a conformational change in the PrP protomer of suPrP (represented here as a sphere) to form infectious and PK-resistant assemblies (represented as stack of torus); (**B**) The conserved differential proteolytic pattern of rfPrPT1-Ov-21K and rfPrPT2-Ov-19K suggests that their respective suPrPs (represented respectively as yellow and red spheres) exhibit distinct conformations. During the refolding step (**C**), two modes of organization contribute to the cohesion within PrP^Sc assemblies. Weak interactions (in blue) are involved in maintaining the overall quaternary structure by stacking suPrPs, when strong interactions are involved in the cohesion of PrP protomers in suPrP oligomers. The weakness of the interactions interlinking suPrP means that PrP^Sc assembly and disassembly are highly dynamic events, even in the absence of a chaotropic agent and free suPrP could exist in equilibrium with infectious assemblies. From [186].

7. Conclusions

Prion strains and species barrier phenomena still remain difficult questions to address. Much knowledge has been gained regarding the characterization of the strains (thanks in particular to the PMCA in vitro methods and to the gradient fractionation techniques). The model proposed by our lab identified some features for the faithful reproduction of prions and proposed a kind of generic polymerization principle where prion fibres exist in equilibrium with one or possibly several sub-units that contain the structural strain determinants. Crossing of species barrier could result from the emergence of one of these sub-assemblies: this hypothesis is in good agreement with the quasi species theory proposed by Weissmann and colleagues. Another non-exclusive hypothesis relies on the ability of the prion inoculum to undergo progressive templating deformation by iterative adjustment of the host PrP structure upon oligomerization. This hypothesis also fits the model proposed by Makarava et al., which focuses on the strain adaptation. It would explain why and how a given strain, adapted to its host, is able to infect a new host, sometimes rapidly and sometimes very slowly. Finally, it is worth emphasizing that the fundamental questions addressed in this topic are supposed to bring high impact on how we understand past outbreaks of transmissible spongiform encephalopathies and potential ones. These threats are linked to hidden circulating prions (and adapting to its human host within the lymphoid organs), or to closer contacts between wild cervids and domestic ruminants (in the context of spreading of CWD in Europe).

Conflicts of Interest: The authors declare no conflict of interest.

References

1. Collinge, J.; Palmer, M.S.; Sidle, K.C.; Hill, A.F.; Gowland, I.; Meads, J.; Asante, E.; Bradley, R.; Doey, L.J.; Lantos, P.L. Unaltered susceptibility to BSE in transgenic mice expressing human prion protein. *Nature* **1995**, *378*, 779–783. [CrossRef] [PubMed]
2. Collinge, J.; Sidle, K.C.; Meads, J.; Ironside, J.; Hill, A.F. Molecular analysis of prion strain variation and the aetiology of "new variant" CJD. *Nature* **1996**, *383*, 685–690. [CrossRef] [PubMed]
3. Gajdusek, D.C.; Gibbs, C.J.; Alpers, M. Experimental transmission of a Kuru-like syndrome to chimpanzees. *Nature* **1966**, *209*, 794–796. [CrossRef] [PubMed]
4. Griffith, J.S. Self-replication and scrapie. *Nature* **1967**, *215*, 1043–1044. [CrossRef] [PubMed]
5. Prusiner, S.B. Novel proteinaceous infectious particles cause scrapie. *Science* **1982**, *216*, 136–144. [CrossRef] [PubMed]
6. Cuillé, J. Chelle La tremblante du mouton est bien inoculable. *CR Séances Acad. Sci. Paris* **1936**, *206*, 78–79.
7. Plummer, P.J. Scrapie—A Disease of Sheep: A Review of the literature. *Can. J. Comp. Med. Vet. Sci.* **1946**, *10*, 49–54. [PubMed]
8. Pattison, I.H.; Gordon, W.S.; Millson, G.C. Experimental Production of Scrapie in Goats. *J. Comp. Pathol. Ther.* **1959**, *69*, 300IN19–312IN20. [CrossRef]
9. Pattison, I.H.; Millson, G.C. Scrapie produced experimentally in goats with special reference to the clinical syndrome. *J. Comp. Pathol.* **1961**, *71*, 101–109. [CrossRef]
10. Hadlow, W.J.; Race, R.E.; Kennedy, R.C. Experimental infection of sheep and goats with transmissible mink encephalopathy virus. *Can. J. Vet. Res.* **1987**, *51*, 135–144. [PubMed]
11. Nonno, R.; Bari, M.A.D.; Cardone, F.; Vaccari, G.; Fazzi, P.; Dell'Omo, G.; Cartoni, C.; Ingrosso, L.; Boyle, A.; Galeno, R.; et al. Efficient Transmission and Characterization of Creutzfeldt–Jakob Disease Strains in Bank Voles. *PLoS Pathog.* **2006**, *2*. [CrossRef] [PubMed]
12. Watts, J.C.; Giles, K.; Patel, S.; Oehler, A.; DeArmond, S.J.; Prusiner, S.B. Evidence That Bank Vole PrP Is a Universal Acceptor for Prions. *PLoS Pathog.* **2014**, *10*. [CrossRef] [PubMed]
13. Orrú, C.D.; Groveman, B.R.; Raymond, L.D.; Hughson, A.G.; Nonno, R.; Zou, W.; Ghetti, B.; Gambetti, P.; Caughey, B. Bank Vole Prion Protein As an Apparently Universal Substrate for RT-QuIC-Based Detection and Discrimination of Prion Strains. *PLoS Pathog.* **2015**, *11*, e1004983. [CrossRef]

14. Robinson, M.M.; Hadlow, W.J.; Knowles, D.P.; Huff, T.P.; Lacy, P.A.; Marsh, R.F.; Gorham, J.R. Experimental infection of cattle with the agents of transmissible mink encephalopathy and scrapie. *J. Comp. Pathol.* **1995**, *113*, 241–251. [CrossRef]

15. Büeler, H.; Aguzzi, A.; Sailer, A.; Greiner, R.-A.; Autenried, P.; Aguet, M.; Weissmann, C. Mice devoid of PrP are resistant to scrapie. *Cell* **1993**, *73*, 1339–1347. [CrossRef]

16. Telling, G.C.; Scott, M.; Hsiao, K.K.; Foster, D.; Yang, S.L.; Torchia, M.; Sidle, K.C.; Collinge, J.; DeArmond, S.J.; Prusiner, S.B. Transmission of Creutzfeldt-Jakob disease from humans to transgenic mice expressing chimeric human-mouse prion protein. *Proc. Natl. Acad. Sci. USA* **1994**, *91*, 9936–9940. [CrossRef] [PubMed]

17. Brandner, S.; Isenmann, S.; Raeber, A.; Fischer, M.; Sailer, A.; Kobayashi, Y.; Marino, S.; Weissmann, C.; Aguzzi, A. Normal host prion protein necessary for scrapie-induced neurotoxicity. *Nature* **1996**, *379*, 339–343. [CrossRef] [PubMed]

18. Fraser, H.; Dickinson, A.G. Scrapie in mice: Agent-strain differences in the distribution and intensity of grey matter vacuolation. *J. Comp. Pathol.* **1973**, *83*, 29–40. [CrossRef]

19. Asher, D.M.; Gibbs, C.J.; Gajdusek, D.C. Pathogenesis of subacute spongiform encephalopathies. *Ann. Clin. Lab. Sci.* **1976**, *6*, 84–103. [PubMed]

20. Lowenstein, D.H.; Butler, D.A.; Westaway, D.; McKinley, M.P.; DeArmond, S.J.; Prusiner, S.B. Three hamster species with different scrapie incubation times and neuropathological features encode distinct prion proteins. *Mol. Cell. Biol.* **1990**, *10*, 1153–1163. [CrossRef] [PubMed]

21. Kimberlin, R.H.; Walker, C.A.; Fraser, H. The Genomic Identity of Different Strains of Mouse Scrapie Is Expressed in Hamsters and Preserved on Reisolation in Mice. *J. Gen. Virol.* **1989**, *70*, 2017–2025. [CrossRef] [PubMed]

22. Kimberlin, R.H.; Walker, C.A. Pathogenesis of mouse scrapie: Dynamics of agent replication in spleen, spinal cord and brain after infection by different routes. *J. Comp. Pathol.* **1979**, *89*, 551–562. [CrossRef]

23. Prusiner, S.B.; Groth, D.F.; Cochran, S.P.; Masiarz, F.R.; McKinley, M.P.; Martinez, H.M. Molecular properties, partial purification and assay by incubation period measurements of the hamster scrapie agent. *Biochemistry* **1980**, *19*, 4883–4891. [CrossRef] [PubMed]

24. Mould, D.L.; Dawson, A.M.; Rennie, J.C. Very early replication of scrapie in lymphocytic tissue. *Nature* **1970**, *228*, 779–780. [CrossRef] [PubMed]

25. Béringue, V.; Andreoletti, O.; Le Dur, A.; Essalmani, R.; Vilotte, J.-L.; Lacroux, C.; Reine, F.; Herzog, L.; Biacabe, A.-G.; Baron, T.; et al. A Bovine Prion Acquires an Epidemic Bovine Spongiform Encephalopathy Strain-Like Phenotype on Interspecies Transmission. *J. Neurosci.* **2007**, *27*, 6965–6971. [CrossRef] [PubMed]

26. Caughey, B.; Raymond, G.J.; Bessen, R.A. Strain-dependent differences in beta-sheet conformations of abnormal prion protein. *J. Biol. Chem.* **1998**, *273*, 32230–32235. [CrossRef] [PubMed]

27. Ostapchenko, V.G.; Sawaya, M.R.; Makarava, N.; Savtchenko, R.; Nilsson, K.P.R.; Eisenberg, D.; Baskakov, I.V. Two amyloid States of the prion protein display significantly different folding patterns. *J. Mol. Biol.* **2010**, *400*, 908–921. [CrossRef] [PubMed]

28. Cobb, N.J.; Apostol, M.I.; Chen, S.; Smirnovas, V.; Surewicz, W.K. Conformational Stability of Mammalian Prion Protein Amyloid Fibrils Is Dictated by a Packing Polymorphism within the Core Region. *J. Biol. Chem.* **2014**, *289*, 2643–2650. [CrossRef] [PubMed]

29. Silva, C.J.; Erickson-Beltran, M.L.; Dynin, I.C. Covalent Surface Modification of Prions: A Mass Spectrometry-Based Means of Detecting Distinctive Structural Features of Prion Strains. *Biochemistry* **2016**, *55*, 894–902. [CrossRef] [PubMed]

30. Langeveld, J.P.; Jacobs, J.G.; Erkens, J.H.; Bossers, A.; van Zijderveld, F.G.; van Keulen, L.J. Rapid and discriminatory diagnosis of scrapie and BSE in retro-pharyngeal lymph nodes of sheep. *BMC Vet. Res.* **2006**, *2*, 19. [CrossRef] [PubMed]

31. Cali, I.; Castellani, R.; Alshekhlee, A.; Cohen, Y.; Blevins, J.; Yuan, J.; Langeveld, J.P.M.; Parchi, P.; Safar, J.G.; Zou, W.-Q.; et al. Co-existence of scrapie prion protein types 1 and 2 in sporadic Creutzfeldt–Jakob disease: Its effect on the phenotype and prion-type characteristics. *Brain* **2009**, *132*, 2643–2658. [CrossRef] [PubMed]

32. Bessen, R.A.; Marsh, R.F. Biochemical and physical properties of the prion protein from two strains of the transmissible mink encephalopathy agent. *J. Virol.* **1992**, *66*, 2096–2101. [PubMed]

33. Korth, C.; Stierli, B.; Streit, P.; Moser, M.; Schaller, O.; Fischer, R.; Schulz-Schaeffer, W.; Kretzschmar, H.; Raeber, A.; Braun, U.; et al. Prion (PrP^{Sc})-specific epitope defined by a monoclonal antibody. *Nature* **1997**, *390*, 36337. [CrossRef] [PubMed]

34. Safar, J.; Wille, H.; Itri, V.; Groth, D.; Serban, H.; Torchia, M.; Cohen, F.E.; Prusiner, S.B. Eight prion strains have PrP(Sc) molecules with different conformations. *Nat. Med.* **1998**, *4*, 1157–1165. [CrossRef] [PubMed]
35. Safar, J.G.; Geschwind, M.D.; Deering, C.; Didorenko, S.; Sattavat, M.; Sanchez, H.; Serban, A.; Vey, M.; Baron, H.; Giles, K.; et al. Diagnosis of human prion disease. *Proc. Natl. Acad. Sci. USA* **2005**, *102*, 3501–3506. [CrossRef] [PubMed]
36. Choi, Y.P.; Peden, A.H.; Gröner, A.; Ironside, J.W.; Head, M.W. Distinct Stability States of Disease-Associated Human Prion Protein Identified by Conformation-Dependent Immunoassay. *J. Virol.* **2010**, *84*, 12030–12038. [CrossRef] [PubMed]
37. Mays, C.E.; Kim, C.; Haldiman, T.; van der Merwe, J.; Lau, A.; Yang, J.; Grams, J.; Di Bari, M.A.; Nonno, R.; Telling, G.C.; et al. Prion disease tempo determined by host-dependent substrate reduction. *J. Clin. Investig.* **2014**, *124*, 847–858. [CrossRef] [PubMed]
38. Neuendorf, E.; Weber, A.; Saalmueller, A.; Schatzl, H.; Reifenberg, K.; Pfaff, E.; Groschup, M.H. Glycosylation Deficiency at Either One of the Two Glycan Attachment Sites of Cellular Prion Protein Preserves Susceptibility to Bovine Spongiform Encephalopathy and Scrapie Infections. *J. Biol. Chem.* **2004**, *279*, 53306–53316. [CrossRef] [PubMed]
39. Moudjou, M.; Chapuis, J.; Mekrouti, M.; Reine, F.; Herzog, L.; Sibille, P.; Laude, H.; Vilette, D.; Andréoletti, O.; Rezaei, H.; et al. Glycoform-independent prion conversion by highly efficient, cell-based, protein misfolding cyclic amplification. *Sci. Rep.* **2016**, *6*. [CrossRef] [PubMed]
40. Srivastava, S.; Katorcha, E.; Daus, M.L.; Lasch, P.; Beekes, M.; Baskakov, I.V. Sialylation controls prion fate in vivo. *J. Biol. Chem.* **2016**. [CrossRef]
41. Katorcha, E.; Makarava, N.; Savtchenko, R.; d'Azzo, A.; Baskakov, I.V. Sialylation of Prion Protein Controls the Rate of Prion Amplification, the Cross-Species Barrier, the Ratio of PrP^Sc Glycoform and Prion Infectivity. *PLoS Pathog.* **2014**, *10*, e1004366. [CrossRef] [PubMed]
42. Telling, G.C.; Parchi, P.; DeArmond, S.J.; Cortelli, P.; Montagna, P.; Gabizon, R.; Mastrianni, J.; Lugaresi, E.; Gambetti, P.; Prusiner, S.B. Evidence for the Conformation of the Pathologic Isoform of the Prion Protein Enciphering and Propagating Prion Diversity. *Science* **1996**, *274*, 2079–2082. [CrossRef] [PubMed]
43. Makarava, N.; Ostapchenko, V.G.; Savtchenko, R.; Baskakov, I.V. Conformational Switching within Individual Amyloid Fibrils. *J. Biol. Chem.* **2009**, *284*, 14386–14395. [CrossRef] [PubMed]
44. Mahal, S.P.; Baker, C.A.; Demczyk, C.A.; Smith, E.W.; Julius, C.; Weissmann, C. Prion strain discrimination in cell culture: The cell panel assay. *Proc. Natl. Acad. Sci. USA* **2007**, *104*, 20908–20913. [CrossRef] [PubMed]
45. Li, J.; Browning, S.; Mahal, S.P.; Oelschlegel, A.M.; Weissmann, C. Darwinian Evolution of Prions in Cell Culture. *Science* **2010**, *327*, 869–872. [CrossRef] [PubMed]
46. Weissmann, C.; Li, J.; Mahal, S.P.; Browning, S. Prions on the move. *EMBO Rep.* **2011**, *12*, 1109–1117. [CrossRef] [PubMed]
47. Weissmann, C. Mutation and Selection of Prions. *PLoS Pathog.* **2012**, *8*, e1002582. [CrossRef] [PubMed]
48. Oelschlegel, A.M.; Weissmann, C. Acquisition of Drug Resistance and Dependence by Prions. *PLoS Pathog.* **2013**, *9*, e1003158. [CrossRef] [PubMed]
49. Collinge, J.; Clarke, A.R. A General Model of Prion Strains and Their Pathogenicity. *Science* **2007**, *318*, 930–936. [CrossRef] [PubMed]
50. Saá, P.; Cervenakova, L. Protein misfolding cyclic amplification (PMCA): Current status and future directions. *Virus Res.* **2015**. [CrossRef] [PubMed]
51. Saborio, G.P.; Permanne, B.; Soto, C. Sensitive detection of pathological prion protein by cyclic amplification of protein misfolding. *Nature* **2001**, *411*, 810–813. [CrossRef] [PubMed]
52. Moudjou, M.; Sibille, P.; Fichet, G.; Reine, F.; Chapuis, J.; Herzog, L.; Jaumain, E.; Laferrière, F.; Richard, C.-A.; Laude, H.; et al. Highly Infectious Prions Generated by a Single Round of Microplate-Based Protein Misfolding Cyclic Amplification. *mBio* **2014**, *5*, e00829-13. [CrossRef] [PubMed]
53. Haldiman, T.; Kim, C.; Cohen, Y.; Chen, W.; Blevins, J.; Qing, L.; Cohen, M.L.; Langeveld, J.; Telling, G.C.; Kong, Q.; et al. Co-existence of Distinct Prion Types Enables Conformational Evolution of Human PrP^Sc by Competitive Selection. *J. Biol. Chem.* **2013**, *288*, 29846–29861. [CrossRef] [PubMed]
54. Castilla, J.; Saá, P.; Hetz, C.; Soto, C. In vitro generation of infectious scrapie prions. *Cell* **2005**, *121*, 195–206. [CrossRef] [PubMed]
55. Castilla, J.; Morales, R.; Saá, P.; Barria, M.; Gambetti, P.; Soto, C. Cell-free propagation of prion strains. *EMBO J.* **2008**, *27*, 2557–2566. [CrossRef] [PubMed]

56. Green, K.M.; Castilla, J.; Seward, T.S.; Napier, D.L.; Jewell, J.E.; Soto, C.; Telling, G.C. Accelerated High Fidelity Prion Amplification Within and Across Prion Species Barriers. *PLOS Pathog.* **2008**, *4*, e1000139. [CrossRef] [PubMed]

57. Munoz-Montesino, C.; Sizun, C.; Moudjou, M.; Herzog, L.; Reine, F.; Chapuis, J.; Ciric, D.; Igel-Egalon, A.; Laude, H.; Béringue, V.; et al. Generating Bona Fide Mammalian Prions with Internal Deletions. *J. Virol.* **2016**, *90*, 6963–6975. [CrossRef] [PubMed]

58. Munoz-Montesino, C.; Sizun, C.; Moudjou, M.; Herzog, L.; Reine, F.; Igel-Egalon, A.; Barbereau, C.; Chapuis, J.; Ciric, D.; Laude, H.; et al. A stretch of residues within the protease-resistant core is not necessary for prion structure and infectivity. *Prion* **2017**, *11*, 25–30. [CrossRef] [PubMed]

59. Deleault, N.R.; Lucassen, R.W.; Supattapone, S. RNA molecules stimulate prion protein conversion. *Nature* **2003**, *425*, 717–720. [CrossRef] [PubMed]

60. Deleault, N.R.; Geoghegan, J.C.; Nishina, K.; Kascsak, R.; Williamson, R.A.; Supattapone, S. Protease-resistant Prion Protein Amplification Reconstituted with Partially Purified Substrates and Synthetic Polyanions. *J. Biol. Chem.* **2005**, *280*, 26873–26879. [CrossRef] [PubMed]

61. Deleault, N.R.; Harris, B.T.; Rees, J.R.; Supattapone, S. Formation of native prions from minimal components in vitro. *Proc. Natl. Acad. Sci. USA* **2007**, *104*, 9741–9746. [CrossRef] [PubMed]

62. Wang, F.; Wang, X.; Yuan, C.-G.; Ma, J. Generating a Prion with Bacterially Expressed Recombinant Prion Protein. *Science* **2010**, *327*, 1132–1135. [CrossRef] [PubMed]

63. Wilham, J.M.; Orrú, C.D.; Bessen, R.A.; Atarashi, R.; Sano, K.; Race, B.; Meade-White, K.D.; Taubner, L.M.; Timmes, A.; Caughey, B. Rapid End-Point Quantitation of Prion Seeding Activity with Sensitivity Comparable to Bioassays. *PLoS Pathog.* **2010**, *6*. [CrossRef] [PubMed]

64. Christina, D.; Orrú, J.M.W. Prion disease blood test using immunoprecipitation and improved quaking-induced conversion. *mBio* **2011**, *2*, e00078-11. [CrossRef]

65. Henderson, D.M.; Davenport, K.A.; Haley, N.J.; Denkers, N.D.; Mathiason, C.K.; Hoover, E.A. Quantitative Assessment of Prion Infectivity in Tissues and Body Fluids by RT-QuIC. *J. Gen. Virol.* **2014**. [CrossRef]

66. Espinosa, J.C.; Nonno, R.; Di Bari, M.; Aguilar-Calvo, P.; Pirisinu, L.; Fernández-Borges, N.; Vanni, I.; Vaccari, G.; Marín-Moreno, A.; Frassanito, P.; et al. PrP^c Governs Susceptibility to Prion Strains in Bank Vole, While Other Host Factors Modulate Strain Features. *J. Virol.* **2016**, *90*, 10660–10669. [CrossRef] [PubMed]

67. Orrú, C.D.; Favole, A.; Corona, C.; Mazza, M.; Manca, M.; Groveman, B.R.; Hughson, A.G.; Acutis, P.L.; Caramelli, M.; Zanusso, G.; et al. Detection and Discrimination of Classical and Atypical L-Type Bovine Spongiform Encephalopathy by Real-Time Quaking-Induced Conversion. *J. Clin. Microbiol.* **2015**, *53*, 1115–1120. [CrossRef] [PubMed]

68. Magnusson, K.; Simon, R.; Sjölander, D.; Sigurdson, C.J.; Hammarström, P.; Nilsson, K.P.R. Multimodal fluorescence microscopy of prion strain specific PrP deposits stained by thiophene-based amyloid ligands. *Prion* **2014**, *8*, 319–329. [CrossRef] [PubMed]

69. Bessen, R.A.; Marsh, R.F. Distinct PrP properties suggest the molecular basis of strain variation in transmissible mink encephalopathy. *J. Virol.* **1994**, *68*, 7859–7868. [PubMed]

70. Bessen, R.A.; Kocisko, D.A.; Raymond, G.J.; Nandan, S.; Lansbury, P.T.; Caughey, B. Non-genetic propagation of strain-specific properties of scrapie prion protein. *Nature* **1995**, *375*, 698–700. [CrossRef] [PubMed]

71. Tixador, P.; Herzog, L.; Reine, F.; Jaumain, E.; Chapuis, J.; Le Dur, A.; Laude, H.; Béringue, V. The Physical Relationship between Infectivity and Prion Protein Aggregates Is Strain-Dependent. *PLoS Pathog.* **2010**, *6*, e1000859. [CrossRef] [PubMed]

72. Laferrière, F.; Tixador, P.; Moudjou, M.; Chapuis, J.; Sibille, P.; Herzog, L.; Reine, F.; Jaumain, E.; Laude, H.; Rezaei, H.; et al. Quaternary Structure of Pathological Prion Protein as a Determining Factor of Strain-Specific Prion Replication Dynamics. *PLoS Pathog.* **2013**, *9*, e1003702. [CrossRef] [PubMed]

73. Baron, T.G.M.; Madec, J.-Y.; Calavas, D. Similar Signature of the Prion Protein in Natural Sheep Scrapie and Bovine Spongiform Encephalopathy-Linked Diseases. *J. Clin. Microbiol.* **1999**, *37*, 3701–3704. [PubMed]

74. Casalone, C.; Zanusso, G.; Acutis, P.; Ferrari, S.; Capucci, L.; Tagliavini, F.; Monaco, S.; Caramelli, M. Identification of a second bovine amyloidotic spongiform encephalopathy: Molecular similarities with sporadic Creutzfeldt-Jakob disease. *Proc. Natl. Acad. Sci. USA* **2004**, *101*, 3065–3070. [CrossRef] [PubMed]

75. Bartz, J.C.; Bessen, R.A.; McKenzie, D.; Marsh, R.F.; Aiken, J.M. Adaptation and Selection of Prion Protein Strain Conformations following Interspecies Transmission of Transmissible Mink Encephalopathy. *J. Virol.* **2000**, *74*, 5542–5547. [CrossRef] [PubMed]

76. Le Dur, A.; Laï, T.L.; Stinnakre, M.-G.; Laisné, A.; Chenais, N.; Rakotobe, S.; Passet, B.; Reine, F.; Soulier, S.; Herzog, L.; et al. Divergent prion strain evolution driven by PrP(c) expression level in transgenic mice. *Nat. Commun.* **2017**, *8*, 14170. [CrossRef] [PubMed]

77. Kimberlin, R.H.; Walker, C.A. Evidence that the Transmission of One Source of Scrapie Agent to Hamsters Involves Separation of Agent Strains from a Mixture. *J. Gen. Virol.* **1978**, *39*, 487–496. [CrossRef] [PubMed]

78. Race, R.; Chesebro, B. Scrapie infectivity found in resistant species. *Nature* **1998**, *392*, 770. [CrossRef] [PubMed]

79. Hill, A.F.; Joiner, S.; Linehan, J.; Desbruslais, M.; Lantos, P.L.; Collinge, J. Species-barrier-independent prion replication in apparently resistant species. *Proc. Natl. Acad. Sci. USA* **2000**, *97*, 10248–10253. [CrossRef] [PubMed]

80. Prusiner, S.B.; Scott, M.; Foster, D.; Pan, K.-M.; Groth, D.; Mirenda, C.; Torchia, M.; Yang, S.-L.; Serban, D.; Carlson, G.A.; et al. Transgenetic studies implicate interactions between homologous PrP isoforms in scrapie prion replication. *Cell* **1990**, *63*, 673–686. [CrossRef]

81. Castilla, J.; Gonzalez-Romero, D.; Saá, P.; Morales, R.; De Castro, J.; Soto, C. Crossing the Species Barrier by PrP^Sc Replication In Vitro Generates Unique Infectious Prions. *Cell* **2008**, *134*, 757–768. [CrossRef] [PubMed]

82. Bruce, M.E.; Will, R.G.; Ironside, J.W.; McConnell, I.; Drummond, D.; Suttie, A.; McCardle, L.; Chree, A.; Hope, J.; Birkett, C.; et al. Transmissions to mice indicate that "new variant" CJD is caused by the BSE agent. *Nature* **1997**, *389*, 498–501. [CrossRef] [PubMed]

83. Scott, M.R.; Will, R.; Ironside, J.; Nguyen, H.-O.B.; Tremblay, P.; DeArmond, S.J.; Prusiner, S.B. Compelling transgenetic evidence for transmission of bovine spongiform encephalopathy prions to humans. *Proc. Natl. Acad. Sci. USA* **1999**, *96*, 15137–15142. [CrossRef] [PubMed]

84. Asante, E.A.; Linehan, J.M.; Desbruslais, M.; Joiner, S.; Gowland, I.; Wood, A.L.; Welch, J.; Hill, A.F.; Lloyd, S.E.; Wadsworth, J.D.F.; et al. BSE prions propagate as either variant CJD-like or sporadic CJD-like prion strains in transgenic mice expressing human prion protein. *EMBO J.* **2002**, *21*, 6358–6366. [CrossRef] [PubMed]

85. Béringue, V.; Herzog, L.; Jaumain, E.; Reine, F.; Sibille, P.; Le Dur, A.; Vilotte, J.-L.; Laude, H. Facilitated Cross-Species Transmission of Prions in Extraneural Tissue. *Science* **2012**, *335*, 472–475. [CrossRef] [PubMed]

86. Williams, E.S.; Young, S. Spongiform encephalopathies in Cervidae. *Rev. Sci. Tech. Int. Off. Epizoot.* **1992**, *11*, 551–567. [CrossRef]

87. Watts, J.C.; Balachandran, A.; Westaway, D. The expanding universe of prion diseases. *PLoS Pathog.* **2006**, *2*, e26. [CrossRef] [PubMed]

88. Benestad, S.L.; Mitchell, G.; Simmons, M.; Ytrehus, B.; Vikøren, T. First case of chronic wasting disease in Europe in a Norwegian free-ranging reindeer. *Vet. Res.* **2016**, *47*. [CrossRef] [PubMed]

89. Sigurdson, C.J.; Manco, G.; Schwarz, P.; Liberski, P.; Hoover, E.A.; Hornemann, S.; Polymenidou, M.; Miller, M.W.; Glatzel, M.; Aguzzi, A. Strain Fidelity of Chronic Wasting Disease upon Murine Adaptation. *J. Virol.* **2006**, *80*, 12303–12311. [CrossRef] [PubMed]

90. Herbst, A.; Velásquez, C.D.; Triscott, E.; Aiken, J.M.; McKenzie, D. Chronic Wasting Disease Prion Strain Emergence and Host Range Expansion. *Emerg. Infect. Dis.* **2017**, *23*, 1598–1600. [CrossRef] [PubMed]

91. Kurt, T.D.; Jiang, L.; Fernández-Borges, N.; Bett, C.; Liu, J.; Yang, T.; Spraker, T.R.; Castilla, J.; Eisenberg, D.; Kong, Q.; et al. Human prion protein sequence elements impede cross-species chronic wasting disease transmission. *J. Clin. Investig.* **2015**, *125*, 1485. [CrossRef] [PubMed]

92. Czub, S.; Schulz-Schaeffer, W.; Stahl-Hennig, C.; Beekes, M.; Schaetzl, H.; Motzkus, D. Chronic Wasting Disease: PRION 2017 CONFERENCE ABSTRACT First Evidence of Intracranial and Peroral Transmission of Chronic Wasting Disease (CWD) into Cynomolgus Macaques: A Work in Progress. Available online: http://chronic-wasting-disease.blogspot.hk/2017/06/prion-2017-conference-abstract-first.html (accessed on 26 December 2017).

93. Di Bari, M.A.; Chianini, F.; Vaccari, G.; Esposito, E.; Conte, M.; Eaton, S.L.; Hamilton, S.; Finlayson, J.; Steele, P.J.; Dagleish, M.P.; et al. The bank vole (*Myodes glareolus*) as a sensitive bioassay for sheep scrapie. *J. Gen. Virol.* **2008**, *89*, 2975–2985. [CrossRef] [PubMed]

94. Peretz, D.; Williamson, R.A.; Legname, G.; Matsunaga, Y.; Vergara, J.; Burton, D.R.; DeArmond, S.J.; Prusiner, S.B.; Scott, M.R. A change in the conformation of prions accompanies the emergence of a new prion strain. *Neuron* **2002**, *34*, 921–932. [CrossRef]

95. Gao, C.; Han, J.; Zhang, J.; Wei, J.; Zhang, B.-Y.; Tian, C.; Zhang, J.; Shi, Q.; Dong, X.-P. Protein Misfolding Cyclic Amplification Cross-Species Products of Mouse-Adapted Scrapie Strain 139A and Hamster-Adapted Scrapie Strain 263K with Brain and Muscle Tissues of Opposite Animals Generate Infectious Prions. *Mol. Neurobiol.* **2017**, *54*, 3771–3782. [CrossRef] [PubMed]

96. Watts, J.C.; Giles, K.; Saltzberg, D.J.; Dugger, B.N.; Patel, S.; Oehler, A.; Bhardwaj, S.; Sali, A.; Prusiner, S.B. Guinea Pig Prion Protein Supports Rapid Propagation of Bovine Spongiform Encephalopathy and Variant Creutzfeldt-Jakob Disease Prions. *J. Virol.* **2016**, *90*, 9558–9569. [CrossRef] [PubMed]

97. Davenport, K.A.; Henderson, D.M.; Bian, J.; Telling, G.C.; Mathiason, C.K.; Hoover, E.A. Insights into Chronic Wasting Disease and Bovine Spongiform Encephalopathy Species Barriers by Use of Real-Time Conversion. *J. Virol.* **2015**, *89*, 9524–9531. [CrossRef] [PubMed]

98. Barria, M.A.; Telling, G.C.; Gambetti, P.; Mastrianni, J.A.; Soto, C. Generation of a New Form of Human PrP^{Sc} in Vitro by Interspecies Transmission from Cervid Prions. *J. Biol. Chem.* **2011**, *286*, 7490–7495. [CrossRef] [PubMed]

99. Marsh, R.F.; Kincaid, A.E.; Bessen, R.A.; Bartz, J.C. Interspecies Transmission of Chronic Wasting Disease Prions to Squirrel Monkeys (*Saimiri sciureus*). *J. Virol.* **2005**, *79*, 13794–13796. [CrossRef] [PubMed]

100. Race, B.; Meade-White, K.D.; Miller, M.W.; Barbian, K.D.; Rubenstein, R.; LaFauci, G.; Cervenakova, L.; Favara, C.; Gardner, D.; Long, D.; et al. Susceptibilities of Nonhuman Primates to Chronic Wasting Disease. *Emerg. Infect. Dis.* **2009**, *15*, 1366–1376. [CrossRef] [PubMed]

101. Scott, M.; Foster, D.; Mirenda, C.; Serban, D.; Coufal, F.; Wälchli, M.; Torchia, M.; Groth, D.; Carlson, G.; DeArmond, S.J.; et al. Transgenic mice expressing hamster prion protein produce species-specific scrapie infectivity and amyloid plaques. *Cell* **1989**, *59*, 847–857. [CrossRef]

102. Race, R.E.; Priola, S.A.; Bessen, R.A.; Ernst, D.; Dockter, J.; Rall, G.F.; Mucke, L.; Chesebro, B.; Oldstone, M.B. Neuron-specific expression of a hamster prion protein minigene in transgenic mice induces susceptibility to hamster scrapie agent. *Neuron* **1995**, *15*, 1183–1191. [CrossRef]

103. Priola, S.A. Prion protein and species barriers in the transmissible spongiform encephalopathies. *Biomed. Pharmacother.* **1999**, *53*, 27–33. [CrossRef]

104. Sarradin, P.; Viglietta, C.; Limouzin, C.; Andréoletti, O.; Daniel-Carlier, N.; Barc, C.; Leroux-Coyau, M.; Berthon, P.; Chapuis, J.; Rossignol, C.; et al. Transgenic Rabbits Expressing Ovine PrP Are Susceptible to Scrapie. *PLoS Pathog.* **2015**, *11*, e1005077. [CrossRef] [PubMed]

105. Tranulis, M.A. Influence of the prion protein gene, *Prnp*, on scrapie susceptibility in sheep. *APMIS* **2002**, *110*, 33–43. [CrossRef] [PubMed]

106. Goldmann, W. PrP genetics in ruminant transmissible spongiform encephalopathies. *Vet. Res.* **2008**, *39*, 1–14. [CrossRef] [PubMed]

107. Chen, C.; Dong, X.-P. Epidemiological characteristics of human prion diseases. *Infect. Dis. Poverty* **2016**, *5*. [CrossRef] [PubMed]

108. Watts, J.C.; Giles, K.; Stöhr, J.; Oehler, A.; Bhardwaj, S.; Grillo, S.K.; Patel, S.; DeArmond, S.J.; Prusiner, S.B. Spontaneous Generation of Rapidly Transmissible Prions in Transgenic Mice Expressing Wild-Type Bank Vole Prion Protein. *Proc. Natl. Acad. Sci. USA* **2012**, *109*, 3498–3503. [CrossRef] [PubMed]

109. Tateishi, J.; Kitamoto, T.; Hoque, M.Z.; Furukawa, H. Experimental transmission of Creutzfeldt-Jakob disease and related diseases to rodents. *Neurology* **1996**, *46*, 532–537. [CrossRef] [PubMed]

110. Palmer, M.S.; Dryden, A.J.; Hughes, J.T.; Collinge, J. Homozygous prion protein genotype predisposes to sporadic Creutzfeldt-Jakob disease. *Nature* **1991**, *352*, 340–342. [CrossRef] [PubMed]

111. Zeidler, M.; Stewart, G.E.; Barraclough, C.R.; Bateman, D.E.; Bates, D.; Burn, D.J.; Colchester, A.C.; Durward, W.; Fletcher, N.A.; Hawkins, S.A.; et al. New variant Creutzfeldt-Jakob disease: Neurological features and diagnostic tests. *Lancet* **1997**, *350*, 903–907. [CrossRef]

112. Fernández-Borges, N.; Espinosa, J.C.; Marín-Moreno, A.; Aguilar-Calvo, P.; Asante, E.A.; Kitamoto, T.; Mohri, S.; Andréoletti, O.; Torres, J.M. Protective Effect of Val129-PrP against Bovine Spongiform Encephalopathy but not Variant Creutzfeldt-Jakob Disease. *Emerg. Infect. Dis.* **2017**, *23*, 1522–1530. [CrossRef] [PubMed]

113. Shibuya, S.; Higuchi, J.; Shin, R.-W.; Tateishi, J.; Kitamoto, T. Protective prion protein polymorphisms against sporadic Creutzfeldt-Jakob disease. *Lancet* **1998**, *351*, 419. [CrossRef]

114. Asante, E.A.; Smidak, M.; Grimshaw, A.; Houghton, R.; Tomlinson, A.; Jeelani, A.; Jakubcova, T.; Hamdan, S.; Richard-Londt, A.; Linehan, J.M.; et al. A naturally occurring variant of the human prion protein completely prevents prion disease. *Nature* **2015**, *522*, 478–481. [CrossRef] [PubMed]

115. Green, K.M.; Browning, S.R.; Seward, T.S.; Jewell, J.E.; Ross, D.L.; Green, M.A.; Williams, E.S.; Hoover, E.A.; Telling, G.C. The elk PRNP codon 132 polymorphism controls cervid and scrapie prion propagation. *J. Gen. Virol.* **2008**, *89*, 598–608. [CrossRef] [PubMed]

116. Cartoni, C.; Schininà, M.E.; Maras, B.; Nonno, R.; Vaccari, G.; Di Bari, M.; Conte, M.; De Pascalis, A.; Principe, S.; Cardone, F.; et al. Quantitative profiling of the pathological prion protein allotypes in bank voles by liquid chromatography–mass spectrometry. *J. Chromatogr. B* **2007**, *849*, 302–306. [CrossRef] [PubMed]

117. Kimberlin, R.H.; Walker, C.A. Competition between strains of scrapie depends on the blocking agent being infectious. *Intervirology* **1985**, *23*, 74–81. [CrossRef] [PubMed]

118. Kimberlin, R.H.; Cole, S.; Walker, C.A. Transmissible mink encephalopathy (TME) in Chinese hamsters: Identification of two strains of TME and comparisons with scrapie. *Neuropathol. Appl. Neurobiol.* **1986**, *12*, 197–206. [CrossRef] [PubMed]

119. Kimberlin, R.H.; Cole, S.; Walker, C.A. Temporary and Permanent Modifications to a Single Strain of Mouse Scrapie on Transmission to Rats and Hamsters. *J. Gen. Virol.* **1987**, *68*, 1875–1881. [CrossRef] [PubMed]

120. Priola, S.A.; Caughey, B.; Race, R.E.; Chesebro, B. Heterologous PrP molecules interfere with accumulation of protease-resistant PrP in scrapie-infected murine neuroblastoma cells. *J. Virol.* **1994**, *68*, 4873–4878. [PubMed]

121. Priola, S.A.; Chesebro, B. A single hamster PrP amino acid blocks conversion to protease-resistant PrP in scrapie-infected mouse neuroblastoma cells. *J. Virol.* **1995**, *69*, 7754–7758. [PubMed]

122. Telling, G.C.; Scott, M.; Mastrianni, J.; Gabizon, R.; Torchia, M.; Cohen, F.E.; DeArmond, S.J.; Prusiner, S.B. Prion propagation in mice expressing human and chimeric PrP transgenes implicates the interaction of cellular PrP with another protein. *Cell* **1995**, *83*, 79–90. [CrossRef]

123. Abid, K.; Morales, R.; Soto, C. Cellular factors implicated in prion replication. *FEBS Lett.* **2010**, *584*, 2409–2414. [CrossRef] [PubMed]

124. Lee, C.I.; Yang, Q.; Perrier, V.; Baskakov, I.V. The dominant-negative effect of the Q218K variant of the prion protein does not require protein X. *Protein Sci. Publ. Protein Soc.* **2007**, *16*, 2166–2173. [CrossRef] [PubMed]

125. Dickinson, A.G.; Fraser, H.; Meikle, V.M.; Outram, G.W. Competition between different scrapie agents in mice. *Nat. New Biol.* **1972**, *237*, 244–245. [CrossRef] [PubMed]

126. Bartz, J.C.; Aiken, J.M.; Bessen, R.A. Delay in onset of prion disease for the HY strain of transmissible mink encephalopathy as a result of prior peripheral inoculation with the replication-deficient DY strain. *J. Gen. Virol.* **2004**, *85*, 265–273. [CrossRef] [PubMed]

127. Shikiya, R.A.; Ayers, J.I.; Schutt, C.R.; Kincaid, A.E.; Bartz, J.C. Coinfecting Prion Strains Compete for a Limiting Cellular Resource. *J. Virol.* **2010**, *84*, 5706–5714. [CrossRef] [PubMed]

128. Bartz, J.C.; Kramer, M.L.; Sheehan, M.H.; Hutter, J.A.L.; Ayers, J.I.; Bessen, R.A.; Kincaid, A.E. Prion Interference Is Due to a Reduction in Strain-Specific PrP^{Sc} Levels. *J. Virol.* **2007**, *81*, 689–697. [CrossRef] [PubMed]

129. Sandberg, M.K.; Al-Doujaily, H.; Sharps, B.; De Oliveira, M.W.; Schmidt, C.; Richard-Londt, A.; Lyall, S.; Linehan, J.M.; Brandner, S.; Wadsworth, J.D.F.; et al. Prion neuropathology follows the accumulation of alternate prion protein isoforms after infective titre has peaked. *Nat. Commun.* **2014**, *5*. [CrossRef]

130. Wiseman, F.K.; Cancellotti, E.; Piccardo, P.; Iremonger, K.; Boyle, A.; Brown, D.; Ironside, J.W.; Manson, J.C.; Diack, A.B. The glycosylation status of PrP^c is a key factor in determining transmissible spongiform encephalopathy transmission between species. *J. Virol.* **2015**, *89*, 4738–4747. [CrossRef] [PubMed]

131. Cancellotti, E.; Wiseman, F.; Tuzi, N.L.; Baybutt, H.; Monaghan, P.; Aitchison, L.; Simpson, J.; Manson, J.C. Altered Glycosylated PrP Proteins Can Have Different Neuronal Trafficking in Brain but Do Not Acquire Scrapie-like Properties. *J. Biol. Chem.* **2005**, *280*, 42909–42918. [CrossRef] [PubMed]

132. Priola, S.A.; Lawson, V.A. Glycosylation influences cross-species formation of protease-resistant prion protein. *EMBO J.* **2001**, *20*, 6692–6699. [CrossRef] [PubMed]

133. Katorcha, E.; Srivastava, S.; Klimova, N.; Baskakov, I.V. Sialylation of Glycosylphosphatidylinositol (GPI) Anchors of Mammalian Prions Is Regulated in a Host-, Tissue- and Cell-specific Manner. *J. Biol. Chem.* **2016**, *291*, 17009–17019. [CrossRef] [PubMed]

134. Srivastava, S.; Makarava, N.; Katorcha, E.; Savtchenko, R.; Brossmer, R.; Baskakov, I.V. Post-conversion sialylation of prions in lymphoid tissues. *Proc. Natl. Acad. Sci. USA* **2015**, *112*, E6654–E6662. [CrossRef] [PubMed]

135. Aguilar-Calvo, P.; Xiao, X.; Bett, C.; Eraña, H.; Soldau, K.; Castilla, J.; Nilsson, K.P.R.; Surewicz, W.K.; Sigurdson, C.J. Post-translational modifications in PrP expand the conformational diversity of prions in vivo. *Sci. Rep.* **2017**, *7*. [CrossRef] [PubMed]

136. McCulloch, L.; Brown, K.L.; Bradford, B.M.; Hopkins, J.; Bailey, M.; Rajewsky, K.; Manson, J.C.; Mabbott, N.A. Follicular Dendritic Cell-Specific Prion Protein (PrP^c) Expression Alone Is Sufficient to Sustain Prion Infection in the Spleen. *PLoS Pathog.* **2011**, *7*, e1002402. [CrossRef] [PubMed]

137. Krautler, N.J.; Kana, V.; Kranich, J.; Tian, Y.; Perera, D.; Lemm, D.; Schwarz, P.; Armulik, A.; Browning, J.L.; Tallquist, M.; et al. Follicular Dendritic Cells Emerge from Ubiquitous Perivascular Precursors. *Cell* **2012**, *150*, 194–206. [CrossRef] [PubMed]

138. Béringue, V.; Demoy, M.; Lasmézas, C.I.; Gouritin, B.; Weingarten, C.; Deslys, J.-P.; Andreux, J.-P.; Couvreur, P.; Dormont, D. Role of spleen macrophages in the clearance of scrapie agent early in pathogenesis. *J. Pathol.* **2000**, *190*, 495–502. [CrossRef]

139. Montrasio, F.; Cozzio, A.; Flechsig, E.; Rossi, D.; Klein, M.A.; Rülicke, T.; Raeber, A.J.; Vosshenrich, C.A.J.; Proft, J.; Aguzzi, A.; et al. B lymphocyte-restricted expression of prion protein does not enable prion replication in prion protein knockout mice. *Proc. Natl. Acad. Sci. USA* **2001**, *98*, 4034–4037. [CrossRef] [PubMed]

140. Prinz, M.; Heikenwalder, M.; Junt, T.; Schwarz, P.; Glatzel, M.; Heppner, F.L.; Fu, Y.-X.; Lipp, M.; Aguzzi, A. Positioning of follicular dendritic cells within the spleen controls prion neuroinvasion. *Nature* **2003**, *425*, 957–962. [CrossRef] [PubMed]

141. Shikiya, R.A.; Langenfeld, K.A.; Eckland, T.E.; Trinh, J.; Holec, S.A.M.; Mathiason, C.K.; Kincaid, A.E.; Bartz, J.C. PrP^Sc formation and clearance as determinants of prion tropism. *PLoS Pathog.* **2017**, *13*, e1006298. [CrossRef] [PubMed]

142. Ironside, J.W.; Head, M.W.; McCardle, L.; Knight, R. Neuropathology of variant Creutzfeldt-Jakob disease. *Acta Neurobiol. Exp.* **2002**, *62*, 175–182. [CrossRef]

143. Brandner, S.; Whitfield, J.; Boone, K.; Puwa, A.; O'Malley, C.; Linehan, J.M.; Joiner, S.; Scaravilli, F.; Calder, I.; Alpers, M.P.; et al. Central and peripheral pathology of kuru: Pathological analysis of a recent case and comparison with other forms of human prion disease. *Philos. Trans. R. Soc. B Biol. Sci.* **2008**, *363*, 3755–3763. [CrossRef] [PubMed]

144. Rubenstein, R.; Chang, B. Re-Assessment of PrP^Sc Distribution in Sporadic and Variant CJD. *PLoS ONE* **2013**, *8*, e66352. [CrossRef] [PubMed]

145. Bons, N.; Lehmann, S.; Nishida, N.; Mestre-Frances, N.; Dormont, D.; Belli, P.; Delacourte, A.; Grassi, J.; Brown, P. BSE infection of the small short-lived primate *Microcebus murinus*. *C. R. Biol.* **2002**, *325*, 67–74. [CrossRef]

146. Herzog, C.; Salès, N.; Etchegaray, N.; Charbonnier, A.; Freire, S.; Dormont, D.; Deslys, J.-P.; Lasmézas, C.I. Tissue distribution of bovine spongiform encephalopathy agent in primates after intravenous or oral infection. *Lancet Lond. Engl.* **2004**, *363*, 422–428. [CrossRef]

147. Lotscher, M.; Recher, M.; Hunziker, L.; Klein, M.A. Immunologically Induced, Complement-Dependent Up-Regulation of the Prion Protein in the Mouse Spleen: Follicular Dendritic Cells Versus Capsule and Trabeculae. *J. Immunol.* **2003**, *170*, 6040–6047. [CrossRef] [PubMed]

148. Heikenwalder, M.; Zeller, N.; Seeger, H.; Prinz, M.; Klöhn, P.-C.; Schwarz, P.; Ruddle, N.H.; Weissmann, C.; Aguzzi, A. Chronic lymphocytic inflammation specifies the organ tropism of prions. *Science* **2005**, *307*, 1107–1110. [CrossRef] [PubMed]

149. Khalifé, M.; Young, R.; Passet, B.; Halliez, S.; Vilotte, M.; Jaffrezic, F.; Marthey, S.; Béringue, V.; Vaiman, D.; Le Provost, F.; et al. Transcriptomic Analysis Brings New Insight into the Biological Role of the Prion Protein during Mouse Embryogenesis. *PLoS ONE* **2011**, *6*, e23253. [CrossRef] [PubMed]

150. Hunter, N.; Houston, F.; Foster, J.; Goldmann, W.; Drummond, D.; Parnham, D.; Kennedy, I.; Green, A.; Stewart, P.; Chong, A. Susceptibility of Young Sheep to Oral Infection with Bovine Spongiform Encephalopathy Decreases Significantly after Weaning. *J. Virol.* **2012**, *86*, 11856–11862. [CrossRef] [PubMed]

151. Avrahami, D.; Gabizon, R. Age-related alterations affect the susceptibility of mice to prion infection. *Neurobiol. Aging* **2011**, *32*, 2006–2015. [CrossRef] [PubMed]

152. Brown, K.L.; Wathne, G.J.; Sales, J.; Bruce, M.E.; Mabbott, N.A. The Effects of Host Age on Follicular Dendritic Cell Status Dramatically Impair Scrapie Agent Neuroinvasion in Aged Mice. *J. Immunol.* **2009**, *183*, 5199–5207. [CrossRef] [PubMed]

153. Brown, K.L.; Mabbott, N.A. Evidence of subclinical prion disease in aged mice following exposure to bovine spongiform encephalopathy. *J. Gen. Virol.* **2014**, *95*, 231–243. [CrossRef] [PubMed]

154. Langevin, C.; Andréoletti, O.; Le Dur, A.; Laude, H.; Béringue, V. Marked influence of the route of infection on prion strain apparent phenotype in a scrapie transgenic mouse model. *Neurobiol. Dis.* **2011**, *41*, 219–225. [CrossRef] [PubMed]

155. Gill, O.N.; Spencer, Y.; Richard-Loendt, A.; Kelly, C.; Dabaghian, R.; Boyes, L.; Linehan, J.; Simmons, M.; Webb, P.; Bellerby, P.; et al. Prevalent abnormal prion protein in human appendixes after bovine spongiform encephalopathy epizootic: Large scale survey. *BMJ* **2013**, *347*, f5675. [CrossRef] [PubMed]

156. Carp, R.I.; Meeker, H.; Sersen, E. Scrapie strains retain their distinctive characteristics following passages of homogenates from different brain regions and spleen. *J. Gen. Virol.* **1997**, *78 Pt 1*, 283–290. [CrossRef] [PubMed]

157. Privat, N.; Levavasseur, E.; Yildirim, S.; Hannaoui, S.; Brandel, J.-P.; Laplanche, J.-L.; Béringue, V.; Seilhean, D.; Haïk, S. Region-specific protein misfolding cyclic amplification reproduces brain tropism of prion strains. *J. Biol. Chem.* **2017**. [CrossRef] [PubMed]

158. Nishida, N.; Harris, D.A.; Vilette, D.; Laude, H.; Frobert, Y.; Grassi, J.; Casanova, D.; Milhavet, O.; Lehmann, S. Successful Transmission of Three Mouse-Adapted Scrapie Strains to Murine Neuroblastoma Cell Lines Overexpressing Wild-Type Mouse Prion Protein. *J. Virol.* **2000**, *74*, 320–325. [CrossRef] [PubMed]

159. Race, R.; Raines, A.; Raymond, G.J.; Caughey, B.; Chesebro, B. Long-Term Subclinical Carrier State Precedes Scrapie Replication and Adaptation in a Resistant Species: Analogies to Bovine Spongiform Encephalopathy and Variant Creutzfeldt-Jakob Disease in Humans. *J. Virol.* **2001**, *75*, 10106–10112. [CrossRef] [PubMed]

160. Race, R.; Meade-White, K.; Raines, A.; Raymond, G.J.; Caughey, B.; Chesebro, B. Subclinical Scrapie Infection in a Resistant Species: Persistence, Replication and Adaptation of Infectivity during Four Passages. *J. Infect. Dis.* **2002**, *186*, S166–S170. [CrossRef] [PubMed]

161. Thackray, A.M.; Hopkins, L.; Lockey, R.; Spiropoulos, J.; Bujdoso, R. Propagation of ovine prions from "poor" transmitter scrapie isolates in ovine PrP transgenic mice. *Exp. Mol. Pathol.* **2012**, *92*, 167–174. [CrossRef] [PubMed]

162. Capobianco, R.; Casalone, C.; Suardi, S.; Mangieri, M.; Miccolo, C.; Limido, L.; Catania, M.; Rossi, G.; Fede, G.D.; Giaccone, G.; et al. Conversion of the BASE Prion Strain into the BSE Strain: The Origin of BSE? *PLoS Pathog.* **2007**, *3*. [CrossRef] [PubMed]

163. Baron, T.; Vulin, J.; Biacabe, A.-G.; Lakhdar, L.; Verchere, J.; Torres, J.-M.; Bencsik, A. Emergence of classical BSE strain properties during serial passages of H-BSE in wild-type mice. *PLoS ONE* **2011**, *6*, e15839. [CrossRef] [PubMed]

164. Plinston, C.; Hart, P.; Hunter, N.; Manson, J.C.; Barron, R.M. Increased susceptibility of transgenic mice expressing human PrP to experimental sheep bovine spongiform encephalopathy is not due to increased agent titre in sheep brain tissue. *J. Gen. Virol.* **2014**, *95*, 1855–1859. [CrossRef] [PubMed]

165. Priem, J.; Langeveld, J.P.M.; van Keulen, L.J.M.; van Zijderveld, F.G.; Andreoletti, O.; Bossers, A. Enhanced Virulence of Sheep-Passaged Bovine Spongiform Encephalopathy Agent Is Revealed by Decreased Polymorphism Barriers in Prion Protein Conversion Studies. *J. Virol.* **2014**, *88*, 2903–2912. [CrossRef] [PubMed]

166. Torres, J.-M.; Espinosa, J.-C.; Aguilar-Calvo, P.; Herva, M.-E.; Relaño-Ginés, A.; Villa-Diaz, A.; Morales, M.; Parra, B.; Alamillo, E.; Brun, A.; et al. Elements Modulating the Prion Species Barrier and Its Passage Consequences. *PLoS ONE* **2014**, *9*. [CrossRef] [PubMed]

167. Chapuis, J.; Moudjou, M.; Reine, F.; Herzog, L.; Jaumain, E.; Chapuis, C.; Quadrio, I.; Boulliat, J.; Perret-Liaudet, A.; Dron, M.; et al. Emergence of two prion subtypes in ovine PrP transgenic mice infected with human MM2-cortical Creutzfeldt-Jakob disease prions. *Acta Neuropathol. Commun.* **2016**, *4*. [CrossRef] [PubMed]

168. Edgeworth, J.A.; Gros, N.; Alden, J.; Joiner, S.; Wadsworth, J.D.F.; Linehan, J.; Brandner, S.; Jackson, G.S.; Weissmann, C.; Collinge, J. Spontaneous generation of mammalian prions. *Proc. Natl. Acad. Sci. USA* **2010**. [CrossRef] [PubMed]

169. Thackray, A.M.; Hopkins, L.; Lockey, R.; Spiropoulos, J.; Bujdoso, R. Emergence of multiple prion strains from single isolates of ovine scrapie. *J. Gen. Virol.* **2011**, *92*, 1482–1491. [CrossRef] [PubMed]

170. Okada, H.; Masujin, K.; Miyazawa, K.; Yokoyama, T. Transmissibility of H-Type Bovine Spongiform Encephalopathy to Hamster PrP Transgenic Mice. *PLoS ONE* **2015**, *10*. [CrossRef] [PubMed]

171. Angers, R.C.; Kang, H.-E.; Napier, D.; Browning, S.; Seward, T.; Mathiason, C.; Balachandran, A.; McKenzie, D.; Castilla, J.; Soto, C.; et al. Prion Strain Mutation Determined by Prion Protein Conformational Compatibility and Primary Structure. *Science* **2010**, *328*, 1154–1158. [CrossRef] [PubMed]

172. Fernández-Borges, N.; de Castro, J.; Castilla, J. In vitro studies of the transmission barrier. *Prion* **2009**, *3*, 220–223. [CrossRef] [PubMed]

173. Gonzalez-Montalban, N.; Lee, Y.J.; Makarava, N.; Savtchenko, R.; Baskakov, I.V. Changes in prion replication environment cause prion strain mutation. *FASEB J. Off. Publ. Fed. Am. Soc. Exp. Biol.* **2013**. [CrossRef] [PubMed]

174. Deleault, N.R.; Walsh, D.J.; Piro, J.R.; Wang, F.; Wang, X.; Ma, J.; Rees, J.R.; Supattapone, S. Cofactor Molecules Maintain Infectious Conformation and Restrict Strain Properties in Purified Prions. *Proc. Natl. Acad. Sci. USA* **2012**. [CrossRef] [PubMed]

175. Ghaemmaghami, S.; Colby, D.W.; Nguyen, H.-O.B.; Hayashi, S.; Oehler, A.; DeArmond, S.J.; Prusiner, S.B. Convergent replication of mouse synthetic prion strains. *Am. J. Pathol.* **2013**, *182*, 866–874. [CrossRef] [PubMed]

176. Sano, K.; Atarashi, R.; Ishibashi, D.; Nakagaki, T.; Satoh, K.; Nishida, N. Conformational Properties of Prion Strains Can Be Transmitted to Recombinant Prion Protein Fibrils in Real-Time Quaking-Induced Conversion. *J. Virol.* **2014**, *88*, 11791–11801. [CrossRef] [PubMed]

177. Sano, K.; Atarashi, R.; Nishida, N. Structural conservation of prion strain specificities in recombinant prion protein fibrils in real-time quaking-induced conversion. *Prion* **2015**, *9*, 237–243. [CrossRef] [PubMed]

178. Makarava, N.; Savtchenko, R.; Baskakov, I.V. Selective Amplification of Classical and Atypical Prions Using Modified Protein Misfolding Cyclic Amplification. *J. Biol. Chem.* **2013**, *288*, 33–41. [CrossRef] [PubMed]

179. Chianini, F.; Fernández-Borges, N.; Vidal, E.; Gibbard, L.; Pintado, B.; De Castro, J.; Priola, S.A.; Hamilton, S.; Eaton, S.L.; Finlayson, J.; et al. Rabbits Are Not Resistant to Prion Infection. *Proc. Natl. Acad. Sci. USA* **2012**, *109*, 5080–5085. [CrossRef] [PubMed]

180. Vidal, E.; Fernández-Borges, N.; Pintado, B.; Eraña, H.; Ordóñez, M.; Márquez, M.; Chianini, F.; Fondevila, D.; Sánchez-Martín, M.A.; Andreoletti, O.; et al. Transgenic Mouse Bioassay: Evidence That Rabbits Are Susceptible to a Variety of Prion Isolates. *PLoS Pathog.* **2015**, *11*, e1004977. [CrossRef] [PubMed]

181. Vidal, E.; Fernández-Borges, N.; Pintado, B.; Ordóñez, M.; Márquez, M.; Fondevila, D.; Torres, J.M.; Pumarola, M.; Castilla, J. Bovine Spongiform Encephalopathy Induces Misfolding of Alleged Prion-Resistant Species Cellular Prion Protein without Altering Its Pathobiological Features. *J. Neurosci. Off. J. Soc. Neurosci.* **2013**, *33*, 7778–7786. [CrossRef] [PubMed]

182. Beck, K.E.; Thorne, L.; Lockey, R.; Vickery, C.M.; Terry, L.A.; Bujdoso, R.; Spiropoulos, J. Strain typing of classical scrapie by transgenic mouse bioassay using protein misfolding cyclic amplification to replace primary passage. *PLoS ONE* **2013**, *8*, e57851. [CrossRef] [PubMed]

183. Cohen, F.E.; Prusiner, S.B. Pathologic Conformations of Prion Proteins. *Annu. Rev. Biochem.* **1998**, *67*, 793–819. [CrossRef] [PubMed]

184. Makarava, N.; Kovacs, G.G.; Savtchenko, R.; Alexeeva, I.; Ostapchenko, V.G.; Budka, H.; Rohwer, R.G.; Baskakov, I.V. A new mechanism for transmissible prion diseases. *J. Neurosci. Off. J. Soc. Neurosci.* **2012**, *32*, 7345–7355. [CrossRef] [PubMed]

185. Makarava, N.; Savtchenko, R.; Alexeeva, I.; Rohwer, R.G.; Baskakov, I.V. New Molecular Insight into Mechanism of Evolution of Mammalian Synthetic Prions. *Am. J. Pathol.* **2016**, *186*, 1006–1014. [CrossRef] [PubMed]

186. Igel-Egalon, A.; Moudjou, M.; Martin, D.; Busley, A.; Knäpple, T.; Herzog, L.; Reine, F.; Lepejova, N.; Richard, C.-A.; Béringue, V.; et al. Reversible unfolding of infectious prion assemblies reveals the existence of an oligomeric elementary brick. *PLoS Pathog.* **2017**, *13*, e1006557. [CrossRef] [PubMed]

MDPI

St. Alban-Anlage 66

4052 Basel

Switzerland

Tel. +41 61 683 77 34

Fax +41 61 302 89 18

www.mdpi.com

Pathogens Editorial Office

E-mail: pathogens@mdpi.com

www.mdpi.com/journal/pathogens

www.ingramcontent.com/pod-product-compliance
Lightning Source LLC
Chambersburg PA
CBHW051848210326
41597CB00033B/5815